NUMERICAL METHODS
FOR EXTERIOR PROBLEMS

Peking University 🏛 Series in Mathematics — Vol. 2

NUMERICAL METHODS FOR EXTERIOR PROBLEMS

Ying Lung-An

Peking University & Xiamen University, China

World Scientific

NEW JERSEY · LONDON · SINGAPORE · BEIJING · SHANGHAI · HONG KONG · TAIPEI · CHENNAI

Published by

World Scientific Publishing Co. Pte. Ltd.

5 Toh Tuck Link, Singapore 596224

USA office: 27 Warren Street, Suite 401-402, Hackensack, NJ 07601

UK office: 57 Shelton Street, Covent Garden, London WC2H 9HE

British Library Cataloguing-in-Publication Data
A catalogue record for this book is available from the British Library.

NUMERICAL METHODS FOR EXTERIOR PROBLEMS
Peking University Series in Mathematics — Vol. 2

ISBN-13 978-981-270-218-0
ISBN-10 981-270-218-0
ISBN-13 978-981-270-526-6 (pbk)
ISBN-10 981-270-526-0 (pbk)

Printed in Singapore

Preface

Partial differential equations play an important role in mathematical physics. Usually some initial conditions or boundary conditions are imposed on the equations, and we are dealing with initial value problems, boundary value problems, or initial-boundary value problems. There are some different points of view to classify the problems: Linear or nonlinear, first order equations or higher order equations, steady problems or evolution problems, interior problems or exterior problems. The first three are according to the properties of the equations, and the last one is according to the properties of the underlying physical domains. For example, the flow problem in a chamber is an interior problem, and the flow problem around an aircraft is an exterior problem.

All partial differential equations possess infinite number of degrees of freedom. To solve a problem numerically, one should approximate the problem so that only a finite number of degrees of freedom are solved in real computation. As a result a truncation error appears in every numerical scheme. This fact is the same for both interior and exterior problems. However when one intends to solve an exterior problem, one faces another kind of difficulty that the domain is infinitely large. Most approaches for the interior problems can not be applied to the exterior problems directly. The importance of exterior problems is obvious, and the challenge is serious.

There are two kinds of approaches to deal with the exterior problems: to truncate the domain, or to solve the problem directly on the infinite domain. The former ones are: the introducing of artificial boundary conditions, and the introducing of perfectly matched layers. The later ones are: the infinite element method, and the spectral method. While in the boundary element method both approaches are applied, depending on the degree of complexity of the domains.

The aim of this monograph is to provide a comprehensive study of different approaches. At the beginning we investigate the mathematical theory of the exterior problems of some typical partial differential equations, which provides the foundation of all numerical methods. Restricted by the space we only state the basic results. The readers can refer to the literature for the details. The main body of the book is on some traditional and new methods. Each of them is effective, and certainly restricted to some particular classes of problems. One omission of the book is that no numerical example is included, because there are too many in the literature, and it is too difficult to choose a few, representing most of the problems and most of the methods. It is also due to the restriction of space. However I am not sure if it is appropriate to do so.

This is an area growing rapidly. Limited by the author's knowledge, some important approaches must have been ignored. There is a proverb in China: " Cast away a brick and attract a jadestone". I will be pleased to listen to all comments.

My friends, Professors Zhang Guan Quan, Guo Ben Yu, Yu De Hao, and Shen Jie kindly provided me their works on this subject. Without their help I could not finish this work. My students, Fang Nengsheng and Liao Caixiu, helped me to proof-read and illustrate the manuscript. I express my sincere thanks to them. Dr. Sim, the editor, helped me a lot for the style file, and Ms. Zhang Ji, the editor, helped me to prepare the final manuscript. I am grateful to them.

Ying Lung-an

Contents

Chapter 1

Exterior Problems of Partial Differential Equations

This chapter is devoted to the theory on well posedness of the exterior problems of some important partial differential equations in mathematical physics. The emphasis is the boundary conditions at the infinity, which play an essential role in the mathematical theory, and differ from different equations and dimensions. An appropriate numerical method must reflect these boundary conditions correctly. In this respect this chapter is the foundation of the other chapters. On the other hand this chapter provides some tools for numerical methods. For example, the potential theory is the basic tool in the boundary element method. Another example is the variational formulation, which is the starting point of the Galerkin method, and the infinite element method and the spectral method belong to this catalogue.

We investigate some typical equations only. No attempt has been made to provide a comprehensive theory on general equations. For example in many cases we consider equations with constant coefficients instead of that with variable coefficients. The reason is that we intend to show the most important facts by means of some simple models. Secondly it is obvious that the complete theory for some general equations may deserve an entire book, which is not our objective. The theory of the Navier-Stokes equations is the subject of a number of outstanding works. We will briefly present some basic results for the exterior problems of the Navier-Stokes equations.

We will consider a bounded domain Ω^c in two dimensional space $\mathbb{R}^2$ or in three dimensional space $\mathbb{R}^3$. Let $\partial\Omega$ be its boundary. For simplicity we assume that $\partial\Omega$ is a connected curve or surface, which is Lipschitz continuous and piecewise smooth. Let Ω be the exterior part of the closed domain $\overline{\Omega^c}$, the complement, then Ω is an unbounded domain. The dimension of the domain Ω is denoted by $\dim(\Omega)$.

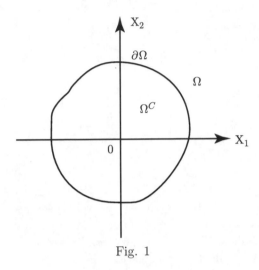

Fig. 1

If the space dimension is one, the domain Ω^c is reduced to an interval (a, b), and the domain Ω is divided into two independent parts: $(-\infty, a)$ and $(b, +\infty)$. We consider the latter one, and it has no harm in assuming $b = 0$, so we will consider the interval $(0, \infty)$.

In the sequel it will be assumed that functions and some related spaces will always be real-valued, unless we explicitly state that some functions are complex-valued in some individual sections. We will always denoted by C a generic constant that may not be the same at different occasions.

1.1 Harmonic equation–potential theory

First of all we investigate a typical exterior boundary value problem: the exterior problem of the Laplace equation:

$$\triangle u = 0, \tag{1.1}$$

where for two dimensional case, $\triangle = \frac{\partial^2}{\partial x_1^2} + \frac{\partial^2}{\partial x_2^2}$, and $x = (x_1, x_2) \in \mathbb{R}^2$, and for three dimensional case, $\triangle = \frac{\partial^2}{\partial x_1^2} + \frac{\partial^2}{\partial x_2^2} + \frac{\partial^2}{\partial x_3^2}$, and $x = (x_1, x_2, x_3) \in \mathbb{R}^3$.
We consider the Dirichlet boundary condition:

$$u = g, \qquad x \in \partial\Omega, \tag{1.2}$$

where g is a continuous function defined on $\partial\Omega$. We also consider the Neŭmann boundary condition:

$$\frac{\partial u}{\partial \nu} = g, \qquad x \in \partial\Omega, \tag{1.3}$$

where g is also a continuous function defined on $\partial\Omega$, and ν is the unit outward normal vector.

Let us consider classical solutions of the problems. A function u is a solution to the Dirichlet problem, if it is continuous on $\overline{\Omega}$, twice continuously differentiable on Ω, and it satisfies the equation (1.1) and the boundary condition (1.2). A function u is a solution to the Neŭmann problem, if it and its first order derivatives are continuous on $\overline{\Omega}$, twice continuously differentiable on Ω, and it satisfies the equation (1.1) and the boundary condition (1.3). However we are interested in well-posed problems, then some boundary conditions at the infinity are required.

$$\begin{cases} u \text{ is a bounded function on } \Omega, \dim(\Omega) = 2, \\ \lim_{|x|\to\infty} u = 0, \qquad\qquad \dim(\Omega) = 3. \end{cases} \tag{1.4}$$

We study uniqueness first. The strong maximum principle is applied here.

Theorem 1. *Let G be a bounded domain, u satisfies the equation (1.1) in G, and u is continuous on the closed domain $\overline{G}$, then if u assumes its maximum (minimum) value at an interior point $x \in G$, then u is a constant on $\overline{G}$.*

Moreover, if the first order derivatives of u are continuous on $\overline{G}$, u is not a constant, and u assumes its maximum (minimum) value at a boundary point $x \in \partial G$, then $\frac{\partial u}{\partial \nu} > 0(< 0)$ at this point.

Theorem 2. *If u_1, u_2 are two solutions satisfying (1.1),(1.2) and (1.4), or if $\dim(\Omega) = 3$ and u_1, u_2 are two solutions satisfying (1.1),(1.3) and (1.4), then $u_1 = u_2$ on Ω.*

Proof.

1. Two dimensional Dirichlet problem

Letting $g = 0$, we are going to verify $u = 0$. We may assume that the origin $O \in \Omega^c$, Then we take a disk $B(O, R_0) \subset \Omega^c$, where O is the center and R_0 is the radius. Consider an auxiliary function $v = u + \varepsilon \ln(r/R_0)$, where the constant $\varepsilon > 0$, then $v|_{\partial\Omega} \geq 0$. v is also a harmonic function. For an arbitrary $x \in \Omega$, let $R_1 > R_0$ be large enough, such that $x \in B(O, R_1)$ and $v|_{r=R_1} > 0$, then by the maximum principle $v \geq 0$ on $\Omega \bigcap B(O, R_1)$.

Therefore $u(x) \geq -\varepsilon \ln(r/R_0)$. Being the same $u(x) \leq \varepsilon \ln(r/R_0)$. But ε is arbitrary, so $u(x) = 0$, and x is arbitrary, so $u = 0$ on Ω.

2. Three dimensional Dirichlet problem

The proof is analogous. We take an auxiliary function $v = u + \varepsilon$, then following the same lines we can prove $v \geq 0$, then get $u = 0$ on Ω.

3. Three dimensional Neŭmann problem

Being the same we take an auxiliary function $v = u + \varepsilon$. Then we consider a ball $B(O, R_1) \supset \Omega^c$. Let R_1 be large enough, such that $v|_{r=R_1} > 0$. If v is a constant on $\Omega \bigcap B(O, R_1)$, then $v \geq 0$ on this domain. If v is not a constant and if v assumes its minimum value at a point $x \in \partial\Omega$, then by the strong maximum principle $\frac{\partial v}{\partial \nu} < 0$, but $\frac{\partial u}{\partial \nu} = 0$ on the boundary $\partial\Omega$, so it is impossible. Therefore v assumes its minimum value at $r = R_1$ and also $v \geq 0$ on $\Omega \bigcap B(O, R_1)$. Then in either case $u(x) \geq -\varepsilon$ on $\Omega \bigcap B(O, R_1)$. Being the same $u(x) \leq \varepsilon$. But ε is arbitrary, so $u = 0$ for $|x| < R_1$. R_1 can be arbitrary large, so $u = 0$ on Ω. $\qquad\square$

Two dimensional Neŭmann problem is more complicated. We have the following result:

Theorem 3. *If $dim(\Omega) = 2$ and (1.1),(1.3) and (1.4) is solvable, then $\int_{\partial\Omega} g(x)\, ds_x = 0$, and if u_1, u_2 are two solutions to this problem, then $u_1 - u_2$ is a constant on Ω.*

Proof.   Let u be a solution to the problem. Let $B(O, R_0) \subset \Omega^c$ as the previous. We define $v(x) = u(\frac{R_0^2}{|x|^2}x)$, and denote by Ω_1 and $\partial\Omega_1$ the image of Ω and $\partial\Omega$ under the mapping $x \to \frac{R_0^2}{|x|^2}x$, then v is also a harmonic function on Ω_1 and satisfies the Neŭmann boundary condition (1.3) on $\partial\Omega_1$. We notice that O is a boundary point of Ω_1, and we will show that it can be removed, that is, if we set $\Omega_2 = \Omega_1 \bigcup\{O\}$ and define a suitable value of $v(O)$, then v is a harmonic function on Ω_2.

We consider a small disk $B(O, R_2) \subset \Omega_2$, and solve a Dirichlet boundary value problem on it with the boundary value $v|_{r=R_2}$. Let the solution be w, which can be expressed in terms of the Poisson integral: (see Section 2.1, for example)

$$w(\rho, \theta) = \frac{R_2^2 - \rho^2}{2\pi} \int_0^{2\pi} \frac{v(\vartheta')}{R_2^2 + \rho^2 - 2R_2\rho\cos(\theta - \vartheta')}\, d\vartheta',$$

where (ρ, θ) are polar coordinates. Using the auxiliary functions $v - w \pm \varepsilon \ln(r/R_2)$ and the same argument as the previous theorem, we can verify that $v = w$ on $B(O, R_2) \setminus \{O\}$. Therefore suffice it to set $v(O) = w(O)$.

Integrating the equation (1.1) on Ω_2 we get $\int_{\partial\Omega_2} \frac{\partial v}{\partial \nu} ds_x = 0$. Returning to the domain Ω, it can be verified that $\int_{\partial\Omega} g(x) ds_x = 0$.

Let u be a solution to the corresponding homogeneous problem. We claim that u is a constant. Because v is the solution to (1.1) on Ω_2 with homogeneous Neŭmann boundary condition, then if v were not a constant then it should assume its maximum value at one point of the boundary $\partial\Omega_1$ and $\frac{\partial v}{\partial \nu} > 0$ at this point, which is impossible. Therefore v is a constant and so is u. □

We notice that the boundary conditions at the infinity (1.4) are different for different dimensions. This fact is essential. For example let $\Omega = \{x; |x| > 1\}$ and we consider the Dirichlet boundary condition $u|_{\partial\Omega} = 1$. We have one solution $u \equiv 1$ in two dimensional case. Since bounded solution is unique, there is no solution satisfying $\lim_{|x|\to\infty} u = 0$. On the other hand bounded solutions are not unique for this example in three dimensional case, because we have two solutions $u \equiv 1$ and $u = 1/r$, where $r = |x|$.

We apply potential theory to study existence. Consider two dimensional case first. For simplicity we assume that $\partial\Omega$ is a smooth curve and we will relax this restriction for the variational formulation later on. It is known that the fundamental solution is $u = \frac{1}{2\pi} \ln \frac{1}{r}$, which is the solution to the equation

$$-\triangle u = \delta, \tag{1.5}$$

where δ is the Dirac δ-function. By this fundamental solution, the solution to the Neŭmann boundary value problem is expressed in terms of a single layer potential:

$$u(x) = \frac{1}{2\pi} \int_{\partial\Omega} \omega(y) \ln \frac{1}{|x-y|} ds_y, \tag{1.6}$$

where s_y is the differential on the curve, and the solution to the Dirichlet boundary value problem is expressed in terms of a double layer potential:

$$u(x) = \frac{1}{2\pi} \int_{\partial\Omega} \sigma(y) \frac{\partial}{\partial \nu_y} \ln \frac{1}{|x-y|} ds_y. \tag{1.7}$$

The functions u satisfy the equation (1.1) on $\mathbb{R}^2 \setminus \partial\Omega$, so we need to determine the densities ω and σ to make the boundary conditions (1.2)-(1.4) satisfied.

The kernel $\frac{1}{2\pi}\ln\frac{1}{|x-y|}$ of the single layer potential is weakly singular, so (1.6) defines a continuous function on $\mathbb{R}^2$. Besides, one useful property of the single layer potential is the following:

Lemma 1. *For the single layer potential (1.6), the limit* $\lim_{|x|\to\infty}u(x)=0$ *if and only if* $\int_{\partial\Omega}\omega(y)\,ds_y=0$, *otherwise the function* u *tends to infinity as* $|x|\to\infty$.

Proof. The single layer potential can be rewritten as

$$u(x)=\frac{1}{2\pi}\int_{\partial\Omega}\omega(y)\ln\frac{|x-y_0|}{|x-y|}\,ds_y+\frac{1}{2\pi}\ln\frac{1}{|x-y_0|}\int_{\partial\Omega}\omega(y)\,ds_y,$$

where $y_0\in\Omega^c$ is a fixed point. Since $\lim_{|x|\to\infty}\ln\frac{|x-y_0|}{|x-y|}=0$ uniformly with respect to $y\in\partial\Omega$ and $\lim_{|x|\to\infty}\ln\frac{1}{|x-y_0|}=-\infty$, the conclusion follows immediately. $\square$

The kernel of the double layer potential is not weakly singular, so it does not define a continuous function on $\mathbb{R}^2$. However one property of the double layer potential is the following:

Lemma 2. *For the double layer potential (1.7) if* σ *is a continuous function on* $\partial\Omega$, *then the following limits exist:*

$$\lim_{\substack{x\to x_0\\ x\in\Omega,\,x_0\in\partial\Omega}}u(x)=\frac{1}{2}\sigma(x_0)+\frac{1}{2\pi}\int_{\partial\Omega}\sigma(y)\frac{\partial}{\partial\nu_y}\ln\frac{1}{|x_0-y|}\,ds_y,$$

$$\lim_{\substack{x\to x_0\\ x\in\Omega^c,\,x_0\in\partial\Omega}}u(x)=-\frac{1}{2}\sigma(x_0)+\frac{1}{2\pi}\int_{\partial\Omega}\sigma(y)\frac{\partial}{\partial\nu_y}\ln\frac{1}{|x_0-y|}\,ds_y.$$

Proof. We consider the case of $\sigma\equiv1$ first. For $x\neq y$ it holds that

$$\frac{\partial}{\partial\nu_y}\ln\frac{1}{|x-y|}=\frac{(x-y)\cdot\nu_y}{|x-y|^2}=\frac{\cos((x-y),\nu_y)}{|x-y|}.$$

For a fixed point $x\in\mathbb{R}^2$, we construct polar coordinates (ρ,θ) with the origin x. Then for $y\in\partial\Omega$

$$d\theta=-\frac{\cos((x-y),\nu_y)}{|x-y|}\,ds_y.$$

Therefore

$$\int_{\partial\Omega} \frac{\partial}{\partial\nu_y} \ln \frac{1}{|x-y|} \, ds_y = \begin{cases} -2\pi, & x \in \Omega^c, \\ -\pi, & x \in \partial\Omega, \\ 0, & x \in \Omega. \end{cases} \tag{1.8}$$

For general cases let $x_0 \in \partial\Omega$ be a fixed point. We set

$$v(x) = \frac{1}{2\pi} \int_{\partial\Omega} (\sigma(y) - \sigma(x_0)) \frac{\partial}{\partial\nu_y} \ln \frac{1}{|x-y|} \, ds_y.$$

Let us show that the function v is continuous at the point x_0, then applying (1.8) we get the desired result. Letting $\varepsilon > 0$, we consider a local neighborhood U of x_0, then take an arc $s \subset \partial\Omega$, $x_0 \in s$, and s is small enough so that

$$\left| \frac{1}{2\pi} \int_s (\sigma(y) - \sigma(x_0)) \frac{\partial}{\partial\nu_y} \ln \frac{1}{|x-y|} \, ds_y \right| < \varepsilon$$

for all $x \in U$. Then we take x close enough to x_0, such that

$$\left| \frac{1}{2\pi} \int_{\partial\Omega \setminus s} (\sigma(y) - \sigma(x_0)) \left\{ \frac{\partial}{\partial\nu_y} \ln \frac{1}{|x-y|} - \frac{\partial}{\partial\nu_y} \ln \frac{1}{|x_0-y|} \right\} \, ds_y \right| < \varepsilon.$$

Then

$$\left| \frac{1}{2\pi} \int_{\partial\Omega} (\sigma(y) - \sigma(x_0)) \left\{ \frac{\partial}{\partial\nu_y} \ln \frac{1}{|x-y|} - \frac{\partial}{\partial\nu_y} \ln \frac{1}{|x_0-y|} \right\} \, ds_y \right| < 3\varepsilon,$$

which implies the continuity of v. □

We turn now to consider the normal derivatives of the single layer potential along the boundary $\partial\Omega$. Since we consider the exterior and interior domains simultaneously, for definiteness, let ν always be the outward normal vector with respect to Ω^c, and $\frac{\partial}{\partial\nu+}$ and $\frac{\partial}{\partial\nu-}$ be referred to the derivatives of functions in Ω and Ω^c respectively. The discontinuity of the derivatives of single layer potentials along $\partial\Omega$ can be characterized as the following:

Lemma 3. *For the single layer potential (1.6) if ω is a continuous function on $\partial\Omega$, then*

$$\frac{\partial u(x_0)}{\partial\nu+} = -\frac{1}{2}\omega(x_0) + \frac{1}{2\pi} \int_{\partial\Omega} \omega(y) \frac{\partial}{\partial\nu_x} \ln \frac{1}{|x_0-y|} \, ds_y, \tag{1.9}$$

$$\frac{\partial u(x_0)}{\partial\nu-} = \frac{1}{2}\omega(x_0) + \frac{1}{2\pi} \int_{\partial\Omega} \omega(y) \frac{\partial}{\partial\nu_x} \ln \frac{1}{|x_0-y|} \, ds_y. \tag{1.10}$$

Proof. For each point $x \in \partial\Omega$ we consider the straight line perpendicular to the curve, then we take derivatives along this line. By this way we extend the normal derivatives $\frac{\partial}{\partial\nu_x}$ to a neighborhood of $\partial\Omega$. Let

$$v(x) = \frac{1}{2\pi} \int_{\partial\Omega} \omega(y) \left\{ \frac{\partial}{\partial\nu_x} \ln \frac{1}{|x-y|} + \frac{\partial}{\partial\nu_y} \ln \frac{1}{|x-y|} \right\} ds_y.$$

For $x_0 \in \partial\Omega$ the kernel

$$\frac{\partial}{\partial\nu_x} \ln \frac{1}{|x-y|} + \frac{\partial}{\partial\nu_y} \ln \frac{1}{|x-y|} = \frac{\cos((y-x),\nu_x) + \cos((x-y),\nu_y)}{|x-y|}$$

is bounded near x_0. Then following the same argument as the previous lemma we can show that v is continuous at x_0, then applying the results of the previous lemma we get the jump results. □

Having expressed the solutions of the Dirichlet and Neŭmann boundary value problems in terms of double layer potentials and single layer potentials, we apply the previous lemmas and get the integral equations for the densities σ and ω:

1. Dirichlet exterior problem:

$$\frac{1}{2}\sigma(x) + \int_{\partial\Omega} \sigma(y) K(x,y) \, ds_y = g(x), \tag{1.11}$$

where $K(x,y) = \frac{1}{2\pi} \frac{\partial}{\partial\nu_y} \ln \frac{1}{|x-y|}$.

2. Dirichlet interior problem:

$$-\frac{1}{2}\sigma(x) + \int_{\partial\Omega} \sigma(y) K(x,y) \, ds_y = g(x). \tag{1.12}$$

3. Neŭmann exterior problem:

$$-\frac{1}{2}\omega(x) + \int_{\partial\Omega} \omega(y) K(y,x) \, ds_y = -g(x). \tag{1.13}$$

4. Neŭmann interior problem:

$$\frac{1}{2}\omega(x) + \int_{\partial\Omega} \omega(y) K(y,x) \, ds_y = g(x). \tag{1.14}$$

$K(x,y)$ is a continuous function on $\partial\Omega \times \partial\Omega$, so it defines a compact operator on the space $C(\partial\Omega)$. The equations (1.11)-(1.14) are the Fredholm integral equations of the second type. The Fredholm alternative theorem is applied here:

Theorem 4. *Let $\mathcal{B}$ be a Banach space, and $K : \mathcal{B} \to \mathcal{B}$ be a compact operator. Then the following alternative holds. Either,*

1. There is a unique solution $u \in \mathcal{B}$ satisfying the equation $u + Ku = f$, for all $f \in \mathcal{B}$.

Or

2. The corresponding homogeneous equation $u + Ku = 0$ admits $k > 0$ linearly independent solutions, its dual equation $v + K^*v = 0$, $v \in \mathcal{B}^*$, also admits k linearly independent solutions, i.e., the dimensions of null spaces are equal, and the equation $u + Ku = f$ is solvable if and only if $vf = 0, \forall v \in \mathcal{B}^*, v + K^*v = 0$.

Theorem 5. *The integral equations (1.12) and (1.13) admit unique solutions.*

Proof. Let ω be a solution to the homogeneous equation associated with (1.13). Then by the equation

$$-\frac{1}{2} \int_{\partial\Omega} \omega(x)\,ds_x + \int_{\partial\Omega}\int_{\partial\Omega} \omega(y)K(y,x)\,ds_y\,ds_x = 0.$$

By exchanging the order of integrals and noting (1.8) we have

$$\int_{\partial\Omega} \omega(x)\,ds_x = 0.$$

With this density ω we consider a single layer potential u defined by (1.6). We notice that ω is a solution to the homogeneous equation. Then by Lemma 3 we get $\frac{\partial u}{\partial \nu^+} = 0$, . By the uniqueness of Neŭmann exterior problems u is a constant on Ω, then by the uniqueness of Dirichlet interior problems u is equal to the same constant on Ω^c, so $\frac{\partial u}{\partial \nu^-} = 0$. We apply Lemma 3 again and get $\omega = 0$. Then by the Fredholm theorem the equations admit unique solutions. $\square$

Theorem 6. *The homogeneous integral equation associated with(1.11)or (1.14) admit one nonzero solution, and any two solutions are linearly dependent.*

Proof. The homogeneous equation associated with (1.11) has a nonzero solution $\sigma \equiv 1$. By the Fredholm theorem the homogeneous equation associated with (1.14) also admits nonzero solutions. Let ω be one of them, then we claim that

$$\int_{\partial\Omega} \omega(x)\,ds_x \neq 0.$$

Otherwise with this density ω we consider s a single layer potential u defined by (1.6), then by Lemma 3 we get $\frac{\partial u}{\partial \nu^-} = 0$. By the uniqueness of Neŭmann interior problems u is a constant on Ω^c, then by Lemma 1 and the uniqueness of Dirichlet exterior problems u is equal to the same constant on Ω, so $\frac{\partial u}{\partial \nu^+} = 0$. We apply Lemma 3 again and get $\omega = 0$, which is not a nonzero solution.

Let ω_1, ω_2 be two solutions to the homogeneous equation associated with (1.14). We consider a linear combination $c_1\omega_1 + c_2\omega_2$, $c_1^2 + c_2^2 \neq 0$, such that $\int_{\partial\Omega}(c_1\omega_1(x) + c_2\omega_2(x))\,ds_x = 0$. Then by the above argument $c_1\omega_1 + c_2\omega_2 \equiv 0$, so they are linearly dependent. $\qquad\square$

Theorem 7. *The problem (1.1),(1.2),(1.4) admits a unique solution.*

Proof. The uniqueness has been given. We prove existence here. We take a constant c such that $\int_{\partial\Omega}(g(x)+c)\omega(x)\,ds_x = 0$, where ω is a nonzero solution to the homogeneous equation associated with (1.14). Then we solve the equation (1.11) with a right hand side $g + c$. Owing to the Fredholm alternative theorem it is solvable. The desired solution to (1.1),(1.2),(1.4) is $u - c$, where u is the double layer potential (1.7). $\qquad\square$

Theorem 8. *The sufficient and necessary condition for (1.1),(1.3),(1.4) admitting a solution is $\int_{\partial\Omega} g(x)\,ds_x = 0$, and the solutions are unique up to adding an arbitrary constant.*

Proof. We solve the equation (1.13), then construct a single layer potential u. If $\int_{\partial\Omega} g(x)\,ds_x = 0$, then one finds that $\int_{\partial\Omega} \omega(x)\,ds_x = 0$ by integrating both sides of (1.13). Then by Lemma 1 u is bounded, so it satisfies the boundary condition (1.4). $\qquad\square$

Next we investigate three dimensional problems. Using single layer potential and double layer potential we also get integral equations (1.11)-(1.14), where $K(x, y) = \frac{1}{4\pi}\frac{\partial}{\partial\nu_y}\frac{1}{|x-y|}$.

Parallel to two dimensional case, we have the following theorems. We will omit most of the proof and only indicate the difference in the argument.

Theorem 9. *The integral equations (1.12) and (1.13) admit unique solutions.*

Theorem 10. *The homogeneous integral equation associated with (1.11) or (1.14) admit one nonzero solution, and any two solutions are linearly dependent.*

Proof. Being the same they admit nonzero solutions.

Let ω_1, ω_2 be two solutions to the homogeneous equation associated with (1.14). Let u_1, u_2 be two single layer potentials with densities ω_1, ω_2. They are constants in Ω^c. We consider a linear combination $c_1\omega_1 + c_2\omega_2$, $c_1^2 + c_2^2 \neq 0$, such that $u = c_1u_1 + c_2u_2 \equiv 0$ in Ω^c. Since $u = 0$ on $\partial\Omega$, by the uniqueness of the Dirichlet problems, $u \equiv 0$ on Ω. Therefore $c_1\omega_1 + c_2\omega_2 \equiv 0$. They are linearly dependent. $\square$

Theorem 11. *The problem (1.1),(1.2),(1.4) admits a unique solution.*

Proof. The uniqueness has been given. We prove existence here. We take a constant c such that $\int_{\partial\Omega}(g(x)+c)\omega(x)\,ds_x = 0$, where ω is a nonzero solution to the homogeneous equation associated with (1.14). Then we solve the equation (1.11) with a right hand side $g + c$. We may assume that the solution u_1 constructed in the proof of Theorem 10 satisfies $u \equiv 1$ in Ω^c. Then the desired solution is $u - cu_1$, where u is the double layer potential (1.7). $\square$

Theorem 12. *The problem (1.1),(1.3),(1.4) admits a unique solution.*

Proof. This is because the corresponding integral equation admits a unique solution. $\square$

The results in this section are collected in the following tables:

Table 1. Boundary value problems

	condition for existence	solutions to homogeneous problems	expressions
2-D Dirichlet	none	0	double layer $-c$
2-D Neŭmann	$\int g\,ds = 0$	constant	single layer $+ C$
3-D Dirichlet	none	0	double layer $-cu_1$
3-D Neŭmann	none	0	single layer

Table 2. Integral equations

	condition for existence	solutions to homogeneous equations
2-D Dirichlet	$\int g\omega\,ds = 0$	constant
2-D Neŭmann	none	0
3-D Dirichlet	$\int g\omega\,ds = 0$	constant
3-D Neŭmann	none	0

1.2   Poisson equations

We investigate the Poisson equation,

$$-\triangle u = f, \qquad x \in \Omega, \tag{1.15}$$

in this section. For simplicity we assume that $f \in C_0(\bar{\Omega})$, that is, f is continuous with a bounded support. The uniqueness of the problem (1.15),(1.2),(1.4) or the problem (1.15),(1.3),(1.4) comes from the results in the previous section directly, so we study existence here.

For two dimensional case the Newton potential is

$$u(x) = \frac{1}{2\pi} \int_{\mathbb{R}^2} f(y) \ln \frac{1}{|x - y|} \, dy, \tag{1.16}$$

where f is extended to the entire space. If f is extended by zero, the integral is in fact taken on Ω. It can be verified that u satisfies the equation (1.15). Analogous to Lemma 1, we have

Lemma 4. *For the Newton potential (1.16), the limit $\lim_{|x|\to\infty} u(x) = 0$ if and only if $\int_{\mathbb{R}^2} f(y)\, dy = 0$, otherwise the function u tends to infinity.*

We extend the function f to Ω^c so that $\int_{\mathbb{R}^2} f(y)\, dy = 0$. Let u_0 be the corresponding Newton potential, then $u - u_0$ satisfies a boundary value problem of the Laplace equation, so the results in the previous section can be applied here. The Dirichlet problem admits a unique solution. As for the Neŭmann problem we notice that

$$\int_\Omega f(x)\, dx = -\int_{\partial\Omega} \frac{\partial u_0}{\partial \nu}\, ds_x.$$

Thus the sufficient and necessary condition of existing a solution is

$$\int_\Omega f(x)\, dx + \int_{\partial\Omega} g(x)\, ds_x = 0,$$

and the solutions are unique up to adding an arbitrary constant.

For three dimensional case the Newton potential is

$$u(x) = \frac{1}{4\pi} \int_{\mathbb{R}^3} f(y) \frac{1}{|x - y|}\, dy. \tag{1.17}$$

We can define $f = 0$ on Ω^c then verify that both the Dirichlet problem and the Neŭmann problem are uniquely solvable.

1.3 Poisson equations–variational formulation

To study the variational formulation and weak solutions the (complex) Lax-Milgram theorem is applied:

Theorem 13. *Let $\mathcal{H}$ be a Hilbert space, $M, \alpha > 0$, and $a(\cdot, \cdot)$ be a sesquilinear form on $\mathcal{H}$ with*
1. $|a(u, v)| \leq M\|u\|\|v\|, \quad \forall u, v \in \mathcal{H}$,
2. $|a(u, u)| \geq \alpha\|u\|^2, \quad \forall u \in \mathcal{H}$.
Let f be a bounded linear functional on $\mathcal{H}$. Then the equation

$$a(u, v) = f(v), \quad \forall v \in \mathcal{H}$$

admits a unique solution $u \in \mathcal{H}$, and the mapping $f \to u$ is bounded.

We list some notations which are commonly used: $W^{m,p}(\Omega)$ and $H^m(\Omega)$, $m \in \mathbb{R}, 1 \leq p \leq \infty$, are Sobolev spaces, and $\| \cdot \|_{m,p}, \| \cdot \|_m$ are norms in these spaces. The semi-norms are denoted by $|\cdot|_{m,p}, |\cdot|_m$ respectively. The L^2 inner product is denoted by $(\cdot, \cdot)$. $W_0^{m,p}(\Omega)$ and $H_0^m(\Omega)$ are the closures of $C_0^\infty(\Omega)$ in $W^{m,p}(\Omega)$ and $H^m(\Omega)$ respectively.

$H^m(\Omega)$ is a Hilbert space which is chosen as the basic space in many variational formulations of partial differential equations. However we have seen two particular solutions in Section 1.1, namely, $u \equiv 1$ for two dimensional exterior problems and $u = 1/r$ for three dimensional exterior problems. They are certainly not in the space $L^2(\Omega)$. So some different spaces have to be introduced. The following lemmas are useful in introducing some new spaces.

Lemma 5. *We assume that dim $(\Omega) = 3$ and $u \in C_0^\infty(\Omega)$, then*

$$\int_\Omega \frac{u^2(x)}{|x - y|^2} \, dx \leq 4 \int_\Omega |\nabla u(x)|^2 \, dx, \tag{1.18}$$

where $y \in \mathbb{R}^3$.

Proof. We have

$$2 \int \sum_{k=1}^3 \frac{\partial u(x)}{\partial x_k} u(x) \frac{x_k - y_k}{|x - y|^2} \, dx$$

$$= \int \sum_{k=1}^3 \frac{\partial u^2(x)}{\partial x_k} \cdot \frac{x_k - y_k}{|x - y|^2} \, dx = - \int \frac{u^2(x)}{|x - y|^2} \, dx.$$

Using the Cauchy inequality to get

$$\int \frac{u^2(x)}{|x-y|^2} \, dx \leq 2 \left(\int \frac{u^2(x)}{|x-y|^2} \sum_{k=1}^{3} \frac{(x_k - y_k)^2}{|x-y|^2} \, dx \right)^{\frac{1}{2}} \cdot \left(\int |\nabla u(x)|^2 \, dx \right)^{\frac{1}{2}}.$$

But $\sum_{k=1}^{3} \frac{(x_k - y_k)^2}{|x-y|^2} = 1$, which gives the result. □

Lemma 6. *We assume that dim* $(\Omega) = 2$, $B(O,1) \subset \Omega^c$ *and* $u \in C_0^\infty(\Omega)$, *then*

$$\int_\Omega \frac{u^2(x)}{|x|^2 \ln^2 |x|} \, dx \leq 4 \int_\Omega |\nabla u(x)|^2 \, dx. \tag{1.19}$$

Proof. We have

$$2 \int_\Omega \sum_{k=1}^{2} \frac{\partial u(x)}{\partial x_k} u(x) \frac{x_k}{|x|^2 \ln |x|} \, dx$$

$$= \int_\Omega \sum_{k=1}^{2} \frac{\partial u^2(x)}{\partial x_k} \cdot \frac{x_k}{|x|^2 \ln |x|} \, dx = - \int_\Omega \frac{u^2(x)}{|x|^2 \ln^2 |x|} \, dx.$$

Then following the same lines we obtain the result. □

Let us define a set $H_0^{1,*}(\Omega)$, the closure of $C_0^\infty(\Omega)$ with respect to the norm $|\cdot|_1$. With this norm it is a Hilbert space. By Lemma 5 and Lemma 6 it is easy to see that

Lemma 7. *If dim* $(\Omega) = 2$ *and* $B(O, R_0) \subset \Omega^c$, *then the norm* $|\cdot|_1$ *in* $H_0^{1,*}(\Omega)$ *is equivalent to*

$$\| \cdot \|_{1,*} = \left(|\cdot|_1^2 + \left\| \frac{\cdot}{|x| \ln |x/R_0|} \right\|_0^2 \right)^{\frac{1}{2}}.$$

Proof. We apply Lemma 6 with a mapping $x \to x/R_0$ and get the result directly. □

Lemma 8. *If dim* $(\Omega) = 3$ *and* $O \in \Omega^c$, *then the norm* $|\cdot|_1$ *in* $H_0^{1,*}(\Omega)$ *is equivalent to*

$$\| \cdot \|_{1,*} = \left(|\cdot|_1^2 + \left\| \frac{\cdot}{|x|} \right\|_0^2 \right)^{\frac{1}{2}}.$$

For simplicity we will always assume that $|x| > R_0$ in Ω later on.

We consider the variational formulation of (1.15),(1.2) first. Let $g \in H^{1/2}(\partial\Omega)$. By the inverse trace theorem of Sobolev spaces there is a function $u_0 \in H^1(\Omega)$, such that $u_0|_{\partial\Omega} = g$. It is easy to make the support of u_0 bounded. Then $u - u_0$ satisfies $-\triangle(u - u_0) = f + \triangle u_0$, and $(u - u_0)|_{\partial\Omega} = 0$. Therefore we will consider homogeneous boundary condition. The variational formulation is: For a given f in the dual space $\left(H_0^{1,*}(\Omega)\right)'$ find $u \in H_0^{1,*}(\Omega)$ such that

$$a(u,v) = <f,v>, \qquad \forall v \in H_0^{1,*}(\Omega), \tag{1.20}$$

where

$$a(u,v) = \int_\Omega \nabla u \cdot \nabla v \, dx.$$

By the Lax-Milgram theorem we get easily:

Theorem 14. *The problem (1.20) admits a unique solution.*

The dual product $< \cdot, \cdot >$ can be realized by Lebesgue integrals. Owing to (1.19) and (1.18), if $|x| \ln |x/R_0| f \in L^2(\Omega)$ for two dimensional problems, we can set $< f, v > = \int_\Omega f v \, dx$, and $|x| f \in L^2(\Omega)$ is suitable for three dimensional problems.

To define the variational formulation of the Neŭmann problems some other spaces are needed. Letting $B(O, R_1) \supset \Omega^c$ and $R_2 > R_1$, we consider a cut-off function $\zeta \in C^\infty$ such that $\zeta(x) \equiv 1$ for $|x| < R_1$, $\zeta(x) \equiv 0$ for $|x| > R_2$, and $0 \leq \zeta(x) \leq 1$ for all x. We define

$$H^{1,*}(\Omega) = \{u; \zeta u \in H^1(\Omega \textstyle\bigcap B(O, R_2)), (1 - \zeta)u \in H_0^{1,*}(\Omega)\}.$$

Equipped with the norm $\| \cdot \|_{1,*}$ the set $H^{1,*}(\Omega)$ is a Hilbert space. The definition of $H^{1,*}(\Omega)$ is independent of the chosen of the cut-off function, because if we take another cut-off function ζ_1, certainly $(\zeta - \zeta_1)u \in H^1$, then the set $H^{1,*}(\Omega)$ keeps the same.

Lemma 9. *For two dimensional domains, $1 \in H^{1,*}(\Omega)$.*

Proof. Let ζ be the function defined as above. Let $\zeta_1(x) = \zeta(x/a)$. $a > 0$ is a constant. Then

$$\int_{\mathbb{R}^2} |\nabla \zeta_1(x)|^2 \, dx = \frac{1}{a^2} \int_{\mathbb{R}^2} |\nabla \zeta(x/a)|^2 \, dx = \int_{\mathbb{R}^2} |\nabla \zeta(x)|^2 \, dx.$$

Let $a \to \infty$. ζ_1 is uniformly bounded in $H^{1,*}(\Omega)$. We extract a subsequence convergent weakly. On the other hand

$$\lim_{a \to \infty} \int_\Omega \frac{1}{|x|^2 \ln^2 |x/R_0|} (\zeta_1(x) - 1)^2 \, dx = 0.$$

Therefore the limit is 1, which is in the space $H^{1,*}(\Omega)$.   □

Letting $g \in H^{-1/2}(\partial\Omega)$, the variational formulation of the problem (1.15)(1.3) is: Find $u \in H^{1,*}(\Omega)$, such that

$$a(u, v) =< f, v >_\Omega + < g, v >_{\partial\Omega}, \qquad \forall v \in H^{1,*}(\Omega). \tag{1.21}$$

Certainly the problem (1.21) is not uniquely solvable for two dimensional problems. Because if u is a solution, $u + C$ is also a solution, where C is an arbitrary constant. We modify the problem as the following: Define

$$V = \left\{ u \in H^{1,*}(\Omega); \int_{\Omega \cap B(O,R_1)} u \, dx = 0 \right\}.$$

Then consider the problem: Find $u \in V$, such that

$$a(u, v) =< f, v >_\Omega + < g, v >_{\partial\Omega}, \qquad \forall v \in V. \tag{1.22}$$

Lemma 10. *The problem (1.22) admits a unique solution.*

Proof.   We show that $|\cdot|_1$ is an equivalent norm in V, then by the Lax-Milgram theorem the conclusion follows. If $|u|_1 = 0$, then u is a constant on Ω. Since $\int_{\Omega \cap B(O,R_1)} u \, dx = 0$, $u \equiv 0$ on Ω. By the closed graph theorem it is an equivalent norm to $\|\cdot\|_{1,*}$. Therefore there is a constant C, such that

$$\|u\|_{1,*}^2 \leq C|u|_1^2 = Ca(u, u), \quad \forall u \in V.$$ □

Returning to the problem (1.21), we have the following:

Theorem 15. *The sufficient and necessary condition for (1.21) admitting a solution is:*

$$< f, 1 >_\Omega + < g, 1 >_{\partial\Omega} = 0. \tag{1.23}$$

Moreover, if u_1 and u_2 are two solutions to (1.21), then $u_1 - u_2$ is a constant in Ω.

Proof.   If u is a solution, we take $v = 1$ in (1.21), then $a(u, 1) = 0$, which shows (1.23) is necessary. To show it is also sufficient, we solve the problem (1.22) first and get a solution $u \in V \subset H^{1,*}(\Omega)$. Letting v be an arbitrary in $H^{1,*}(\Omega)$, we take a constant C, such that $v+C \in V$. Then we take $v+C$ as the test function v in (1.22) to obtain

$$a(u,v) = a(u,v+C)-a(u,C) = a(u,v+C) = <f,v+C>_\Omega + <g,v+C>_{\partial\Omega}.$$

Noting (1.23), (1.21) follows immediately.

Let u_1, u_2 be two solutions to (1.21), then $a(u_1 - u_2, v) = 0$. Let $v = u_1 - u_2$, then we get $|u_1 - u_2|_1 = 0$, which yields that $u_1 - u_2$ is a constant in Ω. $\qquad\square$

The situation for three dimensional problems is different:

Theorem 16. *If dim* $(\Omega) = 3$, *the problem (1.21) admits a unique solution.*

Proof. We need to verify that $|\cdot|_1$ is an equivalent norm. Let $|u|_1 = 0$, then u is a constant C in Ω. By the definition of $\|\cdot\|_{1,*}$,

$$\int_\Omega \frac{C^2}{|x|^2}\, dx < \infty,$$

which gives $u = C = 0$. Thus $|\cdot|_1$ is a norm. By the closed graph theorem, it is equivalent to $\|\cdot\|_{1,*}$. $\qquad\square$

1.4 Helmholtz equations

We consider the wave equation,

$$\frac{\partial^2 u}{\partial t^2} = \triangle u, \tag{1.24}$$

where the variables are normalized to make the wave speed to be 1. Using separation of variables, we set $u(x,t) = X(x)T(t)$, then X and T satisfy

$$\frac{T''}{T} = -\lambda, \qquad \frac{\triangle X}{X} = -\lambda.$$

The general solution T is

$$T = c_1 \cos \omega t + c_2 \sin \omega t, \qquad \lambda = \omega^2.$$

Related to ω the solution u is

$$u = X_1(x) \cos \omega t + X_2(x) \sin \omega t = \text{Re}\,(X(x)e^{-i\omega t}),$$

where $X(x) = X_1(x) + iX_2(x)$.

To solve $X(x)$ we work with complex-valued functions in this section. In general we investigate the exterior problems of the Helmholtz equations,

$$-(\triangle + \lambda)u = f, \qquad \lambda \in \mathbb{C}, \tag{1.25}$$

with the boundary condition,

$$u = 0, \qquad x \in \partial\Omega. \tag{1.26}$$

With respect to the operator $-\triangle$ we define two sets: the spectrum, $\sigma(-\triangle) = \{\lambda \in \mathbb{C}; \operatorname{Im} \lambda = 0, \lambda \geq 0\}$, and the resolvent set, $\rho(-\triangle) = \mathbb{C} \setminus \sigma(-\triangle)$.

Theorem 17. *Let $\lambda \in \rho(-\triangle)$, $f \in L^2(\Omega)$, then the problem (1.25), (1.26) admits a unique solution $u \in H_0^1(\Omega)$.*

Proof.   The assertion is proved by applying the Lax-Milgram theorem. Clearly only the coercive condition of the sesqui-linear form,

$$a(u, v) = \int_\Omega (\nabla u \cdot \overline{\nabla v} - \lambda u \overline{v}) \, dx,$$

is to be verified. Let $\lambda = \lambda_1 + i\lambda_2$. If $\lambda_2 = 0$, then $\lambda_1 < 0$, then

$$|a(u, u)| \geq \min(1, -\lambda)\|u\|_1^2.$$

Then existence and uniqueness follows. If $\lambda_2 \neq 0$, we take a constant $\varepsilon \in (0, 1)$ and make the following estimation:

$$\sqrt{2}|a(u, u)| \geq |\operatorname{Rea}(u, u)| + |\operatorname{Ima}(u, u)| \geq \varepsilon \operatorname{Rea}(u, u) + |\operatorname{Ima}(u, u)|$$
$$= \varepsilon(|u|_1^2 - \lambda_1\|u\|_0^2) + |\lambda_2| \cdot \|u\|_0^2.$$

Let ε be small enough, so that $\alpha = |\lambda_2| - \varepsilon\lambda_1 > 0$, then the right hand side has a lower bound $\min(\alpha, \varepsilon)\|u\|_1$, which yields the conclusion. $\square$

We turn now to consider the case of $\lambda = \omega^2, \omega \geq 0$. The particular case of $\lambda = 0$ is the subject of the previous sections, where we know that $H_0^1(\Omega)$ is not a suitable working space. To study this problem, we derive two particular solutions to the homogeneous equation first. Let $\dim(\Omega) = 3$. If $u = f(r), r = |x|$, then f satisfies the Bessel equation:

$$\frac{d^2 f}{dr^2} + \frac{2}{r}\frac{df}{dr} + \omega^2 f = 0.$$

There are two particular solutions:

$$u_1 = \frac{1}{4\pi r}e^{i\omega r}, \quad u_2 = \frac{1}{4\pi r}e^{-i\omega r}. \tag{1.27}$$

The first one is known as an "outgoing wave", and the second one is an "incoming wave", because when we return to the wave equation (1.24) and consider the solutions of separation of variables, they are

$$u_1 = \frac{1}{4\pi r}e^{-i\omega(t-r)}, \quad u_2 = \frac{1}{4\pi r}e^{-i\omega(t+r)}.$$

The first one represents a sphere wave spreading with a speed of one, and the second one represents a sphere wave converging to the origin, also with a speed of one. It is observed that for these two particular solutions both u and ∇u are not in $L^2(\Omega)$. To choose a suitable space, we consider a weighted norm, $\|\rho\nabla u\|_0 + \|\rho u\|_0$, $\rho = \frac{1}{1+r}$, then the particular solutions are bounded with respect to the norm. However if we want to achieve a result of uniqueness, one of u_1 or u_2 should be excluded in the definition. Relatively the outgoing wave is applied more often, because the scattering problem in scientific computing is: To study the reflecting of an obstacle under a known incoming wave. The outgoing wave satisfies the famous Sommerfield radiation condition:

$$\frac{\partial u}{\partial r} - i\omega u = O(\frac{1}{r^2}),$$

which leads to the inner products,

$$(u,v)_0 = \int_\Omega u \cdot \overline{v}\, dx, \quad (u,v)_{0,\rho} = (\rho u, \rho v)_0,$$

$$(u,v)_V = (u,v)_{0,\rho} + (\nabla u, \nabla v)_{0,\rho} + \left(\left(\frac{\partial u}{\partial r} - i\omega u\right), \left(\frac{\partial v}{\partial r} - i\omega v\right)\right)_0,$$

and the Hilbert space,

$$V = \{u \in H^1_{loc}(\Omega); u|_{\partial\Omega} = 0, \|u\|_V = (u,u)_V^{1/2} < \infty\}.$$

It is easy to see that for the case of $\omega = 0$, $V = H_0^{1,*}(\Omega)$. Let $\sigma(x) = 1 + |x|$ and $L^2_\sigma(\Omega) = \{f; \sigma f \in L^2(\Omega)\}$. The variational problem is: For a given $f \in L^2_\sigma(\Omega)$, find $u \in V$, such that

$$a(u,v) = (f,v)_0, \quad \forall v \in C_0^\infty(\Omega). \tag{1.28}$$

To prove well-posedness of the problem, we start by proving Rellich's estimate:

Lemma 11. *Let $u \in C^2$ be a solution to the equation $\triangle u + \lambda u = 0$ for $|x| > R_1$, $\lambda > 0$. If u is not identity to zero, then there exist constants*

$p > 0$ and $R_2 > R_1$, such that

$$\int_{R_1 < |x| < R} |u(x)|^2 \, dx \geq pR, \quad \forall R > R_2.$$

Proof. We consider sphere coordinates and the sphere harmonic operator on a unit sphere, $\triangle_0$, then

$$\triangle = \frac{\partial^2}{\partial r^2} + \frac{2}{r} \frac{\partial}{\partial r} + \frac{1}{r^2} \triangle_0.$$

Let $S_{n,j}$, $n \in \mathbb{N}, j = 0, 1, \cdots, n$ be the eigenfunctions to $\triangle_0$, then

$$-\triangle_0 S_{n,j} = n(n+1) S_{n,j}.$$

Setting $v = ru$, we see v is not identity to zero either, and satisfies

$$\left(\frac{\partial^2}{\partial r^2} + \frac{1}{r^2} \triangle_0 + \lambda \right) v = 0.$$

There exist a $r_* > R_0$, such that $v|_{r=r_*}$ is not identity to zero, then there is a sphere harmonic function $S_{n,j}$ to make

$$g(r_*) = \int_{S^2} v(r_* x) S_{n,j}(x) \, dx \neq 0.$$

By this we define a function $g(r)$. We have

$$0 = \int_{S^2} \left(\frac{\partial^2}{\partial r^2} + \frac{1}{r^2} \triangle_0 + \lambda \right) v(rx) S_{n,j}(x) \, dx$$

$$= g''(r) + \lambda g(r) - \frac{n(n+1)}{r^2} \int_{S^2} v(rx) S_{n,j}(x) \, dx$$

$$= g''(r) + \left(\lambda - \frac{n(n+1)}{r^2} \right) g(r).$$

Therefore g is a solution to the Bessel equation,

$$g(r) = \sqrt{r} (c_1 J_\nu(\sqrt{\lambda} r) + c_2 Y_\nu(\sqrt{\lambda} r)),$$

where $\nu = n + \frac{1}{2}$, J_ν, Y_ν are Bessel functions, and c_1 and c_2 are two constants. By the fact that $g(r_*) \neq 0$, we have $|c_1| + |c_2| > 0$. The asymptotic behavior of the Bessel functions as $r \to \infty$ is:

$$\sqrt{r} J_\nu(r) = \sqrt{\frac{2}{\pi}} \cos(r - a_n) + O\left(\frac{1}{r} \right),$$

$$\sqrt{r}Y_\nu(r) = \sqrt{\frac{2}{\pi}}\sin(r - a_n) + O\left(\frac{1}{r}\right),$$

where $a_n = \pi(n+1)/2$. Using this relation and

$$\int_{S^2} |u(rx)|^2 r^2\, dx = \int_{S^2} |v(rx)|^2\, dx \geq |g(r)|^2,$$

we get

$$\int_{R_1 < |x| < R} |u(x)|^2\, dx \geq \int_{R_1}^{R} |g(r)|^2\, dr$$

$$\geq \frac{2}{\pi}\int_{R_1}^{R} (|c_1|^2 \cos^2(\sqrt{\lambda}(r - a_n)) + |c_2|^2 \sin^2(\sqrt{\lambda}(r - a_n)))\, dr$$

$$+ \frac{2}{\pi}\int_{R_1}^{R} (c_1\overline{c_2} + c_2\overline{c_1}) \cos(\sqrt{\lambda}(r - a_n)) \sin(\sqrt{\lambda}(r - a_n))\, dr + O(\log r)$$

$$= \frac{1}{\pi}\int_{R_1}^{R} (|c_1|^2 + |c_2|^2)\, dr + \frac{1}{\pi}\int_{R_1}^{R} (|c_1|^2 - |c_2|^2) \cos(2\sqrt{\lambda}(r - a_n))\, dr$$

$$+ \frac{2}{\pi}\int_{R_1}^{R} (c_1\overline{c_2} + c_2\overline{c_1}) \cos(\sqrt{\lambda}(r - a_n)) \sin(\sqrt{\lambda}(r - a_n))\, dr + O(\log r).$$

We notice that the second term and the third term are bounded. Thus the conclusion follows. □

Lemma 12. *If u is a solution to the homogeneous problem (1.28), then $u = 0$.*

Proof.   The case of $\lambda = 0$ has been investigated in Section 1.3. It remains to study the case of $\lambda > 0$.

Let $B(O, R_1) \supset \Omega^c$. We take a cut-off function $\zeta \in C_0^\infty(\mathbb{R}^3)$ such that $\zeta \equiv 1$ on $B(O, R_1)$. By the properties of solutions to elliptic differential equations, u is an analytic function, then we have $a(u, \zeta u) = 0$. Let $\Omega_1 = \Omega \bigcap B(O, R_1)$. We notice that $\triangle u + \lambda u = 0$ and apply the Green's formula to get

$$a(u, \zeta u) = \int_{\Omega} (\nabla u \cdot \overline{\nabla(\zeta u)} \quad \lambda u \overline{\zeta u})\, dx$$

$$= \int_{\Omega_1} (\nabla u \cdot \overline{\nabla(\zeta u)} - \lambda u \overline{\zeta u})\, dx + \int_{|x| > R_1} (\nabla u \cdot \overline{\nabla(\zeta u)} - \lambda u \overline{\zeta u})\, dx$$

$$= \int_{\Omega_1} (\nabla u \cdot \overline{\nabla(\zeta u)} - \lambda u \overline{\zeta u})\, dx + \int_{|x| = R_1} \nabla u \cdot \nu \overline{u}\, ds.$$

The first integral is real, thus

$$0 = \text{Im}\, a(u, \zeta u) = \text{Im} \int_{|x|=R_1} \nabla u \cdot \nu \bar{u} \, ds.$$

Then it holds that

$$\int_{|x|=R_1} \left| \frac{\partial u}{\partial r} - i\omega u \right|^2 ds = \int_{|x|=R_1} \left(\left| \frac{\partial u}{\partial r} \right|^2 + \omega^2 |u|^2 \right) ds,$$

which yields

$$\omega^2 \int_{|x|>R_1} |u|^2 \, dx \leq \left\| \frac{\partial u}{\partial r} - i\omega u \right\|_0^2 < \infty.$$

Then the Rellich's estimate yields $u \equiv 0$ for $r > R_1$, but u is an analytic function, so $u \equiv 0$ on Ω. $\qquad\square$

To study existence we consider the Newton potential,

$$u(x) = \int_{\mathbb{R}^3} f(y) K(x - y) \, dy, \tag{1.29}$$

first, where $K(x - y) = \frac{1}{4\pi} \frac{1}{|x-y|} e^{i\sqrt{\lambda}|x-y|}$, $\lambda \in \mathbb{C}$.

Lemma 13. *If $|\lambda| < \Lambda$, $u \in C^2(\mathbb{R}^3)$, $\|u\|_V^2 = \|u\|_{0,\rho}^2 + \|\nabla u\|_{0,\rho}^2 + \|\frac{\partial u}{\partial r} - i\sqrt{\lambda}u\|_0^2 < \infty$, $f \in L^2(\mathbb{R}^3)$, f is compactly supported, and $-(\triangle + \lambda)u = f$, then u can be expressed in terms of (1.29), and the estimate,*

$$\|u\|_V < C\|f\|_{0,\sigma} \tag{1.30}$$

holds, with a constant C, depending on Λ.

Proof. We take a cut-off function $\zeta \in C^\infty(\mathbb{R}^3)$ with supp $\zeta \subset B(O, 2)$, and $\zeta(x) \equiv 1$ for $|x| < 1$. For a given $x \in \mathbb{R}^3$ let $a > |x|$. Then

$$\zeta(x/a)u(x) = \int \zeta(y/a)u(y)\delta(x - y) \, dy$$

$$= -\int \zeta(y/a)u(y)(\triangle + \lambda)K(x - y) \, dy$$

$$= -\int \{\triangle(\zeta(y/a)u(y))K(x - y) + \lambda\zeta(y/a)u(y)K(x - y)\} \, dy, \tag{1.31}$$

$$\triangle(\zeta(y/a)u(y)) = \zeta(y/a)\triangle u(y) + u(y)\triangle\zeta(y/a) + 2\nabla\zeta(y/a) \cdot \nabla u(y)).$$

We see that

$$\int (u(y)\triangle\zeta(y/a) + 2\nabla\zeta(y/a) \cdot \nabla u(y))K(x-y)\, dy$$

$$= \int \nabla\zeta(y/a) \cdot ((\nabla u(y))K(x-y) - u(y)\nabla K(x-y))\, dy$$

$$= \int \nabla\zeta(y/a) \cdot ((\nabla u(y) - i\sqrt{\lambda}u(y))K(x-y) - u(y)(\nabla - i\sqrt{\lambda})K(x-y))\, dy.$$

Since $u \in V$, the right hand side tends to zero as $a \to \infty$. Then letting $a \to \infty$ in (1.31), the expression (1.29) follows.

Next let us verify the estimate (1.30), which is a direct consequence of the Hölder inequality. For example

$$\left\| \frac{\partial u}{\partial r} - i\omega u \right\|_0^2$$

$$= \int \left| \frac{1}{4\pi} \int f(y) \frac{1}{|x-y|^2} e^{i\sqrt{\lambda}|x-y|} \frac{\partial|x-y|}{\partial r}\, dy \right|^2 dx$$

$$\leq C \int \left(\int \frac{|f(y)|^2(1+|y|)^2|y|^{1/2}}{|x-y|^{3/2}}\, dy \right) \left(\int \frac{dy}{|x-y|^{5/2}(1+|y|)^2|y|^{1/2}} \right) dx,$$

where

$$\int \frac{dy}{|x-y|^{5/2}(1+|y|)^2|y|^{1/2}} \leq \int \frac{dy}{|x-y|^{5/2}|y|^{5/2}} \leq \frac{C}{|x|^2}.$$

We get

$$\left\| \frac{\partial u}{\partial r} - i\omega u \right\|_0^2 \leq C \int\int \frac{|f(y)|^2(1+|y|)^2|y|^{1/2}}{|x-y|^{3/2}|x|^2}\, dx\, dy$$

$$\leq C \int |f(y)|^2(1+|y|)^2\, dy = C\|f\|_{0,\sigma}^2.$$

The other two terms can be estimated in a similar way. $\qquad\square$

Lemma 14. *If f is compactly supported, $|\lambda| < \Lambda$, then the solution u to the problem (1.28) satisfies the estimate,*

$$\|u\|_V < C(\|u\|_{0,\Omega \cap B(O,R)} + \|f\|_{0,\sigma}), \qquad (1.32)$$

where $R > R_2 > R_1$, $B(O,R_1) \supset \Omega^c$, and the constant C depends on Λ and R.

Proof. We take a cut-off function $\zeta \in C^\infty(\mathbb{R}^3)$ with supp $\zeta \subset B(O,R)$, $0 \leq \zeta \leq 1$, and $\zeta \equiv 1$ for $|x| < R_2$. The estimate of ζu is the standard

H^1-norm estimate for elliptic equations.

$$\begin{aligned}
|\zeta u|_1^2 &= (\nabla(\zeta u), \nabla(\zeta u)) \\
&= (\nabla u, \nabla(\zeta^2 u)) + ((\nabla\zeta)u, \nabla(\zeta u)) - (\nabla u, (\nabla\zeta)\zeta u) \\
&= (f, \nabla(\zeta^2 u)) + \lambda(u, \zeta^2 u) + ((\nabla\zeta)u, \nabla(\zeta u)) - (\nabla(\zeta u), (\nabla\zeta)u) \\
&\quad + \|(\nabla\zeta)u\|_0^2,
\end{aligned}$$

which gives

$$|u|_{1,\Omega \cap B(O,R_2)} \leq |\zeta u|_{1,\Omega \cap B(O,R)} \leq C(\|u\|_{0,\Omega \cap B(O,R)} + \|f\|_{0,\Omega \cap B(O,R)}). \tag{1.33}$$

To estimate the exterior part we notice that the Lemma 13 also holds for all $u \in V$, which is obtained by approaching u by a sequence of smooth functions. We take another function $\zeta_1 \in C^\infty(\mathbb{R}^3)$, $0 \leq \zeta_1 \leq 1$, $\zeta_1 \equiv 1$ for $|x| > R_2$ and $\zeta_1 \equiv 0$ for $|x| < R_1$. Extend u by zero to Ω^c, then u is defined on $\mathbb{R}^3$. $\zeta_1 u$ satisfies

$$-(\triangle + \lambda)(\zeta_1 u) = \zeta_1 f - (\triangle\zeta_1)u - 2(\nabla\zeta_1) \cdot \nabla u.$$

By Lemma 13 it holds that

$$\|\zeta_1 u\|_V \leq C(\|u\|_{0,\Omega \cap B(O,R)} + \|f\|_{0,\sigma} + \|\nabla u\|_{0,B(O,R_2)\backslash B(O,R_1)}).$$

Then by (1.33) the inequality (1.32) follows. $\qquad\square$

Lemma 15. *If f is compactly supported, $|\lambda| < \Lambda$, then the solution u to the problem (1.28) satisfies the estimate,*

$$\|u\|_V < C\|f\|_{0,\sigma}, \tag{1.34}$$

where the constant C depends on Λ and the domain Ω.

Proof. If it were not the case, then there would be a sequence $\{\lambda_n, f_n, u_n\}$, such that

$$\|u_n\|_V = 1, \qquad \|f_n\|_{0,\sigma} < 1/n.$$

We take a sub-sequence, still denoted by $\{\lambda_n, f_n, u_n\}$, such that $\lambda_n \to \lambda$, $u_n \rightharpoonup u$ in V. Once more by Rellich's selection theorem there exists a sub-sequence of this sub-sequence, such that $\|u_n - u\|_{0,\Omega \cap B(O,R)} \to 0$. By (1.32) $u_n \to u$ strongly in V, and u satisfies

$$a(u, v) = 0, \qquad \forall v \in C_0^\infty(\Omega).$$

By uniqueness $u = 0$. On the other hand passing to the limit we get $\|u\|_V = 1$. This is a contradiction. $\qquad\square$

Finally we prove the existence theorem.

Theorem 18. *The problem (1.28) admits a unique solution.*

Proof. We prove existence. First we assume that f is compactly supported. We take a constant $\varepsilon > 0$ and set $\lambda_\varepsilon = \lambda + i\varepsilon$. By Theorem 17 the problem (1.25),(1.26) admits a unique solution $u_\varepsilon \in H_0^1(\Omega)$ with respect to λ_ε. Let $\varepsilon_n \to +0$, then $\sqrt{\lambda_{\varepsilon_n}} \to \omega$. By Lemma15 there is a sub-sequence $\{u_{\varepsilon_n}\}$, converging weakly to u in V. Taking the limit in (1.28), we verify that u is a solution.

For general f we take a sequence $\{f_n\}$ with compact support and converging to f in $L_\sigma^2(\Omega)$. Then by Lemma 15 the corresponding u_n is a Cauchy sequence in V. Taking the limit in (1.28), we get the desired solution. $\qquad\square$

1.5 Linear elasticity

Let the displacement in an isotropic body be $u = (u_1, u_2, u_3)^T$, where by T we denote transpose of a vector or a matrix. Let the strains be

$$\epsilon_{ij}(u) = \epsilon_{ji}(u) = \frac{1}{2}\Big(\frac{\partial u_i}{\partial x_j} + \frac{\partial u_j}{\partial x_i}\Big), \quad 1 \leq i,j \leq 3.$$

Let $\sigma_{ij}(u)$, $1 \leq i,j \leq 3$, be stresses, then the Hooke's law reads

$$\sigma_{ij}(u) = \sigma_{ji}(u) = \lambda \left(\sum_{k=1}^{3} \epsilon_{kk}(u)\right) \delta_{ij} + 2\mu\epsilon_{ij}(u),$$

where $\lambda > 0, \mu > 0$ are the Lamé coefficients and δ_{ij} is the Kronecker's symbol. Let the Young's modulus and the Poisson ratio of the material be E and $\tilde{\nu}$, $E > 0$, $0 < \tilde{\nu} < \frac{1}{2}$, then the relations between them and the Lamé coefficients are the following:

$$E = \frac{\mu(3\lambda + 2\mu)}{\lambda + \mu}, \qquad \tilde{\nu} = \frac{\lambda}{2(\lambda + \mu)}.$$

Let the external force be $f = (f_1, f_2, f_3)^T$, then the equilibrium equation is

$$-\sum_{j=1}^{3} \frac{\partial}{\partial x_j} \sigma_{ij}(u) = f_i, \quad 1 \leq i \leq 3.$$

The above formulas lead to the Lamé equation for u:

$$-\mu \triangle u - (\lambda + \mu)\text{grad div} u = f, \quad x \in \Omega. \tag{1.35}$$

We consider the Dirichlet boundary condition,

$$u = 0, \qquad x \in \partial\Omega, \tag{1.36}$$

first, which is associated with a fixed boundary. We will work in the space $V = \left(H_0^{1,*}(\Omega)\right)^3$, and the corresponding bilinear form is

$$a(u,v) = \int_\Omega \sum_{i,j=1}^3 \sigma_{ij}(u)\epsilon_{ij}(v) \, dx$$

$$= \int_\Omega \left\{ \lambda \text{div} u \text{div} v + 2\mu \sum_{i,j=1}^3 \epsilon_{ij}(u)\epsilon_{ij}(v) \right\} dx.$$

Let V' be the dual space of V. We assume that $f \in V'$, for example, $f \in (L_{loc}^2(\Omega))^3$ and $\int_\Omega r^2 |f|^2 \, dx < \infty$. The weak formulation of the problem (1.35),(1.36) is: Find $u \in V$, such that

$$a(u,v) = < f, v >, \qquad \forall v \in V. \tag{1.37}$$

To study the existence and uniqueness, we need Korn's inequality:

Theorem 19. *It holds that*

$$\sum_{i,j=1}^3 \int_\Omega \left(\frac{\partial u_i}{\partial x_j}\right)^2 dx \leq 4 \sum_{i,j=1}^3 \int_\Omega \epsilon_{ij}^2(u) \, dx, \tag{1.38}$$

for all $u \in V$.

Proof. It would be sufficient to consider $u \in (C_0^\infty(\Omega))^3$ only. Let

$$\frac{\partial u_i}{\partial x_j} = \epsilon_{ij}(u) + r_{ij}, \tag{1.39}$$

where

$$r_{ij} = \frac{1}{2}(\frac{\partial u_i}{\partial x_j} - \frac{\partial u_j}{\partial x_i}),$$

then we have

$$\frac{\partial u_i}{\partial x_j} \cdot \frac{\partial u_j}{\partial x_i} = \epsilon_{ij}^2 - r_{ij}^2. \tag{1.40}$$

Applying the Green's formula and noting that $u|_{\partial\Omega} = 0$, we get

$$\int_\Omega \frac{\partial u_i}{\partial x_j} \cdot \frac{\partial u_j}{\partial x_i}\, dx = \int_\Omega \frac{\partial u_i}{\partial x_i} \cdot \frac{\partial u_j}{\partial x_j}\, dx. \tag{1.41}$$

By (1.40) and (1.41) we get

$$\sum_{i,j=1}^{3} \int_\Omega r_{ij}^2(u)\, dx = \sum_{i,j=1}^{3} \int_\Omega \epsilon_{ij}^2(u)\, dx - \int_\Omega (\mathrm{div}\, u)^2\, dx \le \sum_{i,j=1}^{3} \int_\Omega \epsilon_{ij}^2(u)\, dx. \tag{1.42}$$

By (1.39) we also have

$$\left(\frac{\partial u_i}{\partial x_j} \right)^2 \le 2(\epsilon_{ij}^2(u) + r_{ij}^2).$$

The above two inequalities imply (1.38). □

Applying the Lax-Milgram theorem, Theorem 19, and Lemma 5 we obtain the following immediately:

Theorem 20. *The problem (1.37) admits a unique solution.*

Next let us consider the Neümann boundary condition:

$$\sum_{j=1}^{3} \sigma_{ij}(u)\nu_j = g_i, \qquad x \in \partial\Omega, 1 \le i \le 3, \tag{1.43}$$

which is associated with a distributed force g applied on the boundary $\partial\Omega$. Let $V = \left(H^{1,*}(\Omega)\right)^3$. We assume that $f \in V'$ and $g \in (L^2(\partial\Omega))^3$, then the problem is: Find $u \in V$, such that

$$a(u,v) = < f, v >_\Omega + (g,v)_{\partial\Omega}. \tag{1.44}$$

For the well-posedness of this problem we need the following Korn's inequality:

Theorem 21. *We assume that Ω_0 is a bounded domain, then there exists a constant C, depending on Ω_0, such that*

$$\|u\|_{1,\Omega_0} \le C \left(\sum_{i,j=1}^{3} \int_{\Omega_0} \epsilon_{ij}^2(u)\, dx + \|u\|_{0,\Omega_0}^2 \right)^{1/2}, \tag{1.45}$$

for all $u \in (H^1(\Omega_0))^3$.

This inequality is not trivial. The proof of it can be found in [Leis, R. (1986)]. Applying Theorem 21 we have the following:

Theorem 22. *The problem (1.44) admits a unique solution.*

Proof.   Owing to the Lax-Milgram theorem we need to prove

$$a(u,u) \geq \alpha\|u\|_{1,*}^2, \quad \forall u \in V, \quad \alpha > 0.$$

It is proved by contradiction. If there exists a sequence $\{u_n\} \subset V$, such that $\|u_n\|_{1,*} = 1$ and $a(u_n, u_n) \leq 1/n$, let $B(O, R)$ be a ball with large R, such that $\partial\Omega \subset B(O, R)$, then $\{u_n\}$ is bounded in $(H^1(\Omega_1))^3$, $\Omega_1 = \Omega \cap B(O, R)$. Then there is a subsequence, still denoted by $\{u_n\}$, such that u_n convergence strongly in $(L^2(\Omega_1))^3$ and weakly in $(H^1(\Omega_1))^3$. Let the limit be u. Since R is arbitrary, u is defined on Ω. Because $a(u_n, u_n) \to 0$ as $n \to \infty$, $\epsilon_{ij}(u_n)$ converge to zero in $L^2(\Omega)$, $1 \leq i, j \leq 3$, so $\epsilon_{ij}(u) = 0$, which implies u is a rigid body motion, that is, a linear function. Since $\|u_n\|_{1,*,\Omega_1} \leq 1$, we have $\|u\|_{1,*,\Omega_1} \leq 1$, and thus $\|u\|_{1,*} \leq 1$. But u is linear, hence $u = 0$. Owing to the Korn's inequality (1.45) u_n converges strongly in $(H^1(\Omega_1))^3$. We define a cut-off function $\zeta \in C_0^\infty(\Omega)$ such that $0 \leq \zeta \leq 1$ and $\zeta \equiv 1$ if $|x|$ is large enough. Then $a(\zeta u_n, \zeta u_n) \to 0$. Owing to the Korn's inequality (1.38) ζu_n converges with respect to the norm $|\cdot|_1$, which means ζu_n converges in $H_0^{1,*}(\Omega)$. We already know that u_n converges strongly in $(H^1(\Omega_1))^3$ for any large R. Therefore u_n converges strongly in $H^{1,*}(\Omega)$. Then $\|u\|_{1,*} = 1$, which contradicts $u = 0$. $\qquad\square$

We turn now to two dimensional problems. There are two cases:

1. Plane stress. $\sigma_{11}, \sigma_{22}, \sigma_{12}$ depend on x_1, x_2 only, and $\sigma_{13} = \sigma_{23} = \sigma_{33} = 0$.

We eliminate $\epsilon_{13}, \epsilon_{23}, \epsilon_{33}$ in the Hooke's law, then expresses the stresses as

$$\sigma_{ij}(u) = \sigma_{ji}(u) = \frac{\tilde{\nu}E}{1 - \tilde{\nu}^2}\left(\sum_{k=1}^2 \epsilon_{kk}(u)\right)\delta_{ij} + \frac{E}{1 + \tilde{\nu}}\epsilon_{ij}(u),$$

for $i, j = 1, 2$. Therefore if we denote

$$\lambda = \frac{\tilde{\nu} E}{1 - \tilde{\nu}^2}, \qquad \mu = \frac{E}{2(1 + \tilde{\nu})},$$

the equation for u keeps the same as (1.35). Only the parameters are different.

2. Plane strain. $\epsilon_{11}, \epsilon_{22}, \epsilon_{12}$ depend on x_1, x_2 only, and $\epsilon_{13} = \epsilon_{23} = \epsilon_{33} = 0$.

The Hooke's law is

$$\sigma_{ij}(u) = \sigma_{ji}(u) = \lambda \left(\sum_{k=1}^{2} \epsilon_{kk}(u) \right) \delta_{ij} + 2\mu \epsilon_{ij}(u),$$

for $i, j = 1, 2$. The parameters in the equation keeps the same as the three dimensional case.

The argument for the existence and uniqueness of the Dirichlet and Neŭmann problems is analogous to that in Section 1.3. We only give the results here. The Dirichlet problem admits a unique solution. The sufficient and necessary condition for the Neŭmann problem admitting a solution is: $< f, 1 >_\Omega + < g, 1 >_{\partial\Omega} = 0$, and the solutions are not unique. An arbitrary constant vector is a solution to the homogeneous Neŭmann problem, and if u_1 and u_2 are two solutions, then $u_1 - u_2$ is a constant vector in Ω.

1.6 Bi-harmonic equations

We investigate two dimensional bi-harmonic equations,

$$\triangle^2 u = f, \qquad (1.46)$$

in this section. One background of this equation is the bending problem of thin plates, where u is the deflection, the displacement perpendicular to the plates. Another background is plane elasticity. If the stresses are expressed by $\sigma_{11} = \frac{\partial^2 A}{\partial x_2^2}$, $\sigma_{12} = \frac{\partial^2 A}{\partial x_1 \partial x_2}$, $\sigma_{22} = \frac{\partial^2 A}{\partial x_1^2}$, where A is the Airy stress function, then A satisfies (1.46).

First of all let us derive the corresponding bilinear form. Suppose that $u \in C^4(\overline{\Omega})$ and we take $v \in C_0^\infty(\overline{\Omega})$. Integrating by parts we get

$$\int_\Omega \triangle^2 u \cdot v \, dx$$

$$= \int_{\partial\Omega} \frac{\partial \triangle u}{\partial \nu} \cdot v \, ds - \int_\Omega \nabla \triangle u \cdot \nabla v \, dx$$

$$= \int_{\partial\Omega} \frac{\partial \triangle u}{\partial \nu} \cdot v \, ds - \int_\Omega \tilde{\nu} \nabla \triangle u \cdot \nabla v \, dx$$

$$- \int_\Omega (1 - \tilde{\nu}) \left(\frac{\partial^3 u}{\partial x_1^3} \frac{\partial v}{\partial x_1} + \frac{\partial^3 u}{\partial x_1^2 \partial x_2} \frac{\partial v}{\partial x_2} + \frac{\partial^3 u}{\partial x_1 \partial x_2^2} \frac{\partial v}{\partial x_1} + \frac{\partial^3 u}{\partial x_2^3} \frac{\partial v}{\partial x_2} \right) dx$$

$$= \int_{\partial\Omega} \frac{\partial \triangle u}{\partial \nu} \cdot v \, ds - \int_{\partial\Omega} \tilde{\nu} \triangle u \cdot \frac{\partial v}{\partial \nu} \, ds$$

$$- \int_{\partial\Omega} (1 - \tilde{\nu}) \left(\frac{\partial^2 u}{\partial x_1^2} \frac{\partial v}{\partial x_1} \nu_1 + \frac{\partial^2 u}{\partial x_1 \partial x_2} \frac{\partial v}{\partial x_2} \nu_1 + \frac{\partial^2 u}{\partial x_1 \partial x_2} \frac{\partial v}{\partial x_1} \nu_2 + \frac{\partial^2 u}{\partial x_2^2} \frac{\partial v}{\partial x_2} \nu_2 \right) ds$$

$$+ \int_\Omega \tilde{\nu} \triangle u \cdot \triangle v \, dx + \int_\Omega (1 - \tilde{\nu}) \left(\frac{\partial^2 u}{\partial x_1^2} \frac{\partial^2 v}{\partial x_1^2} + 2 \frac{\partial^2 u}{\partial x_1 \partial x_2} \frac{\partial^2 v}{\partial x_1 \partial x_2} + \frac{\partial^2 u}{\partial x_2^2} \frac{\partial^2 v}{\partial x_2^2} \right) dx$$

$$= \int_{\partial\Omega} \frac{\partial \triangle u}{\partial \nu} \cdot v \, ds - \int_{\partial\Omega} \tilde{\nu} \triangle u \cdot \frac{\partial v}{\partial \nu} \, ds$$

$$- \int_{\partial\Omega} (1 - \tilde{\nu}) \left(\frac{\partial^2 u}{\partial x_1^2} \nu_1^2 + 2 \frac{\partial^2 u}{\partial x_1 \partial x_2} \nu_1 \nu_2 + \frac{\partial^2 u}{\partial x_2^2} \nu_2^2 \right) \frac{\partial v}{\partial \nu} \, ds$$

$$- \int_{\partial\Omega} (1 - \tilde{\nu}) \left(-\frac{\partial^2 u}{\partial x_1^2} \nu_1 \nu_2 + \frac{\partial^2 u}{\partial x_1 \partial x_2} (\nu_1^2 - \nu_2^2) + \frac{\partial^2 u}{\partial x_2^2} \nu_1 \nu_2 \right) \frac{\partial v}{\partial s} \, ds$$

$$+ \int_\Omega \tilde{\nu} \triangle u \cdot \triangle v \, dx + \int_\Omega (1 - \tilde{\nu}) \left(\frac{\partial^2 u}{\partial x_1^2} \frac{\partial^2 v}{\partial x_1^2} + 2 \frac{\partial^2 u}{\partial x_1 \partial x_2} \frac{\partial^2 v}{\partial x_1 \partial x_2} + \frac{\partial^2 u}{\partial x_2^2} \frac{\partial^2 v}{\partial x_2^2} \right) dx$$

$$= \int_{\partial\Omega} \frac{\partial \triangle u}{\partial \nu} \cdot v \, ds - \int_{\partial\Omega} \tilde{\nu} \triangle u \cdot \frac{\partial v}{\partial \nu} \, ds$$

$$- \int_{\partial\Omega} (1 - \tilde{\nu}) \left(\frac{\partial^2 u}{\partial x_1^2} \nu_1^2 + 2 \frac{\partial^2 u}{\partial x_1 \partial x_2} \nu_1 \nu_2 + \frac{\partial^2 u}{\partial x_2^2} \nu_2^2 \right) \frac{\partial v}{\partial \nu} \, ds$$

$$+ \int_{\partial\Omega} (1 - \tilde{\nu}) \frac{\partial}{\partial s} \left(-\frac{\partial^2 u}{\partial x_1^2} \nu_1 \nu_2 + \frac{\partial^2 u}{\partial x_1 \partial x_2} (\nu_1^2 - \nu_2^2) + \frac{\partial^2 u}{\partial x_2^2} \nu_1 \nu_2 \right) v \, ds$$

$$+ \int_\Omega \tilde{\nu} \triangle u \cdot \triangle v \, dx + \int_\Omega (1 - \tilde{\nu}) \left(\frac{\partial^2 u}{\partial x_1^2} \frac{\partial^2 v}{\partial x_1^2} + 2 \frac{\partial^2 u}{\partial x_1 \partial x_2} \frac{\partial^2 v}{\partial x_1 \partial x_2} + \frac{\partial^2 u}{\partial x_2^2} \frac{\partial^2 v}{\partial x_2^2} \right) dx,$$

where $\nu = (\nu_1, \nu_2)$ is the unit outward normal vector, and s is the unit clockwise tangential vector.

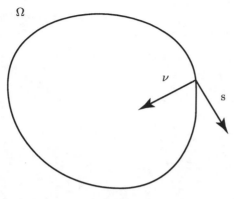

Fig. 2

For plane elasticity problems, $\tilde{\nu}$ is the Poisson ratio $0 < \tilde{\nu} \le \frac{1}{2}$, and for the Stokes problem in fluid dynamics, $\tilde{\nu} = 1$. We define

$$
a(u,v) = \int_\Omega \tilde{\nu} \triangle u \cdot \triangle v \, dx
$$
$$
+ \int_\Omega (1 - \tilde{\nu}) \left(\frac{\partial^2 u}{\partial x_1^2} \frac{\partial^2 v}{\partial x_1^2} + 2 \frac{\partial^2 u}{\partial x_1 \partial x_2} \frac{\partial^2 v}{\partial x_1 \partial x_2} + \frac{\partial^2 u}{\partial x_2^2} \frac{\partial^2 v}{\partial x_2^2} \right) dx,
\tag{1.47}
$$

the lateral shearing force,

$$
Tu = -\frac{\partial \triangle u}{\partial \nu} + (1 - \tilde{\nu}) \frac{\partial}{\partial s} \left(\frac{\partial^2 u}{\partial x_1^2} \nu_1 \nu_2 - \frac{\partial^2 u}{\partial x_1 \partial x_2} (\nu_1^2 - \nu_2^2) - \frac{\partial^2 u}{\partial x_2^2} \nu_1 \nu_2 \right),
\tag{1.48}
$$

and the bending moment load,

$$
Mu = \tilde{\nu} \triangle u + (1 - \tilde{\nu}) \left(\frac{\partial^2 u}{\partial x_1^2} \nu_1^2 + 2 \frac{\partial^2 u}{\partial x_1 \partial x_2} \nu_1 \nu_2 + \frac{\partial^2 u}{\partial x_2^2} \nu_2^2 \right),
\tag{1.49}
$$

then

$$
a(u,v) - \int_{\partial\Omega} Tu \cdot v \, ds - \int_{\partial\Omega} Mu \cdot \frac{\partial v}{\partial \nu} \, ds = \int_\Omega fv \, dx.
\tag{1.50}
$$

For plate problems there are several kind of boundary conditions, namely,

1. Dirichlet boundary condition, $u = g_1, \frac{\partial u}{\partial \nu} = g_2$;
2. Neŭmann boundary condition, $Tu = g_1, Mu = g_2$;
3. simply supported boundary condition, $u = g_1, Mu = g_2$.

To study these boundary value problems we define some spaces with respect to the exterior domain Ω, and at first we prove a lemma.

Lemma 16. *We assume that $B(O, 1) \subset \Omega^c$ and $u \in C_0^\infty(\Omega)$, then*

$$\int_\Omega \frac{u^2(x)}{|x|^4 \ln^4 |x|} \, dx \leq \frac{4}{9} \int_\Omega \frac{|\nabla u(x)|^2}{|x|^2 \ln^2 |x|} \, dx. \tag{1.51}$$

Proof.   We have

$$2 \int_\Omega \sum_{k=1}^2 \frac{\partial u(x)}{\partial x_k} u(x) \frac{x_k}{|x|^4 \ln^3 |x|} \, dx$$

$$= \int_\Omega \sum_{k=1}^2 \frac{\partial u^2(x)}{\partial x_k} \cdot \frac{x_k}{|x|^4 \ln^3 |x|} \, dx$$

$$= \int_\Omega u^2(x) \left\{ \frac{2}{|x|^4 \ln^3 |x|} + \frac{3}{|x|^4 \ln^4 |x|} \right\} \, dx$$

$$\geq 3 \int_\Omega \frac{u^2(x)}{|x|^4 \ln^4 |x|} \, dx.$$

Using the Cauchy inequality to get

$$3 \int \frac{u^2(x)}{|x|^4 \ln^4 |x|} \, dx \leq 2 \left(\int \frac{u^2(x)}{|x|^4 \ln^4 |x|} \sum_{k=1}^2 \frac{(x_k)^2}{|x|^2} \, dx \right)^{\frac{1}{2}} \cdot \left(\int \frac{|\nabla u(x)|^2}{|x|^2 \ln^2 |x|} \, dx \right)^{\frac{1}{2}}.$$

But $\sum_{k=1}^2 \frac{(x_k)^2}{|x|^2} = 1$, which gives the result.   □

Let us define a space $H_0^{2,*}(\Omega)$, the closure of $C_0^\infty(\Omega)$ with respect to the norm $|\cdot|_2$. Equipped with this norm it is a Hilbert space. By Lemma 6 and Lemma 16 it is easy to see that

Lemma 17. *If $B(O, R_0) \subset \Omega^c$, then the norm $|\cdot|_2$ in $H_0^{2,*}(\Omega)$ is equivalent to*

$$\| \cdot \|_{2,*} = \left(|\cdot|_2^2 + \left\| \frac{\nabla \cdot}{|x| \ln |x/R_0|} \right\|_0^2 + \left\| \frac{\cdot}{|x|^2 \ln^2 |x/R_0|} \right\|_0^2 \right)^{\frac{1}{2}}. \tag{1.52}$$

Letting $B(O, R_1) \supset \Omega^c$ and $R_2 > R_1$, we consider a cut-off function $\zeta \in C^\infty$ such that $\zeta(x) \equiv 1$ for $|x| < R_1$, $\zeta(x) \equiv 0$ for $|x| > R_2$, and $0 \leq \zeta(x) \leq 1$ for all x. We define

$$H^{2,*}(\Omega) = \{u; \zeta u \in H^2(\Omega \textstyle\bigcap B(O, R_2)), (1 - \zeta)u \in H_0^{2,*}(\Omega)\}.$$

Equipped with the norm $\| \cdot \|_{2,*}$ the set $H^{2,*}(\Omega)$ is a Hilbert space. Being analogous to the space $H^{1,*}(\Omega)$, the definition of $H^{2,*}(\Omega)$ is independent of the chosen of the cut-off function ζ.

Lemma 18. *Let p be a polynomial, the order of which is not greater than one, denoted by $p \in P_1(\Omega)$, then $p \in H^{2,*}(\Omega)$.*

Proof. Let ζ be the cut-off function defined as above. Let $\zeta_1(x) = \zeta(x/a)$. $a > 0$ is a constant. Then

$$\frac{\partial^2(\zeta_1 p)}{\partial x_i x_j} = \frac{\partial^2 \zeta_1}{\partial x_i x_j} p + \frac{\partial \zeta_1}{\partial x_i} \frac{\partial p}{\partial x_j} + \frac{\partial \zeta_1}{\partial x_j} \frac{\partial p}{\partial x_i}.$$

By Lemma 9 the semi-norm $|\cdot|_1$ of the second and the third terms are uniformly bounded with respect to a. For the first term we have

$$\int_{\mathbb{R}^2} \left\{ \frac{\partial^2 \zeta_1}{\partial x_i x_j} p \right\}^2 (x)\, dx = \frac{1}{a^4} \int_{\mathbb{R}^2} \left\{ \frac{\partial^2 \zeta}{\partial x_i x_j}(x/a) p(x) \right\}^2 dx$$

$$= \frac{1}{a^2} \int_{\mathbb{R}^2} \left\{ \frac{\partial^2 \zeta}{\partial x_i x_j}(x) p(ax) \right\}^2 dx,$$

which is also uniformly bounded. Let $a \to \infty$, then $\zeta_1 p$ is uniformly bounded in $H^{2,*}(\Omega)$. We extract a subsequence converging weakly, the limit of which is p, which is in the space $H^{2,*}(\Omega)$. □

The variational formulation of the Dirichlet problem is: Given $f \in \left(H_0^{2,*}(\Omega)\right)'$, $g_1 \in H^{3/2}(\partial\Omega)$, and $g_2 \in H^{1/2}(\partial\Omega)$, find $u \in H^{2,*}(\Omega)$, such that $u|_{\partial\Omega} = g_1$, $\frac{\partial u}{\partial \nu}|_{\partial\Omega} = g_2$, and

$$a(u, v) = \langle f, v \rangle, \qquad \forall v \in H_0^{2,*}(\Omega), \tag{1.53}$$

where $\left(H_0^{2,*}(\Omega)\right)'$ is the dual space of $H_0^{2,*}(\Omega)$.

Theorem 23. *The problem (1.53) admits a unique solution.*

Proof. By the inverse trace theorem there exists $u_0 \in H^{2,*}(\Omega)$ satisfying the boundary condition, so $u - u_0$ satisfies homogeneous boundary conditions. Applying the Green's formula we get

$$\int_\Omega \tilde{\nu} \triangle u \cdot \triangle u\, dx = \int_\Omega \tilde{\nu} \left\{ \left(\frac{\partial^2 u}{\partial x_1^2}\right)^2 + 2\left(\frac{\partial^2 u}{\partial x_1 \partial x_2}\right)^2 + \left(\frac{\partial^2 u}{\partial x_2^2}\right)^2 \right\} dx = \tilde{\nu}|u|_2^2$$

in $H_0^{2,*}(\Omega)$. Then applying the Lax-Milgram theorem the existence and uniqueness follows. □

The variational formulation of the Neumann problem is: Given $f \in \left(H_0^{2,*}(\Omega)\right)'$, $g_1 \in H^{-3/2}(\partial\Omega)$, and $g_2 \in H^{-1/2}(\partial\Omega)$, find $u \in H^{2,*}(\Omega)$, such that

$$a(u, v) = \langle g_1, v \rangle_{\partial\Omega} + \langle g_2, \frac{\partial v}{\partial \nu} \rangle_{\partial\Omega} + \langle f, v \rangle_\Omega, \qquad \forall v \in H^{2,*}(\Omega). \tag{1.54}$$

Theorem 24. *We assume that $\tilde{\nu} < 1$ then the sufficient and necessary condition for (1.54) admitting a solution is:*

$$< g_1, v >_{\partial \Omega} + < g_2, \frac{\partial v}{\partial \nu} >_{\partial \Omega} + < f, v >_\Omega = 0, \quad \forall v \in P_1(\Omega).$$

Moreover, if u_1 and u_2 are two solutions to (1.54), then $u_1 - u_2 \in P_1(\Omega)$.

Proof. It is easy to see that any $p \in P_1(\Omega)$ is a solution to the corresponding homogeneous problem. We define a space $V = \{u \in H^{2,*}(\Omega); \int_{\Omega \cap B(O,R_1)} pu \, dx = 0, \forall p \in P_1(\Omega)\}$. $|\cdot|_2$ is a norm on V, because if $|u|_2 = 0$, then $u \in P_1(\Omega)$, then by the definition of V we get $u = 0$ by taking $p = u$. We replace the space $H^{2,*}(\Omega)$ in (1.54) by V. By the definition (1.47) $a(u, u) \geq (1 - \tilde{\nu})|u|_2^2$. Owing to the Lax-Milgram theorem there is a unique solution. If $v \in H^{2,*}(\Omega)$, we divide it into two parts $v = v_1 + v_2$, such that $v_1 \in V$, and $v_2 \in P_1(\Omega)$. It is easy to see that (1.54) holds either for $v = v_1$ or for $v = v_2$. Therefore u is also a solution to the problem (1.54). $\qquad\square$

To define the variational formulation of the simply supported boundary value problem we set $V = \{u \in H^{2,*}(\Omega); u|_{\partial \Omega} = 0\}$, then the problem is; Given $f \in V'$, $g_1 \in H^{3/2}(\partial \Omega)$, and $g_2 \in H^{-1/2}(\partial \Omega)$, find $u \in H^{2,*}(\Omega)$, such that $u|_{\partial \Omega} = g_1,$, and

$$a(u, v) = < g_2, \frac{\partial v}{\partial \nu} >_{\partial \Omega} + < f, v >_\Omega, \qquad \forall v \in V. \tag{1.55}$$

Theorem 25. *We assume that $\tilde{\nu} < 1$. The problem (1.55) admits a unique solution.*

Proof. It has no harm in assuming $g_1 = 0$, because by the inverse trace theorem we can find $u_0 \in H^{2,*}(\Omega)$ such that $u_0|_{\partial \Omega} = g_1$, then consider the problem for $u - u_0$. We only need to verify that $|\cdot|_2$ is a norm in V. If not, then there would be a sequence $\{u_n\}$ in V such that $|u_n|_2 \to 0$ and

$$\left(\left\| \frac{\nabla u_n}{|x| \ln |x/R_0|} \right\|_0^2 + \left\| \frac{u_n}{|x|^2 \ln^2 |x/R_0|} \right\|_0^2 \right)^{\frac{1}{2}} = 1. \tag{1.56}$$

Then by the compact imbedding, there is a subsequence, still denoted by $\{u_n\}$, converging strongly in $H^1(\Omega \cap B(O, R_1))$ and converging weakly in $H^{2,*}(\Omega)$. Since $|u_n|_2 \to 0$, it also converges in $H^2(\Omega \cap B(O, R_1))$, then u, the limit of $\{u_n\}$, is a polynomial in P_1. By the boundary condition, $u = 0$ on $\Omega \cap B(O, R_1)$. R_1 is arbitrary, so we can extract a diagonal subsequence such that $u_n \to 0$, in $H^2(\Omega \cap B(O, R_1))$ for any R_1. Then by the definition of the space $H^{2,*}(\Omega)$, $u = 0$, then $\|u\|_{2,*} = 0$, which contradicts (1.56). $\quad\square$

1.7 Steady Navier-Stokes equations -linearized problems

1.7.1 *Navier-Stokes equations*

The motion of viscous incompressible fluids is governed by the system of Navier-Stokes equations,

$$\frac{\partial u}{\partial t} + u \cdot \nabla u + \nabla p = \tilde{\nu} \triangle u + f, \tag{1.57}$$

$$\nabla \cdot u = 0, \tag{1.58}$$

where u is the velocity, p is the kinematic pressure, f is the external body force, and the constant $\tilde{\nu}$ represents the kinematic viscosity. For the steady case $\frac{\partial u}{\partial t} = 0$. Introducing the dimensionless Reynolds number

$$Re = \frac{VL}{\tilde{\nu}},$$

where L and V are reference length and velocity respectively, the equation (1.57) becomes

$$Re \cdot u \cdot \nabla u + \nabla p = \triangle u + f, \tag{1.59}$$

where u, p and f are new nondimensional quantities. For low Reynolds number flow the first term on the left hand side is small comparing with the first term on the right hand side, so it can be neglected. The equation (1.59) becomes

$$- \triangle u + \nabla p = f, \tag{1.60}$$

which is the Stokes equation. Let us consider a typical problem: Let Ω^c be a ball $B(O, 1)$, which is a rigid body moving in a viscous incompressible fluid with a constant speed. We fix the coordinate system on this body, then the equation (1.60) is combined with the boundary conditions

$$u = 0, \qquad x \in \partial B(O, 1), \tag{1.61}$$

$$\lim_{|x| \to \infty} u = u_\infty, \tag{1.62}$$

where $-u_\infty$ is the speed of the body. Stokes has derived the analytic solution to the problem (1.60)-(1.62):

$$u(r) = u_\infty - \frac{3}{4} \nabla \times \left\{ |x|^2 \nabla \times \left(\frac{u_\infty}{|x|} \right) \right\} - \frac{1}{4} \nabla \times \nabla \times \left(\frac{u_\infty}{|x|} \right),$$

$$p(x) = -\frac{3}{2} u_\infty \cdot \nabla \left(\frac{1}{|x|} \right).$$

We are going to investigate the boundary value problems for general exterior domains.

1.7.2   Stokes equations

We consider the equation (1.60) with homogeneous boundary condition

$$u = 0, \qquad x \in \partial\Omega \qquad\qquad (1.63)$$

first. The dimension of the domain Ω is n, $n = 2, 3$. For the time being we don't study the behavior of the solution at the infinity, and leave the problem to later on. We introduce a Hilbert space,

$$V = \{ u \in \left(H_0^{1,*}(\Omega) \right)^n ; \nabla \cdot u = 0 \}, \quad n = 2, 3$$

and a bilinear form

$$a(u, v) = \int_\Omega \nabla u : \nabla v \, dx.$$

Assume that f belongs to V', the dual space of V, then the weak formulation of (1.60),(1.63) is: Find $u \in V$, such that

$$a(u, v) = < f, v >, \quad \forall v \in V. \qquad\qquad (1.64)$$

Theorem 26. *The problem (1.64) admits a unique solution.*

Proof.   By Lemma 5 and Lemma 6 the bilinear form $a(\cdot, \cdot)$ is coercive in V, then owing to the Lax-Milgram theorem existence and uniqueness follows.   □

To investigate the problems with inhomogeneous boundary conditions and the existence of p in (1.60), we need some lemmas.

Lemma 19. *Let g be a function in $\left(H^{1/2}(\partial\Omega) \right)^n$ that satisfies $\int_{\partial\Omega} g \cdot \nu \, ds = 0$. Then there exists a function $u \in \left(H^{1,*}(\Omega) \right)^n$ such that*

$$\nabla \cdot u = 0, \qquad u|_{\partial\Omega} = g,$$

and

$$\|u\|_{1,*} \le C \|g\|_{\frac{1}{2}}.$$

Proof.   For the sake of simplicity we consider the case of $n = 2$ and $\partial\Omega$ is sufficiently smooth only. For the general case see [Galdi, G.P. (1994)].

If $g \cdot \nu = 0$, then by the inverse trace theorem of Sobolev spaces there is $\psi \in H^2(\Omega)$ with a compact support, such that

$$\psi = 0. \qquad \frac{\partial\psi}{\partial\nu} = -g \cdot \tau$$

on $\partial\Omega$, where τ is the unit clockwise tangential vector, and $\|\psi\|_2 \le C\|g\|_{\frac{1}{2}}$. Let $u = \operatorname{curl} \psi$, then

$$u \cdot \nu = \frac{\partial\psi}{\partial\tau} = 0 = g \cdot \nu,$$

and

$$u \cdot \tau = -\frac{\partial\psi}{\partial\nu} = g \cdot \tau,$$

so u is the desired one.

When $g \cdot \nu$ does not vanish on $\partial\Omega$, let p be the solution to the Neŭmann problem

$$\triangle p = 0, \quad \left.\frac{\partial p}{\partial\nu}\right|_{\partial\Omega} = g \cdot \nu.$$

By Theorem 15 the solution p exists and belongs to the space $H^{1,*}(\Omega)$ and $|p|_1 \le C\|g\|_{\frac{1}{2}}$. Moreover, since $g \in \left(H^{1/2}(\partial\Omega)\right)^n$, by the theory of elliptic equations, $p \in H^2$ locally. Then by the expression of p in terms of single layer potentials, we see that $|p|_2 \le C\|g\|_{\frac{1}{2}}$. We set $u_1 = \nabla p$, then $\nabla \cdot u_1 = 0$, and $u_1 \cdot \nu = g \cdot \nu$ on $\partial\Omega$. $u - u_1$ can be defined as the previous case. $\qquad\square$

Lemma 20. *For any $f \in L^2(\Omega)$, there exists $u \in \left(H_0^{1,*}(\Omega)\right)^n$, such that $\nabla \cdot u = f$, and $\|u\|_{1,*} \le C\|f\|_0$ with a constant C depending on Ω.*

Proof. Let $\{f_m\} \subset C_0^\infty(\Omega)$ be a sequence approximating f in $L^2(\Omega)$. We define functions ψ_m to solve $\triangle\psi_m = f_m$ on $\mathbb{R}^n$. ψ_m can be expressed in terms of the Newton potential. Owing to the theory of singular integrals (see [Calderon, A.P. and Zygmund, A. (1957)]),

$$|\psi_m|_2 \le C\|f_m\|_0. \tag{1.65}$$

Let us consider the case of dim $(\Omega) = 3$ first. Since f_m is compactly supported, applying the expression of the Newton potential we deduce that $\nabla\psi_m = O(|x|^{-2})$ as $|x| \to \infty$. We consider a cut-off function $\zeta_1 \in C^\infty(\mathbb{R})$, such that $\zeta_1(\xi) \equiv 1$ for $\xi < 1$, $\zeta(\xi) \equiv 0$ for $\xi > 2$, and $0 \le \zeta_1(\xi) \le 1$ for all ξ. Let $\zeta(x) = \zeta_1(|x|/R)$, then

$$\left|\frac{\partial}{\partial x_i}(\zeta\nabla\psi_m)\right| \le \left|\zeta\frac{\partial}{\partial x_i}(\nabla\psi_m)\right| + \left|\nabla\psi_m\frac{\partial}{\partial x_i}\zeta\right| \le \left|\frac{\partial}{\partial x_i}(\nabla\psi_m)\right| + \frac{C}{R}|\nabla\psi_m|.$$

Therefore $|\zeta\nabla\psi_m|_1$ is bounded and $\nabla\psi_m$ can be approximated by functions in $(C_0^\infty(\mathbb{R}^3))^3$ with respect to $|\cdot|_1$. Consequently $\nabla\psi_m \in \left(H_0^{1,*}(\mathbb{R}^3)\right)^3$. Applying Lemma 5 and (1.65) we obtain $\|\nabla\psi_m\|_{1,*} \le C\|f_m\|_0$. Then by the trace theorem of Sobolev spaces, $\|\nabla\psi_m\|_{\frac{1}{2},\partial\Omega} \le C\|f_m\|_0$.

For the case of dim $(\Omega) = 2$, we take R large enough such that $B(O, R) \supset \Omega^c$, then take constant vectors c_m, such that

$$\int_{B(O,R)} (\nabla \psi_m + c_m)\, dx = 0.$$

Owing to the Poincaré inequality

$$\|\nabla \psi_m + c_m\|_{0,B(O,R)} \leq C |\nabla \psi_m + c_m|_{1,B(O,R)} \leq C |\psi_m|_2.$$

By the definition of $H^{1,*}$ we see that $\nabla \psi_m + c_m \in \left(H^{1,*}(\Omega)\right)^2$. We also have $\|\nabla \psi_m + c_m\|_{1,*} \leq C \|f_m\|_0$, and $\|\nabla \psi_m + c_m\|_{\frac{1}{2},\partial\Omega} \leq C \|f_m\|_0$.

ψ_m is the solution to the Laplace equation in Ω^c, so $\int_{\partial\Omega} \frac{\partial \psi_m}{\partial \nu}\, ds = 0$. Let $g = -\nabla \psi_m$, or $g = -\nabla \psi_m - c_m$ on $\partial\Omega$. We apply Lemma 19 to get a function in $(H^{1,*}(\Omega))^n$, denoted by w_m. Set

$$u_m = \nabla \psi_m + w_m, \text{ or } u_m = \nabla \psi_m + c_m + w_m.$$

We get $u_m \in \left(H_0^{1,*}(\Omega)\right)^n$, $\nabla \cdot u_m = f_m$, and $\|u_m\|_{1,*} \leq C\|f_m\|_0$. Let $m \to \infty$, then the conclusion follows. $\square$

Lemma 20 leads to the inequality: There exists a constant $\beta > 0$ such that

$$\inf_{p\in L^2(\Omega),p\neq 0}\ \sup_{u\in(H_0^{1,*}(\Omega))^n, u\neq 0} \frac{(p, \nabla \cdot u)}{\|u\|_{1,*}\|p\|_0} \geq \beta, \qquad (1.66)$$

which is the inf-sup condition, or Babuska-Brezzi condition, for this problem. We will make use of this condition later on for different problems.

For the inhomogeneous boundary condition

$$u = g, \qquad x \in \partial\Omega,$$

where g is a prescribed velocity field on the boundary, with $g \in \left(H^{1/2}(\partial\Omega)\right)^n$ but in general $\int_{\partial\Omega} g \cdot \nu\, ds$ may not be zero, by the inverse trace theorem of Sobolev spaces, there is a function $u_1 \in \left(H^1(\Omega)\right)^n$ with compact support on $\bar{\Omega}$ and satisfying $u_1|_{\partial\Omega} = g$. Then by Lemma 20 there exists $u_2 \in \left(H_0^{1,*}(\Omega)\right)^n$, such that $\nabla \cdot u_2 = -\nabla \cdot u_1$. Let $u - u_1 - u_2$ and p be the new variables, then they satisfy the Stokes problem with a homogeneous boundary condition, and the the external force term becomes $f + \triangle(u_1 + u_2)$. It is easy to see that it belongs to the space V'. By this way, the problem is reduced to (1.64).

We turn now to the existence of p. Let us define $b(u, p) = (p, \nabla \cdot u)$, then we deduce from the equation (1.60) that the equation for p in weak form is

$$b(v, p) = a(u, v) - <f, v>, \qquad \forall v \in \left(H_0^{1,*}(\Omega)\right)^n. \qquad (1.67)$$

Theorem 27. *The equation (1.67) admits a unique solution $p \in L^2(\Omega)$, and it holds that*

$$\|p\|_0 \leq C(\|f\|_{V'} + |u|_1). \tag{1.68}$$

Proof. The functional

$$F(v) = a(u, v) - <f, v>$$

is bounded in $\left(H_0^{1,*}(\Omega)\right)^n$ and vanishes in V. Consider the operator A : $v \in \left(H_0^{1,*}(\Omega)\right)^n \to \nabla \cdot v \in L^2(\Omega)$. A is linear and bounded. By Lemma 20 its range is $L^2(\Omega)$. Then $\ker(A)^\perp = R(A^*)$, where ker is the kernel, $*$ is the adjoint, and $\perp$ is annihilator. Now $\ker(A) = V$, therefore $R(A^*) = V^\perp$. Thus all functionals F are in the range of A^*. On the other hand, by definition, every element in $R(A^*)$ is of the form $\mathcal{L}(Av)$, where $\mathcal{L}$ is a functional on $L^2(\Omega)$. By the Riesz representation theorem there is a unique $p \in L^2(\Omega)$, such that

$$F(v) = \mathcal{L}(Av) = \int_\Omega pAv \, dx = \int_\Omega p\nabla \cdot v \, dx.$$

Thus the proof is complete. $\qquad \square$

1.7.3 *Behavior of solutions at the infinity*

We need some lemmas. First of all let us prove the Wirtinger inequality.

Lemma 21. *Let S_3 be the unit sphere in $\mathbb{R}^3$, $S_3 = \partial B(O, 1)$. If $f \in H^1(S_3)$ and $\bar{f}$ is the average of f, then*

$$\|f - \bar{f}\|_0 \leq C\|\nabla^* f\|_0, \tag{1.69}$$

where ∇^* is the projection of ∇ on S_3.

Proof. We extend f by constant along all lines perpendicular to S_3 to the domain Ω_0, the spherical shell of radii $1/2$ and 1. Then $|\nabla^* f|^2 = r^2 |\nabla f|^2$, $r = |x|$. By the Poincare inequality we have

$$\|f - \bar{f}\|_{0,\Omega_0} \leq C\|\nabla f\|_{0,\Omega_0}.$$

On the other hand we have

$$\|f - \bar{f}\|_{0,S_3}^2 \leq C\|f - \bar{f}\|_{0,\Omega_0}^2.$$

Thus (1.69) follows. $\qquad \square$

Lemma 22. *Let $\Omega \subset \mathbb{R}^3$ and $u \in H^{1,*}(\Omega)$, then*

$$\int_{S_3} |u(R,s)|^2 \, ds \leq \frac{1}{R} \int_{|x|>R} |\nabla u|^2 \, dx,$$

where $s \in S_3$ and $u = u(\rho, s)$, $\rho = |x|$.

Proof. It would be sufficient to prove the inequality for smooth u. Let

$$D_r(R) = \int_R^r \int_{S_3} |\nabla u(\rho, s)|^2 \rho^2 \, ds \, d\rho,$$

then by the Wirtinger inequality we obtain

$$D_r(R) = \int_{S_3} \int_R^r \left|\frac{\partial u}{\partial \rho}\right|^2 \rho^2 \, d\rho \, ds + \int_R^r \int_{S_3} |\nabla^* u|^2 \rho^2 \, ds \, d\rho$$

$$\geq \int_{S_3} \left|\int_R^r \frac{\partial u}{\partial \rho} \, d\rho\right|^2 \left(\int_R^r \rho^{-2} \, d\rho\right)^{-1} ds + C^{-1} \int_R^r \int_{S_3} |u - \bar{u}|^2 \rho^2 \, ds \, d\rho.$$

Therefore

$$D_r(R) \geq R \int_{S_3} |u(r) - u(R)|^2 \, ds + C^{-1} \int_R^r \int_{S_3} |u - \bar{u}|^2 \rho^2 \, ds \, d\rho. \quad (1.70)$$

Disregarding, for the moment, the second term on the right hand side and letting $R, r \to \infty$, we deduce that u converges strongly in $L^2(S_3)$ to some function u^*. Set $u_0 = \bar{u}^*$, and $w = u - u_0$. Then

$$\lim_{|x|\to\infty} \int_{S_3} w(x) \, ds = 0. \quad (1.71)$$

Rewriting (1.70) with w instead of u, we recover the existence of a sequence $\{r_m\}$, with $r_m \to \infty$, such that

$$\lim_{m\to\infty} \int_{S_3} |w(r_m) - \bar{w}(r_m)|^2 \, ds = 0,$$

which, because of (1.71) implies

$$\lim_{m\to\infty} \int_{S_3} |w(r_m)|^2 \, ds = 0.$$

Inserting it into (1.70) written with w in place of u and letting $r \to \infty$ yields

$$\int_{S_3} |w(R,s)|^2 \, ds \leq \frac{1}{R} \int_{|x|>R} |\nabla u|^2 \, dx. \quad (1.72)$$

Finally we show that $u_0 = 0$. We have

$$(r-R)|u_0|^2 \int_{S_3} ds = \int_{S_3} \int_R^r \frac{|u_0|^2}{|x|^2} \, dx \leq 2 \int_{S_3} \int_R^r \frac{|u|^2}{|x|^2} \, dx + 2 \int_{S_3} \int_R^r \frac{|w|^2}{|x|^2} \, dx.$$

By Lemma 5 the first term on the right hand side is bounded, and by the estimate (1.72) the second term is bounded by $O(\log r)$ as $r \to \infty$, so $u_0 = 0$. $\square$

The above result shows that the space $H^{1,*}(\Omega)$ in the formulation (1.64) indeed implies a zero boundary condition at the infinity for three dimensional problems.

1.7.4 Stokes paradox

For three dimensional problems, the result can be extended to nonhomogeneous boundary conditions at the infinity:

$$\lim_{|x| \to \infty} u = u_\infty,$$

where u_∞ is a given constant vector. Because if the boundary condition on $\partial\Omega$ is $u = g, x \in \partial\Omega$, we first solve the problem with the boundary condition

$$\lim_{|x| \to \infty} u = 0, \quad u|_{\partial\Omega} = g - u_\infty,$$

to get a solution u_1. Then $u_1 + u_\infty$ is the desired solution.

The two dimensional case is different, because the unique solution to the boundary condition $u|_{\partial\Omega} = -u_\infty$ and $f = 0$, is $u \equiv -u_\infty$, which is not zero at the infinity if $u_\infty \neq 0$. As a result, the problem with boundary condition

$$\lim_{|x| \to \infty} u = u_\infty, \quad u|_{\partial\Omega} = 0,$$

and $f = 0$ is not solvable. This phenomenon is called "Stokes paradox", first discovered by Stokes.

1.7.5 Oseen flow

Another approach to linearize the Navier-Stokes equation is the Oseen flow,

$$u_0 \cdot \nabla u + \nabla p = \triangle u + f, \tag{1.73}$$

$$\nabla \cdot u - 0, \tag{1.74}$$

where u_0 is a constant vector. For simplicity we assume that $f \in \left(L^2(\Omega)\right)^n$ with a compact support. Taking a positive constant ε we consider the following approximate problem:

$$u_0 \cdot \nabla u + \nabla p + \varepsilon u = \triangle u + f, \tag{1.75}$$

$$\nabla \cdot u = 0, \tag{1.76}$$

$$\lim_{|x| \to \infty} u = 0, \quad u|_{\partial\Omega} = 0. \tag{1.77}$$

We define a space $V_\varepsilon = \{u \in \left(H_0^1(\Omega)\right)^n; \nabla \cdot u = 0\}$ and a bilinear form,

$$a_\varepsilon(u, v) = (\nabla u, \nabla v) + (u_0 \cdot \nabla u, v) + \varepsilon(u, v).$$

Then we have

Lemma 23. *The problem: Find $u \in V_\varepsilon$, such that*

$$a_\varepsilon(u, v) = (f, v), \qquad \forall v \in V_\varepsilon,$$

admits a unique solution.

Proof. Since $(u_0 \cdot \nabla u, u) = 0$, we have

$$a_\varepsilon(u, u) = |u|_1^2 + \varepsilon \|u\|_0^2 \geq \min(1, \varepsilon) \|u\|_1^2. \tag{1.78}$$

Owing to the Lax-Milgram theorem uniqueness and existence follows. □

To investigate the limit as $\varepsilon \to 0$, we define two spaces $V = \{u \in \left(H_0^{1,*}(\Omega)\right)^n; \nabla \cdot u = 0\}$ and $\mathcal{V} = \{u \in (C_0^\infty(\Omega))^n; \nabla \cdot u = 0\}$. Then we obtain

Lemma 24. *The problem: Find $u \in V$, such that*

$$(\nabla u, \nabla v) + (u_0 \cdot \nabla u, v) = (f, v), \quad \forall v \in \mathcal{V},$$

admits at least a solution.

Proof. Let a series of ε tend to zero. By (1.78)

$$|u|_1^2 + \varepsilon \|u\|_0^2 \leq \|f\|_{V'} \|u\|_V \leq C \|f\|_{V'} |u|_1,$$

so the solutions are uniformly bounded in V. Therefore we can subtract a subseries converging weakly in V. Applying Lemma 5 and Lemma 6 we get the strong convergence in $L_{loc}^2(\Omega)$. Passing to the limit the conclusion follows. □

We remark that $\mathcal{V}$ can not be replaced by V in the weak formulation, otherwise $(u_0 \cdot \nabla u, v)$ makes no sense.

Moreover we remark that the space $H_0^{1,*}(\Omega)$ in the formulation implies a zero boundary condition at the infinity for three dimensional problems. Using the same approach as for the Stokes equation, we can easily extend the results to more general boundary conditions,

$$\lim_{|x| \to \infty} u = u_\infty, \quad u|_{\partial\Omega} = g. \tag{1.79}$$

The two dimensional case is different. However we still have the following:

Lemma 25. *We assume that dim $(\Omega) = 2$ and $u_0 \neq 0$. Then the solution satisfies $\|u\|_{0,r} < C$ with $r > 3$.*

Proof.   We define a cut-off function $\zeta \in C^\infty(\mathbb{R}^2)$ such that $\zeta(x) \equiv 1$ for $|x| < R$, $\zeta(x) \equiv 0$ for $|x| > 2R$, and $0 \leq \zeta(x) \leq 1$ for all x with R large enough. Let $w = (1 - \zeta)u$, $\phi = (1 - \zeta)p$, and (u, p) be the solution to (1.75)-(1.77). Then (w, ϕ) solves

$$u_0 \cdot \nabla w + \nabla \phi + \varepsilon w = \triangle w + f_1,$$

$$\nabla \cdot w = f_2,$$

where $f_1 = -(u_0 \cdot \nabla \zeta)u - (\nabla \zeta)p + (\triangle \zeta)u + 2\nabla \zeta \cdot \nabla u + (1 - \zeta)f$, $f_2 = -\nabla \zeta \cdot u$. Noting that $|\nabla \zeta| < \frac{1}{R}$ for $R < |x| < 2R$, we have $\|f_1\|_{0,q} < C$, $\|f_2\|_{0,q} < C$, $1 < q < \infty$, uniformly with respect to R, and f_1, f_2 compactly supported. Let the Fourier transform of $w(x), \phi(x)$ be $\hat{w}(\xi), \hat{\phi}(\xi)$, then they satisfy

$$iu_0 \cdot \xi \hat{w} + i\xi \hat{\phi} + \varepsilon \hat{w} = -|\xi|^2 \hat{w} + \hat{f}_1,$$

$$i\xi \cdot \hat{w} = \hat{f}_2.$$

After some calculation we get

$$\hat{w} = \frac{|\xi|^2 \hat{f}_1 - (\xi \cdot \hat{f}_1)\xi}{|\xi|^2(iu_0 \cdot \xi + |\xi|^2 + \varepsilon)} - \frac{i\xi}{|\xi|^2}\hat{f}_2.$$

Applying a proposition stated below we notice that $|\xi_1^{\frac{2}{3}}| \cdot |\xi_2^{\frac{2}{3}}| \leq |u_0|^{-1}|u_0 \cdot \xi| + |\xi|^2$ and find $\beta = \frac{2}{3}$ for the first term, and the second term is typical corresponding to the Laplace operator, where $\hat{f}_2(0) = \int_\Omega f_2 \, dx = 0$. Thus we have $\|w\|_{0,r} < C$ with $r > 3$, where the constant C is independent of ε. Letting $R \to \infty$ and passing to the limit, we obtain the final result.   □

The following proposition can be found in [Lizorkin, P.I. (1963)].

Proposition 1. *Let $\hat{u}(\xi), \hat{w}(\xi)$ be the Fourier transform of $u(x), w(x)$, and $\hat{w}(\xi) = \Phi(\xi)\hat{u}(\xi)$, and let Φ be continuous together with the derivative*

$$\frac{\partial^n \Phi}{\partial \xi_1 \cdots \partial \xi_n}$$

and all preceding derivatives for $|\xi_i| > 0$, $i = 1, \cdots n$. Then if for some $\beta \in [0, 1)$ and $M > 0$

$$|\xi_1|^{\kappa_1 + \beta} \cdots |\xi_n|^{\kappa_n + \beta} \left| \frac{\partial^\kappa \Phi}{\partial \xi_1^{\kappa_1} \cdots \partial \xi_n^{\kappa_n}} \right| \leq M,$$

where κ_i is zero or one and $\kappa = \sum_{i=1}^{n} \kappa_i = 0, 1, \cdots, n$, the transform $u \to w$ defines a bounded linear operator from $L^q(\mathbb{R}^n)$ into $L^r(\mathbb{R}^n)$, $1 < q < \infty$, $1/r = 1/q - \beta$, and we have $\|w\|_{0,r} \leq C\|u\|_{0,q}$, with a constant C depending only on $M, r,$ and q.

Therefore the solution u also tends to zero at the infinity in the sense of this lemma. The results can be extended to more general boundary conditions (1.79). This is the essential difference between the Oseen flow and the Stokes equation, for which we already know that the boundary condition at the infinity is impossible for two dimensional problems.

Uniqueness for the boundary value problems of the Oseen flow holds. For details, see [Galdi, G.P. (1994)].

1.8   Steady Navier-Stokes equations

We investigate three dimensional Navier-Stokes equations

$$u \cdot \nabla u + \nabla p = \triangle u + f, \tag{1.80}$$

$$\nabla \cdot u = 0, \tag{1.81}$$

with the homogeneous boundary codition

$$\lim_{|x| \to \infty} u = 0, \quad u|_{\partial\Omega} = 0 \tag{1.82}$$

first. This is a nonlinear problem, and to prove existence we need the Brouwer fixed point theorem:

Theorem 28. *Let B denote a non-void, convex and compact subset of a finite dimensional space, and let F be a continuous mapping from B to B. Then F has at least one fixed point.*

Corollary 1. *Let H be a finite dimensional Hilbert space whose scalar product is denoted by $(\cdot, \cdot)$ and corresponding norm by $|\cdot|$. Let F be a continuous mapping from H into H with the following property: There exists $R > 0$ such that*

$$(F(u), u) > 0, \qquad \forall u \in H \text{ with } |u| = R. \tag{1.83}$$

Then there exists an element u in H such that $F(u) = 0$ and $|u| \leq R$.

Proof. The proof proceeds by contradiction. Suppose that $F(u) \neq 0$ in the Ball $B = \{u \in H; |u| \leq R\}$. Then the mapping $u \to -R\frac{F(u)}{|F(u)|}$ is continuous from B into B. As the dimension of H is finite, and since B is non-void, convex and compact, we can apply the Brouwer theorem asserting that there exists u in B such that $u = -R\frac{F(u)}{|F(u)|}$. Thus $|u| = R$ and $(F(u), u) = -R|F(u)| < 0$, since $u \in B$. This contradicts (1.83). $\square$

We define two spaces $V = \{u \in \left(H_0^{1,*}(\Omega)\right)^3 ; \nabla \cdot u = 0\}$ and $\mathcal{V} = \{u \in (C_0^\infty(\Omega))^3 ; \nabla \cdot u = 0\}$. Assume that $f \in V'$, the dual space of V. Then the weak formulation of (1.80)- (1.82) is: Find $u \in V$, such that

$$(\nabla u, \nabla v) + (u \cdot \nabla u, v) =\ <f, v>, \quad \forall v \in \mathcal{V}. \tag{1.84}$$

Let us introduce some lemmas first.

Lemma 26. *There exists a denumerable set of functions $\{\varphi_k\}$ whose linear hull is dense in V and the set has the following properties:*

(1) $\varphi_k \in \mathcal{V}$, for all $k = 1, 2 \cdots$.

(2) $(\nabla \varphi_k, \nabla \varphi_j) = \delta_{kj}$, for all $k, j = 1, 2, \cdots$.

(3) Given $\varphi \in \mathcal{V}$, for any $\varepsilon > 0$ there exists a linear combination of $\varphi_k' s$, such that

$$\left\| \varphi - \sum_{k=1}^{m} c_k \varphi_k \right\|_{1,*} < \varepsilon.$$

Proof. Since V is separable, $\mathcal{V}$, as a subset of V, is also separable. Let $\{\varphi_k\}$ be a basis in V of functions from $\mathcal{V}$. We orthonormalise $\{\varphi_k\}$ in V by the Schmidt procedure, to obtain another denumerable set $\{\varphi_k\}$, whose linear hull is still dense in V. $\square$

Lemma 27. *Let $v \in (H^{1,*}(\Omega))^3$ and $\nabla \cdot v = 0$. Then*

$$(v \cdot \nabla u, u) = 0, \quad \forall u \in (C_0^\infty(\Omega))^3, \tag{1.85}$$

and

$$(v \cdot \nabla u, w) = -(v \cdot \nabla w, u), \quad \forall u, w \in (C_0^\infty(\Omega))^3. \tag{1.86}$$

Proof. Property (1.86) is a direct consequence of (1.85) when we replace the u in it with $u + w$. To prove (1.85) we have

$$\int_\Omega (v \cdot \nabla u) \cdot u \, dx = \frac{1}{2} \int_\Omega v \cdot \nabla |u|^2 \, dx = 0.$$

$\square$

Theorem 29. *The problem (1.84) admits at least one solution.*

Proof.   We use the Galerkin method to prove existence. Let $\{\varphi_k\}$ be the set given in Lemma 26, then we set

$$u_m = \sum_{k=1}^{m} c_k \varphi_k,$$

where c_k satisfy

$$(\nabla u_m, \nabla \varphi_k) + (u_m \cdot \nabla u_m, \varphi_k) = < f, \varphi_k >, \quad \forall k = 1, \cdots, m. \qquad (1.87)$$

To prove the existence of c_k, we make some estimates first. Multiplying (1.87) by c_k and summing up with respect to k, we obtain

$$(\nabla u_m, \nabla u_m) + (u_m \cdot \nabla u_m, u_m) = < f, u_m > .$$

By (1.85) the second term on the left hand side vanishes. The right hand side is bounded by $\|f\|_{V'}\|u_m\|_V$. Let M be a constant such that $M > \|f\|_{V'}$, then we get

$$(\nabla u_m, \nabla u_m) + (u_m \cdot \nabla u_m, u_m) > < f, u_m >, \quad \forall |u_m|_1 = M.$$

Owing to the Corollary 1 of the Brouwer theorem, the solution to (1.87) exists, and $|u_m|_1 \leq \|f\|_{V'}$.

Letting $m \to \infty$, we can extract a subsequence of u_m, converging weakly in V, and by Lemma 5 the subsequence converges strongly in $L^2_{loc}(\Omega)$. Let u be the limit. For the second term on the left hand side of (1.87) we have

$$(u \cdot \nabla u, \varphi_k) - (u_m \cdot \nabla u_m, \varphi_k) = (u \cdot \nabla(u - u_m), \varphi_k) + ((u - u_m) \cdot \nabla u_m, \varphi_k).$$

Noting that φ_k is compactly supported, we find the limit of the above expression is zero. Then we obtain

$$(\nabla u, \nabla \varphi_k) + (u \cdot \nabla u, \varphi_k) = < f, \varphi_k >, \quad \forall k = 1, \cdots, \infty.$$

Since the linear hull of $\{\varphi_k\}$ is dense in $\mathcal{V}$, the conclusion follows. $\square$

We turn now to the inhomogeneous boundary conditions,

$$\lim_{|x| \to \infty} u = u_\infty, \quad u|_{\partial\Omega} = g, \qquad (1.88)$$

and assume that $g \in \left(H^{1/2}(\partial\Omega)\right)^3$. To prove the existence of (1.80),(1.81), (1.88) we need the Sobolev inequality. Let us prove it for a special case, the inequality for three dimensional domains.

Lemma 28. *For all* $u \in C_0^\infty(\Omega)$

$$\|u\|_{0,\frac{3}{2}} \leq \frac{1}{2\sqrt{3}}\|\nabla u\|_{0,1}. \qquad (1.89)$$

Proof. We have

$$|u(x)| \leq \frac{1}{2} \int_{-\infty}^{\infty} \left| \frac{\partial u}{\partial x_1} \right| \, dx_1 \equiv \frac{1}{2} F_1(x_2, x_3)$$

and similar estimates for x_2 and x_3. With the obvious meaning of the symbols we then deduce

$$|2u(x)|^{\frac{3}{2}} \leq (F_1(x_2, x_3) F_2(x_1, x_3) F_3(x_1, x_2))^{\frac{1}{2}}.$$

Integrating over x_1, and using the Schwarz inequality,

$$\int_{-\infty}^{\infty} |2u(x)|^{\frac{3}{2}} \, dx_1$$

$$\leq (F_1(x_2, x_3))^{\frac{1}{2}} \left(\int_{-\infty}^{\infty} F_2(x_1, x_3) \, dx_1 \right)^{\frac{1}{2}} \left(\int_{-\infty}^{\infty} F_3(x_1, x_2) \, dx_1 \right)^{\frac{1}{2}}.$$

Integrating it successively over x_2 and x_3 and applying the same procedure, we find

$$2\|u\|_{0,\frac{3}{2}} \leq \left(\prod_{i=1}^{3} \int_{\mathbb{R}^3} \left| \frac{\partial u}{\partial x_i} \right| \, dx \right)^{\frac{1}{3}} \leq \frac{1}{3} \sum_{i=1}^{3} \int_{\mathbb{R}^3} \left| \frac{\partial u}{\partial x_i} \right| \, dx,$$

which, in turn, after employing the inequality

$$(a_1 + a_2 + a_3)^2 \leq 3(a_1^2 + a_2^2 + a_3^2)$$

gives (1.89). $\square$

Lemma 29. *For all $u \in C_0^{\infty}(\Omega)$*

$$\|u\|_{0,6} \leq \frac{2}{\sqrt{3}} |u|_1. \tag{1.90}$$

Proof. Replacing u with u^4 in (1.89) and then estimate the right hand side by the Hölder inequality. We have

$$\|u\|_{0,6}^4 \leq \frac{2}{\sqrt{3}} \|u\|_{0,6}^3 \|\nabla u\|_0.$$

Then (1.90) follows. $\square$

By the inverse trace theorem of Sobolev spaces there exists $u_1 \in (H^1(\Omega))^3$ with a compact support on $\bar{\Omega}$, such that $\|u_1\|_1 \leq \|g - u_\infty\|_{\frac{1}{2}}$, and $u_1 = g - u_\infty$ on $\partial\Omega$. Let R be large enough such that $B(O, R) \supset$ supp u_1, and let $\Omega_1 = \Omega \cap B(O, R)$. We apply Lemma 20 to get a function $u_2 \in (H_0^1(\Omega_1))^3$ such that $\nabla \cdot u_2 = -\nabla \cdot u_1$ and $\|u_2\|_1 \leq C\|\nabla \cdot u_1\|_0$. The Lemma 20 is about an exterior domain, however it can be verified that it is also applied to bounded domains. Let us extend the function u_2 to zero to the exterior of $B(O, R)$, then $u_2 \in \left(H_0^1(\Omega) \right)^3$. We define $u_0 = u_1 + u_2 + u_\infty$, then u_0 satisfies the boundary condition, and $\nabla \cdot u_0 = 0$, and we notice that the derivatives of u_0 are compactly supported.

Let $w = u - u_0$, then the weak formulation for (1.80),(1.81),(1.88) is: Find $u \in V + u_0$, such that (1.84) holds.

Theorem 30. *There is a positive constant $\delta > 0$, such that if $\|g - u_\infty\|_{\frac{1}{2}} < \delta$, then the weak solution to (1.80),(1.81),(1.88) exists.*

Proof. Being analogous to Theorem 29, we set the approximation to $w = u - u_0$ to be

$$w_m = \sum_{k=1}^{m} c_k \varphi_k,$$

then c_k satisfy

$$
\begin{aligned}
(\nabla w_m, \nabla \varphi_k) + (w_m \cdot \nabla w_m, \varphi_k) + (w_m \cdot \nabla u_0, \varphi_k) & \\
+ (u_0 \cdot \nabla w_m, \varphi_k) + (u_0 \cdot \nabla u_0, \varphi_k) & \qquad (1.91) \\
= -(\nabla u_0, \nabla \varphi_k) + <f, \varphi_k>, \quad \forall k = 1, \cdots, m. &
\end{aligned}
$$

Some terms in (1.91) are the same as those in (1.87). Let us estimate the other ones. By the Hölder inequality,

$$|(w_m \cdot \nabla u_0, w_m)| \leq \|w_m\|_{0,6}^2 \|\nabla u_0\|_{0,\frac{3}{2}}.$$

Since ∇u_0 is compactly supported we get

$$|(w_m \cdot \nabla u_0, w_m)| \leq C |w_m|_1^2 \|\nabla u_0\|_{0,\Omega_1},$$

where we have applied Lemma 29. Being the same

$$|(u_0 \cdot \nabla u_0, w_m)| \leq C \|u_0\|_{1,\Omega_1} \|\nabla u_0\|_0 |w_m|_1,$$

where we have applied the Sobolev embedding theorem to get $\|u_0\|_{0,6,\Omega_1} \leq C \|u_0\|_{1,\Omega_1}$. By (1.85) we have

$$(u_0 \cdot \nabla w_m, w_m) = 0.$$

Finally we have

$$|(\nabla u_0, \nabla w_m)| \leq |u_0|_{1,\Omega_1} |w_m|_1.$$

By the definition of u_0, there is a constant C_1, such that $C \|\nabla u_0\|_{0,\Omega_1} \leq C_1 \|g - u_\infty\|_{\frac{1}{2}}$. We take δ and M such that $\delta < 1/C_1$ and $(1 - \delta C_1)M > \|f\|_{V'} + C \|u_0\|_{1,\Omega_1} \|\nabla u_0\|_0 + |u_0|_{1,\Omega_1}$. By the assumption of the theorem, $\|g - u_\infty\|_{\frac{1}{2}} < \delta$, then if $|w_m|_1 = M$ the combination of the above inequalities can be summarized as

$$(\nabla w_m, \nabla w_m) + (w_m \cdot \nabla w_m, w_m) + (w_m \cdot \nabla u_0, w_m)$$
$$+(u_0 \cdot \nabla w_m, w_m)(u_0 \cdot \nabla u_0, w_m) > -(\nabla u_0, \nabla w_m) + < f, w_m > .$$

The remaining part of the proof is the same as that in Theorem 29 □

The condition for g can be reduced to $|\int_{\partial\Omega} g \cdot \nu \, ds| < \delta$, see [Galdi, G.P. (1994)]. Then for a particular case, $\int_{\partial\Omega} g \cdot \nu \, ds = 0$, there is no restriction on the size of data. Generally speaking the solutions are not unique. An example is given in [Temam, R. (1984)]. Some results for the two dimensional case can be found in [Galdi, G.P. (1994)].

1.9 Heat equation

To investigate evolution problems, some results of the theory of linear operators are needed. Let us introduce some definitions and some results first.

Let H be a Hilbert space and $A : D(A) \to H$ be a linear operator with domain $D(A) \subset H$. Let $\overline{D(A)} = H$. Then the adjoint operator A^* is defined by

$$D(A^*) = \{u \in H; \exists h \in H, (u, Av) = (h, v), \forall v \in D(A)\}, \quad A^*u = h.$$

A is symmetric if and only if $A \subset A^*$. A is self-adjoint if and only if $A = A^*$. Let A be a closed symmetric operator, then the Cayley transform of A is defined by

$$U_A = (A - iI)(A + iI)^{-1},$$

where I is the identity. A bounded linear operator $A : H \to H$ is isometric, if and only if A keeps the inner product invariant:

$$(Au, Av) = (u, v), \qquad \forall u, v \in H.$$

In particular if the range of A, $R(A) = H$, then A is called a unitary operator.

A family of orthogonal projection operators $E(\lambda)$, $-\infty < \lambda < \infty$, in a Hilbert space H is called a resolution of the identity if

$$E(\lambda)E(\mu) = E(\min(\lambda, \mu)),$$

$$E(-\infty) = 0, E(+\infty) = I,$$

where

$$E(-\infty)u = \underset{\lambda \to -\infty}{\text{s-}\lim}\, E(\lambda)u, \quad E(+\infty)u = \underset{\lambda \to +\infty}{\text{s-}\lim}\, E(\lambda)u,$$

$$E(\lambda + 0) = E(\lambda), \qquad \text{where } E(\lambda + 0)u = \underset{\mu \to \lambda+0}{\text{s-}\lim}\, E(\mu)u,$$

and s-lim is the strong limit.

Proposition 2. *For any $u, v \in H$, $(E(\lambda)u, v)$ is a function of bounded variation with respect to λ.*

Proposition 3. *If $f(\lambda)$ is a complex continuous function defined on $(-\infty, \infty)$, and $u \in H$, then for $-\infty < \alpha < \beta < \infty$ we can define*

$$\int_\alpha^\beta f(\lambda)\, dE(\lambda)u,$$

which is the strong limit of the Riemann sum

$$\sum_j f(\chi_j) E(\lambda_j, \lambda_{j+1}]u, \quad \alpha = \lambda_1 < \lambda_2 < \cdots < \lambda_n = \beta, \chi_j \in (\lambda_j, \lambda_{j+1}]$$

as $\max |\lambda_{j+1} - \lambda_j| \to 0$, where $E(\lambda_j, \lambda_{j+1}] = E(\lambda_{j+1}) - E(\lambda_j)$.

Theorem 31. *For a given $u \in H$ the following three conditions are equivalent:*

(1) $\int_{-\infty}^\infty f(\lambda)\, dE(\lambda)u$ exists,
(2) $\int_{-\infty}^\infty |f(\lambda)|^2\, d|E(\lambda)u|^2 < \infty$,
(3) $F(v) = \int_{-\infty}^\infty f(\lambda)\, d(E(\lambda)u, v)$ defines a bounded linear functional.

Theorem 32. *If $f(\lambda)$ is a real continuous function, then by*

$$(Au, v) = \int_{-\infty}^\infty f(\lambda)\, d(E(\lambda)u, v)$$

determines a self-adjoint operator A with $D(A) = D$, where

$$D = \left\{ u; \int_{-\infty}^\infty |f(\lambda)|^2\, d|E(\lambda)u|^2 < \infty \right\},$$

$u \in D$ *and v is arbitrary. Moreover, $AE(\lambda) \sqsupseteq E(\lambda)A$.*

In particular, if $f(\lambda) = \lambda$, we have

$$(Au, v) = \int_{-\infty}^\infty \lambda\, d(E(\lambda)u, v), \qquad \forall u \in D(A), v \in H,$$

which is denoted by

$$A = \int_{-\infty}^\infty \lambda\, dE(\lambda).$$

It is called the spectral resolution of the self-adjoint operator A.

Theorem 33. *A self-adjoint operator in a Hilbert space possesses a uniquely determined spectral resolution.*

Theorem 34. *For a symmetric operator A there exists a closed symmetric extension A^{**}. A closed symmetric operator A possesses a uniquely determined spectral resolution if and only if A is self-adjoint, and A is self-adjoint if and only if its Cayley transform is unitary.*

Let $A : D(A) \subset H \to H$ be a linear operator. The resolvent set of A is defined by
$$\rho(A) =$$
$$\{\lambda \in C; \overline{R(A - \lambda I)} = H, (A - \lambda I)\text{one} - \text{to} - \text{one}, (A - \lambda I)^{-1}\text{continuous}\}.$$
The spectrum of A is defined by $\sigma(A) = C \setminus \rho(A)$, which is further divided into three subsets. The point spectrum is
$$\sigma_p(A) = \{\lambda \in \sigma(A); A - \lambda I \text{ is not one} - \text{to} - \text{one}\}.$$
The continuous spectrum is
$$\sigma_c(A) = \{\lambda \in \sigma(A); \overline{R(A - \lambda I)} = H, (A - \lambda I)\text{one} - \text{to} - \text{one},$$
$$(A - \lambda I)^{-1}\text{is not continuous}\}.$$
The residual spectrum is
$$\sigma_r(A) = \{\lambda \in \sigma(A); \overline{R(A - \lambda I)} \neq H, (A - \lambda I)\text{one} - \text{to} - \text{one}\}.$$

Theorem 35. *Let A be a self-adjoint operator. Then $\lambda \in \sigma_p(A)$ if and only if $E(\lambda) - E(\lambda - 0) \neq 0$, $\sigma_r(A)$ is a null set, and the necessary and sufficient condition for $\lambda \in \sigma(A)$ is that $\forall \varepsilon > 0$, $E(I_*) \neq 0$, where $I_* = (\lambda - \varepsilon, \lambda + \varepsilon)$.*

Let us now investigate the initial boundary value problem of the heat equation. We remark that although the above definitions and results are all for complex numbers, we will assume the functions are real in the remaining part of this section, and it is easy to see that the following result is valid for both real and complex cases. The problem reads
$$\frac{\partial u}{\partial t} = \triangle u + f, \tag{1.92}$$
$$u = 0, \qquad x \in \partial\Omega, \tag{1.93}$$
$$u(x, 0) = u_0(x). \tag{1.94}$$
We assume that $\dim(\Omega) = 3$, but the results are true for two and one dimensional problems as well. We define $H = L^2(\Omega)$, and operator $A = -\triangle$ with domain
$$D(A) = \{u \in H_0^1(\Omega); \triangle u \in H\}.$$
Lemma 30. $A : D(A) \to H$ *is a self-adjoint operator .*

Proof. Clearly A is closed and symmetric. We consider the Cayley transform U_A. Let $v = (A + iI)^{-1}u$, then by Theorem 17 $u \to v$ and $u \to Av$ are bounded in H, therefore U_A is bounded in H. Besides we have $U_A^* U_A = I$, so U_A keeps inner product invariant. From $U_A^* = U_A^{-1}$ we get $R(U_A) = D(U_A^{-1}) = D(U_A^*) = H$, so U_A is a unitary operator. Then A is self-adjoint.   □

We assume that $u_0 \in L^2(\Omega)$, and $f \in L^2((0,T); L^2(\Omega))$ for $T > 0$. Let us write the problem (1.92)-(1.94) in the operator form: Find $u : (0,T] \to D(A)$, such that

$$\frac{du}{dt} + Au = f, \qquad \lim_{t \to +0} u(t) = u_0. \tag{1.95}$$

Theorem 36. *The problem (1.95) admits a unique solution $u(t)$, and $\frac{du}{dt} \in L^2((0,T), L^2(\Omega))$.*

Proof. Formally we get

$$u(t) = e^{-At}u_0 + \int_0^t e^{-At} e^{A\tau} f(\tau)\, d\tau.$$

By Theorem 17 we have $\sigma(A) \subset [0, \infty)$. Then in the notations of spectral resolution the solution to (1.95) can be expressed as

$$u = \int_0^\infty e^{-\lambda t}\, dE(\lambda)u_0 + \int_0^t \int_0^\infty e^{-\lambda(t-\tau)}\, dE(\lambda)\, f(\tau)\, d\tau.$$

Let us verify it is the desired solution. Differential of it gives

$$\frac{du}{dt} = -\int_0^\infty \lambda e^{-\lambda t}\, dE(\lambda)u_0 + f(t) - \int_0^t \int_0^\infty \lambda e^{-\lambda(t-\tau)}\, dE(\lambda)\, f(\tau)\, d\tau.$$

Here we notice that $\lambda e^{-\lambda t}$ is bounded for positive t, so the above expression makes sense. For the first term we see that

$$A \int_0^\infty e^{-\lambda t}\, dE(\lambda)u_0 = \int_0^\infty e^{-\lambda t}\, d\lambda \int_0^\infty \mu\, d_\mu E(\lambda)E(\mu)u_0$$

$$= \int_0^\infty e^{-\lambda t}\, d\lambda \int_0^\lambda \mu\, d_\mu E(\mu)u_0 = \int_0^\infty \lambda e^{-\lambda t}\, dE(\lambda)u_0.$$

The second term can be deduced in the same way. Thus u satisfies the equation. It is obvious that the initial condition is satisfied.

To prove uniqueness we consider a final boundary value problem

$$-\frac{dv}{dt} + Av = g, \qquad \lim_{t \to T-0} v(t) = 0.$$

Then by the above argument v exists. Let u be the solution to (1.95) associated with $f = 0$ and $u_0 = 0$. We have

$$\int_0^T (u(t), g(t))\, dt = \int_0^T (u, -\frac{dv}{dt} + Av)\, dt$$

$$= \int_0^T (\frac{du}{dt} + Au, v)\, dt - (v(T), u(T)) + (u(0), v(0)) = 0.$$

Since g is arbitrary, we obtain $u(t) = 0$. Thus the proof is complete.   □

1.10   Wave equation

We investigate the following initial boundary value problem of the wave equation (1.24) in this section. The initial and boundary conditions are

$$u = 0, \qquad x \in \partial\Omega, \tag{1.96}$$

$$u(x, 0) = u_0(x), \qquad \frac{\partial u}{\partial t}(x, 0) = u_1(x). \tag{1.97}$$

Let $u_0, u_1 \in L^2(\Omega) = H$. We are looking for a weak solution $u \in C([0, \infty); H)$, which satisfies

$$\int_0^\infty \int_\Omega u \left(\frac{\partial^2 v}{\partial t^2} - \Delta v \right) dx dt + \int_\Omega u_0(x) \frac{\partial v}{\partial t}(x, 0)\, dx$$
$$- \int_\Omega u_1(x) v(x, 0)\, dx = 0, \quad \forall v \in \mathcal{V}, \tag{1.98}$$

where

$$\mathcal{V} = C_0([0, \infty); D(A)) \cap C^2([0, \infty); H),$$

and the operator A is the same as that in the previous section.

Theorem 37. *The problem (1.98) admits a unique solution.*

Proof. The equation (1.24) can be expressed as

$$\frac{d^2 u}{dt^2} + Au = 0.$$

The solution to the problem can be formally expressed as

$$u(t) = \cos(A^{1/2} t) u_0 + A^{-1/2} \sin(A^{1/2} t) u_1.$$

Then in the notations of spectral resolution the solution to (1.98) can be expressed as

$$u(t) = \int_0^\infty \cos(\sqrt{\lambda}t)\, dE(\lambda)u_0 + \int_0^\infty \frac{\sin(\sqrt{\lambda}t)}{\sqrt{\lambda}}\, dE(\lambda)u_1. \qquad (1.99)$$

Let us verify that $u(t)$, defined by (1.99), is the desired weak solution. $u(t) \in H$ is well defined because

$$|\cos(\sqrt{\lambda})| \le 1, \qquad \left|\frac{\sin(\sqrt{\lambda}t)}{\sqrt{\lambda}}\right| \le t.$$

In order to prove u is continuous in t, we estimate

$$\|u(t) - u(t_0)\|^2$$
$$\le 2\left(\int_0^\infty |\cos(\sqrt{\lambda}t) - \cos(\sqrt{\lambda}t_0)|^2\, d\|E(\lambda)u_0\|^2\right.$$
$$\left. + \int_0^\infty \frac{1}{\lambda}|\sin(\sqrt{\lambda}t) - \sin(\sqrt{\lambda}t_0)|^2\, d\|E(\lambda)u_1\|^2\right)$$
$$\le 2\left(\lambda_0|t - t_0|^2 \int_0^{\lambda_0} d\|E(\lambda)u_0\|^2\right.$$
$$\left. + 4\int_{\lambda_0}^\infty d\|E(\lambda)u_0\|^2 + |t - t_0|^2 \int_0^{\lambda_0} d\|E(\lambda)u_1\|^2\right)$$
$$\le 2|t - t_0|^2(\lambda_0\|u_0\|^2 + \|u_1\|^2) + 8(\|u_0\|^2 - \|E(\lambda_0)u_0\|^2).$$

We first take λ_0 large enough, then take $|t - t_0|$ small. Then $\|u(t) - u(t_0)\|$ is arbitrary small, which gives $u \in C([0,\infty);H)$. To verify the weak formulation (1.98) we take $v \in \mathcal{V}$, then use Lebesgue's dominated convergence theorem to obtain

$$-\int_\Omega u_0(x)\frac{\partial v}{\partial t}(x,0)\, dx = -\int_0^\infty d(E(\lambda)u_0, \frac{\partial v}{\partial t}(\cdot,0))$$
$$= \int_0^\infty \frac{d}{dt}\int_0^\infty \cos(\sqrt{\lambda}t)\, d(E(\lambda)u_0, \frac{\partial v}{\partial t}(\cdot,t))\, dt$$
$$= \int_0^\infty \int_0^\infty \left(\cos(\sqrt{\lambda}t)\, d(E(\lambda)u_0, \frac{\partial^2 v}{\partial t^2}(\cdot,t))\right.$$
$$\left. - \sqrt{\lambda}\sin(\sqrt{\lambda}t)\, d(E(\lambda)u_0, \frac{\partial v}{\partial t}(\cdot,t))\right)\, dt,$$

and analogously

$$0 = \int_0^\infty \frac{d}{dt} \int_0^\infty \sqrt{\lambda} \sin(\sqrt{\lambda}t)\, d(E(\lambda)u_0, v(\cdot, t))\, dt$$

$$= \int_0^\infty \int_0^\infty (\lambda \cos(\sqrt{\lambda}t)\, d(E(\lambda)u_0, v(\cdot, t))$$

$$+\sqrt{\lambda}\sin(\sqrt{\lambda}t)\, d(E(\lambda)u_0, \frac{\partial v}{\partial t}(\cdot, t)))\, dt.$$

Thus we obtain

$$\int_0^\infty \int_0^\infty \cos(\sqrt{\lambda}t)\, d(E(\lambda)u_0, \frac{\partial^2 v}{\partial t^2}(\cdot, t) + Av(\cdot, t))\, dt = - \int_\Omega u_0(x)\frac{\partial v}{\partial t}(x, 0)\, dx.$$

Similarly we get

$$\int_0^\infty \int_0^\infty \frac{\sin(\sqrt{\lambda}t)}{\sqrt{\lambda}}\, d(E(\lambda)u_1, \frac{\partial^2 v}{\partial t^2}(\cdot, t) + Av(\cdot, t))\, dt = - \int_\Omega u_1(x)v(x, 0)\, dx.$$

The proof of existence is thus complete. To prove uniqueness we take an arbitrary $g \in C_0^\infty(\Omega \times (0, \infty))$ and solve the equation

$$\frac{d^2 v}{dt^2} - Av = g,$$

to get v satisfying the homogeneous boundary condition and vanishing for sufficiently large t. Let u be a solution to the problem (1.98) with homogeneous initial data, then by (1.98)

$$\int_0^\infty \int_\Omega u \left(\frac{\partial^2 v}{\partial t^2} - \Delta v \right)\, dx dt = \int_0^\infty \int_\Omega ug\, dx dt = 0.$$

Since C_0^∞ is dense in L^2, we obtain $u \equiv 0$, which implies uniqueness. To prove the existence of v, we write down the formal expression

$$v(t) = - \int_t^\infty A^{-1/2} \sin(A^{1/2}(t - s))g(s)\, ds.$$

We define a cut-off function $\zeta \in C^\infty(\mathbb{R})$, such that $\zeta \geq 0$, $\zeta(t) \equiv 1$ for $t > 0$, and $\zeta(t) \equiv 0$ for $t < \tau$, $\tau < 0$ fixed. Then we set

$$v(t) = -\zeta(t) \int_t^\infty \int_0^\infty \frac{\sin(\sqrt{\lambda}(t - s))}{\sqrt{\lambda}}\, d(E(\lambda)g)(\cdot, s)\, ds. \tag{1.100}$$

Let us verify that $v(t)$ expressed by (1.100) is the desired solution. Following the same lines we can see $v(t)$ is the weak solution, so the only problem is the regularity of v. For a bounded interval $I \subset [0, \infty)$, $AE(I)$ is a bounded operator and

$$AE(I)v(t) = -\zeta(t) \int_t^\infty \int_{\mu \in I} \mu \, dE(\mu) \int_0^\infty \frac{\sin(\sqrt{\lambda}(t-s))}{\sqrt{\lambda}} \, d(E(\lambda)g)(\cdot, s) \, ds.$$

Since $E(\mu)E(\lambda) = E(\min(\mu, \lambda))$, we have

$$dE(\mu)dE(\lambda) = \begin{cases} 0 \cdot dE(\lambda) = 0, \lambda < \mu, \\ dE(\mu), \qquad \lambda = \mu, \\ 0 \cdot dE(\mu) = 0, \lambda > \mu. \end{cases}$$

Therefore

$$AE(I)v(t) = -\zeta(t) \int_t^\infty \int_{\mu \in I} \frac{\sin(\sqrt{\mu}(t-s))}{\sqrt{\mu}} \, d(E(\mu)Ag)(\cdot, s) \, ds.$$

Letting $I \to [0, \infty)$, one gets

$$\|Av(t)\|_0 \leq \zeta(t) \int_t^\infty |t-s| \|Ag(\cdot, s)\|_0 \, ds < \infty,$$

so $v(t) \in D(A)$. Moreover

$$\|Av(t) - Av(t_0)\|_0$$

$$\leq \zeta(t_0)|t - t_0| \int_{t_0}^\infty \|Ag(\cdot, s)\|_0 \, ds + |\zeta(t) - \zeta(t_0)| \int_t^\infty |t-s| \|Ag(\cdot, s)\|_0 \, ds$$

$$+ \zeta(t_0) \left| \int_{t_0}^t |t-s| \|Ag(\cdot, s)\|_0 \, ds \right|,$$

which implies $v \in C_0([0, \infty); D(A))$. v is also second order continuously differentiable with respect to t, because

$$\frac{d^2}{dt^2} \int_t^\infty \int_0^\infty \frac{\sin(\sqrt{\lambda}(t-s))}{\sqrt{\lambda}} \, d(E(\lambda)g)(\cdot, s) \, ds$$

$$= -g(\cdot, t) - A \int_t^\infty \int_0^\infty \frac{\sin(\sqrt{\lambda}(t-s))}{\sqrt{\lambda}} \, d(E(\lambda)Ag)(\cdot, s) \, ds.$$

Therefore $v \in \mathcal{V}$. $\qquad \square$

1.11 Maxwell equations

We work with complex-valued functions in this section.

The Maxwell equations in vacuum are:

$$\mathrm{div}\,E = \frac{\rho}{\varepsilon}, \tag{1.101}$$

$$\mathrm{curl}E = -\frac{\partial B}{\partial t}, \tag{1.102}$$

$$\mathrm{div}\,B = 0, \tag{1.103}$$

$$\mathrm{curl}B = \mu\left(\varepsilon\frac{\partial E}{\partial t} + j\right), \tag{1.104}$$

where $E = (E_1, E_2, E_3)^T$ is the electric field, $B = (B_1, B_2, B_3)^T$ is the magnetic flux density, ε is the dielectric constant, μ is the permeability, ρ is the electric charge density, and j is the current density. The relation between ρ and j is

$$\frac{\partial \rho}{\partial t} + \mathrm{div}j = 0. \tag{1.105}$$

Let $\hat{H} = (H_1, H_2, H_3)^T$ be the magnetic field, then $B = \mu\hat{H}$. For given ρ and j there are six unknowns, E and B, in the system (1.101)-(1.104). It seems the system is over determined with eight equations. However (1.101) and (1.103) are not independent. If (1.103) is satisfied at time $t = 0$, then applying the operator div to the both sides of (1.102), we find that (1.103) is satisfied at all time t. Being the same, applying div to the both sides of (1.104) and noting (1.105) we can get

$$\frac{\partial}{\partial t}\left(\mathrm{div}\,E - \frac{\rho}{\varepsilon}\right) = 0.$$

Therefore if (1.101) is satisfied at the initial time, then it is satisfied at all times.

Let us consider the total reflection boundary condition, $E \times \nu|_{\partial\Omega} = 0$, $\frac{\partial}{\partial t}B \cdot \nu|_{\partial\Omega} = 0$, and the initial conditions $E|_{t=0} = E_0$, $\hat{H}|_{t=0} = H_0$. To investigate the initial boundary value problem, we introduce some Hilbert spaces first. Define

$$H(\mathrm{curl};\Omega) = \{v \in (L^2(\Omega))^3; \mathrm{curl}\,v \in (L^2(\Omega))^3\},$$

equipped with norm

$$\|v\|_{H(\mathrm{curl};\Omega)} = \{\|v\|_0^2 + \|\mathrm{curl}\,v\|_0^2\}^{1/2}.$$

Lemma 31. *The mapping $v \to v \times \nu|_{\partial\Omega}$, defined on $(C^\infty(\overline{\Omega}))^3$ can be extended by continuity to a linear continuous mapping from $H(\mathrm{curl};\Omega)$ onto $(H^{-1/2}(\partial\Omega))^3$.*

Then we define

$$H_0(\mathrm{curl};\Omega) = \{v \in H(\mathrm{curl};\Omega); v \times \nu|_{\partial\Omega} = 0\}.$$

We choose

$$H = (L^2(\Omega))^3 \times (L^2(\Omega))^3$$

with the weight matrix

$$M = \begin{pmatrix} \varepsilon I & 0 \\ 0 & \mu I \end{pmatrix}$$

such that $(u,v)_H = (u,Mv)$ and $\|u\|_H = (u,u)_H^{\frac{1}{2}}$. We define

$$u = \begin{pmatrix} E \\ \hat{H} \end{pmatrix}, \quad A = -iM^{-1}\begin{pmatrix} 0 & -\bigtriangledown \times \\ \bigtriangledown \times & 0 \end{pmatrix},$$

with domain

$$D(A) = H_0(\mathrm{curl};\Omega) \times H(\mathrm{curl};\Omega).$$

The initial boundary value problem can be written in the form of

$$\frac{du}{dt} + iAu = f, \qquad u(0) = u_0. \tag{1.106}$$

Lemma 32. $A : D(A) \to H$ *is a self-adjoint operator.*

Proof.   clearly A is closed and symmetric. By definition

$$D(A^*) = \{v \in H; \exists g \in H, (Au,v)_H = (u,g)_H, \forall u \in D(A)\}.$$

We denote $v = (v_1, v_2)^T$ and $g = (g_1, g_2)^T$. Letting $\hat{H} = 0$ in u, we have

$$-i\mu^{-1}(\bigtriangledown \times E, v_2) = (E, g_1), \quad \forall E \in H_0(\mathrm{curl};\Omega).$$

Therefore $g_1 = i\mu\mathrm{curl}\, v_2$ and the domain for v_2 is $H(\mathrm{curl};\Omega)$. Similarly letting $E = 0$, we have

$$i\mu^{-1}(\bigtriangledown \times \hat{H}, v_1) = (\hat{H}, g_2), \quad \forall \hat{H} \in H(\mathrm{curl};\Omega).$$

Therefore $g_2 = -i\varepsilon\mathrm{curl}\, v_1$ and the domain for v_1 is $H_0(\mathrm{curl};\Omega)$. Hence

$$D(A^*) = H_0(\mathrm{curl};\Omega) \times H(\mathrm{curl};\Omega) = D(A).$$

$\square$

We assume that $f \in C([0,\infty); H)$. The formulation of a weak solution to (1.106) reads: Find $u \in C([0,\infty); H)$ such that

$$\int_0^\infty \int_\Omega \left\{ u\left(-\frac{\overline{d\varphi}}{dt} + iA\overline{\varphi}\right) - f\overline{\varphi} \right\} dx dt - \int_\Omega u_0(x)\overline{\varphi(x,0)}\, dx = 0, \quad (1.107)$$

for all $\varphi \in C_0([0,\infty); D(A)) \cap C^1([0,\infty); H)$.

The weak formulation (1.107) is equivalent to the classical formulation (1.101)-(1.105) with the total reflection boundary condition and the initial condition. Just like the previous argument claiming that (1.101) and (1.103) are not independent, we take $\phi = \zeta(t)(\nabla\varphi_1, \nabla\varphi_2)^T$, where $\varphi_1 \in C_0^\infty(\Omega)$, $\varphi_2 \in C^\infty(\overline{\Omega})$, and $\zeta \in C_0^\infty([0,\infty))$. Then by the definition of A we find $iuA\overline{\varphi} = 0$, and consequently

$$\int_0^\infty \int_\Omega \left\{ u\left(-\frac{\overline{d\phi}}{dt}\right) - f\overline{\phi} \right\} dx dt - \int_\Omega u_0(x)\overline{\phi(x,0)}\, dx = 0.$$

Since φ_1, φ_2, and ζ are arbitrary, the equations (1.101), (1.103) and the boundary condition for B are satisfied. We turn to prove the existence and uniqueness of a weak solution.

Theorem 38. *The problem (1.107) admits a unique solution $u \in C^1([0,\infty); H)$, and the estimate*

$$\|u(t)\|_H \leq \|u_0\|_H + \int_0^t \|f(\tau)\|_H\, d\tau \quad (1.108)$$

holds.

Proof. Formally we get

$$u(t) = e^{-iAt}u_0 + \int_0^t e^{-iA(t-\tau)} f(\tau)\, d\tau.$$

In the notations of spectral resolution the solution can be expressed as

$$u = \int_{-\infty}^\infty e^{-i\lambda t}\, dE(\lambda)u_0 + \int_0^t \int_{-\infty}^\infty e^{-i\lambda(t-\tau)}\, dE(\lambda)\, f(\tau)\, d\tau.$$

Let us verify that it is the desired solution. We have

$$
\begin{aligned}
(u_0, \varphi(0))_H &= \int_{-\infty}^{\infty} d(E(\lambda)u_0, \varphi(0))_H \\
&= -\int_0^{\infty} \frac{d}{dt} \int_{-\infty}^{\infty} e^{-i\lambda t} \, d(E(\lambda)u_0, \varphi(t))_H \, dt \\
&= \int_0^{\infty} \int_{-\infty}^{\infty} i\lambda e^{-i\lambda t} \, d(E(\lambda)u_0, \varphi(t))_H \, dt \\
&\quad - \int_0^{\infty} \int_{-\infty}^{\infty} e^{-i\lambda t} \, d(E(\lambda)u_0, \frac{d\varphi(t)}{dt})_H \, dt \\
&= \int_0^{\infty} \int_{-\infty}^{\infty} i e^{-i\lambda t} \, d(E(\lambda)u_0, A\varphi(t))_H \, dt \\
&\quad - \int_0^{\infty} \int_{-\infty}^{\infty} e^{-i\lambda t} \, d(E(\lambda)u_0, \frac{d\varphi(t)}{dt})_H \, dt.
\end{aligned}
$$

Similarly we have

$$
\begin{aligned}
(f(\tau), \varphi(\tau))_H &= \int_{\tau}^{\infty} \int_{-\infty}^{\infty} i e^{-i\lambda(t-\tau)} \, d(E(\lambda)f(\tau), A\varphi(t))_H \, dt \\
&\quad - \int_{\tau}^{\infty} \int_{-\infty}^{\infty} e^{-i\lambda(t-\tau)} \, d(E(\lambda)f(\tau), \frac{d\varphi(t)}{dt})_H \, dt.
\end{aligned}
$$

Then (1.107) follows by noting $\int_0^{\infty} d\tau \int_{\tau}^{\infty} dt = \int_0^{\infty} dt \int_0^t d\tau$. It is easy to see that u is continuous with respect to t. Then by (1.107) $\frac{du}{dt}$ exists in the sense of distributions and (1.106) holds, therefore u is continuously differentiable. Taking the inner product of (1.106) with u we get

$$
\frac{1}{2} \frac{d\|u\|_H^2}{dt} + i(Au, u)_H = (f, u)_H.
$$

For the real part we have

$$
\frac{1}{2} \frac{d\|u\|_H^2}{dt} = \mathrm{Re}(f, u)_H \leq \|f\|_H \|u\|_H.
$$

Then (1.108) follows, which implies uniqueness. $\square$

Up to now we have investigated three kinds of evolution problems. Applying some similar argument, the results for some other problems, for example the evolution elasticity problem, can also be obtained.

1.12 Darwin model

The Darwin model is a simplified model [Hewett, D.W. and Nielson, C. (1978)][Li, T.-t. and Qin, T. (1997)] for the Maxwell equations (1.101)-(1.105). It was shown in [Degond, P. and Raviart, P.A. (1992)] that the Darwin model approximates the Maxwell equations up to the second order for B and to the third order for E, provided $\eta = \bar{v}/c$ is small, where $\bar{v}$ is a characteristic velocity, and c is the light velocity. Let us introduce the Darwin model for bounded domains first. For a domain Ω_0 let Γ_i, $0 \leq i \leq m$, be the connected components of the boundary $\partial\Omega_0$, and Γ_0 be the outer boundary. Note that the electric field E can be written as the sum of a transverse component E_T and a longitudinal component E_L, such that $E = E_T + E_L$, and

$$\nabla \cdot E_T = 0, \qquad \nabla \times E_L = 0.$$

The equation (1.104) becomes

$$\nabla \times B = \mu \left(\varepsilon \frac{\partial E_L}{\partial t} + \varepsilon \frac{\partial E_T}{\partial t} + j \right).$$

In order to get the Darwin model, we neglect the term for E_T and get

$$\mathrm{curl} B = \mu \left(\varepsilon \frac{\partial E_L}{\partial t} + j \right). \tag{1.109}$$

Then the system (1.101)-(1.103),(1.109) with the total reflection boundary condition and the initial condition will be reduced to
(a) $E_L = -\nabla\phi$, and ϕ satisfies

$$-\triangle \phi = \frac{\rho}{\varepsilon},$$

$$\phi|_{\Gamma_i} = \alpha_i, \quad 0 \leq i \leq m.$$

$\alpha = \{\alpha_i\}$ satisfies

$$\mathcal{C} \frac{d\alpha}{dt} = \frac{1}{\varepsilon} \int_\Omega j \cdot \nabla\chi \, dx,$$

$$\alpha(0) = \alpha_0,$$

where α_0 depends on E_0, $\mathcal{C} = \{c_{ij}\}$ is the capacitance matrix. Here $c_{ij} = \int_{\Gamma_j} \frac{\partial \chi_i}{\partial \nu}\, ds$, and $\chi = \{\chi_i\}$ is the solution of

$$\triangle \chi_i = 0,$$

$$\chi_i|_{\Gamma_j} = \delta_{ij}.$$

(b) B satisfies

$$-\triangle B = \mu \nabla \times j, \tag{1.110}$$

$$\nabla \cdot B = 0, \tag{1.111}$$

$$B \cdot \nu|_{\partial \Omega_0} = B_0 \cdot \nu|_{\partial \Omega_0}, \tag{1.112}$$

$$(\nabla \times B) \times \nu|_{\partial \Omega_0} = \mu j \times \nu|_{\partial \Omega_0}, \tag{1.113}$$

where $B_0 = \mu H_0$.

(c) E_T satisfies

$$\triangle E_T = \frac{\partial}{\partial t} \nabla \times B, \tag{1.114}$$

$$\nabla \cdot E_T = 0, \tag{1.115}$$

$$E_T \times \nu|_{\partial \Omega_0} = 0, \tag{1.116}$$

$$\int_{\Gamma_i} E_T \cdot \nu\, ds = 0, \quad 1 \le i \le m. \tag{1.117}$$

Now we turn to the exterior domain Ω and assume that $\dim(\Omega) = 2$. We investigate problem (c) as an example, the method to problem (b) is similar. In order to figure out the suitable functional setting for the problem, we introducing some spaces. The space

$$H(\text{curl}, \text{div}; \Omega_0) = \{u \in (L^2(\Omega_0))^2; \nabla \cdot u, \nabla \times u \in L^2(\Omega_0)\},$$

provided with the norm

$$\|u\|_{0,\text{curl},\text{div}} = \left(\|u\|_0^2 + \|\nabla \cdot u\|_0^2 + \|\nabla \times u\|_0^2\right)^{\frac{1}{2}},$$

is a Hilbert space. We take a cut-off function $\zeta \in C^\infty(\Omega)$, which satisfies: $\zeta \equiv 1$ near the boundary $\partial\Omega$, $\zeta \equiv 0$ near the infinity and $0 \leq \zeta \leq 1$. Let $\mathcal{D}'$ be the space of distributions. Define

$$H_{0c}(\Omega) = \{u \in \mathcal{D}'; \|\zeta u\|_{0,\mathrm{curl,div}} < \infty, (1-\zeta)u \in (H_0^{1,*}(\Omega))^2, u \times \nu|_{\partial\Omega} = 0\},$$

and

$$\|u\|_* = \{\|\nabla \times u\|_0^2 + \|\nabla \cdot u\|_0^2 + < u \cdot \nu, 1 >_{\partial\Omega}^2\}^{\frac{1}{2}}.$$

However $\|\cdot\|_*$ does not provide a norm in $H_{0c}(\Omega)$, and actually we have the following result:

Lemma 33. *Let* $V_0 = \{u \in H_{0c}(\Omega); \|u\|_* = 0\}$, *then*

$$V_0 = \{u = (u_1, u_2); u_1 = \frac{\partial\phi}{\partial x_2}, u_2 = -\frac{\partial\phi}{\partial x_1},$$
$$\phi = a(x_2 + f(x)) + b(x_1 + g(x)), (a, b) \in \mathbb{R}^2\},$$

where $f, g \in H^{1,*}(\Omega)$ *satisfy*

$$-\triangle \psi = 0, \quad x \in \Omega,$$

$$\frac{\partial\psi}{\partial\nu}\bigg|_{\partial\Omega} = -\frac{\partial\chi}{\partial\nu},$$

where $\chi(x_1, x_2) = x_2, x_1$ *separately.*

Proof. We take any $u \in V_0$, then it will satisfy

$$\nabla \cdot u = 0, \tag{1.118}$$

$$\nabla \times u = 0, \tag{1.119}$$

$$u \times \nu|_{\partial\Omega} = 0, \tag{1.120}$$

$$\int_{\partial\Omega} u \cdot \nu \, ds = 0. \tag{1.121}$$

From (1.119)-(1.121), $\int_\Gamma -u_2 \, dx_1 + u_1 \, dx_2 = 0$ is true for any closed curve Γ in Ω. Thus we can define

$$\phi(x) = \int_{x_0}^x (-u_2 \, dx_1 + u_1 \, dx_2),$$

where x_0 can be any but fixed point in Ω. Notice now $u_1 = \frac{\partial\phi}{\partial x_2}, u_2 = -\frac{\partial\phi}{\partial x_1}$.

From (1.119),(1.120), we can get

$$- \triangle \phi = \nabla \times \left(\frac{\partial \phi}{\partial x_2}, - \frac{\partial \phi}{\partial x_1} \right) = \nabla \times u = 0,$$

and

$$\frac{\partial \phi}{\partial \nu} \bigg|_{\partial \Omega} = \left(\frac{\partial \phi}{\partial x_2}, - \frac{\partial \phi}{\partial x_1} \right) \times \nu |_{\partial \Omega} = u \times \nu |_{\partial \Omega} = 0.$$

Thus starting from every $u \in V_0$, we will end up with a problem: find a function ϕ, satisfying

$$\begin{aligned} - \triangle \phi &= 0, \qquad x \in \Omega \\ \frac{\partial \phi}{\partial \nu} \bigg|_{\partial \Omega} &= 0. \end{aligned} \qquad (1.122)$$

Besides, u is a bounded harmonic function, which can be developed in a neighborhood of the infinity as follows:

$$u_1 - iu_2 = \sum_{k=0}^{\infty} \frac{z_{0k}}{z^k},$$

where $z_{0k} = a_k + ib_k$, $(a_k, b_k) \in \mathbb{R}^2$, $z = x + iy = re^{i\theta}$. That is

$$(u_1, u_2) = \left(\sum_{k=0}^{\infty} \frac{a_k \cos k\theta + b_k \sin k\theta}{r^k}, \sum_{k=0}^{\infty} \frac{a_k \sin k\theta - b_k \cos k\theta}{r^k} \right).$$

From (1.118),(1.121) we know $a_1 = 0$, so

$$u_1 - iu_2 = \left(a_0 + \frac{b_1 \sin k\theta}{r}, -b_0 - \frac{b_1 \cos k\theta}{r} \right) + \sum_{k=2}^{\infty} \frac{z_{0k}}{z^k}.$$

Hence the asymptotic expansion of $\phi(x)$ near the infinity will be

$$\phi(x) \to a_0(x_2 - x_{20}) + b_0(x_1 - x_{10}) + b_1(\ln r - \ln r_0) + O(\frac{1}{r}) \cong h(x) + c + O(\frac{1}{r}),$$

where $h(x) = a_0 x_2 + b_0 x_1 + b_1 \ln r$.

Introduce a new function ψ, such that $\phi = \psi + h$. Then ψ will satisfy

$$\begin{aligned} - \triangle \psi &= 0, \qquad x \in \Omega \\ \frac{\partial \psi}{\partial \nu} \bigg|_{\partial \Omega} &= - \frac{\partial h}{\partial \nu}. \end{aligned} \qquad (1.123)$$

Notice that it suffices to consider the solution in $H^{1,*}(\Omega)/\mathbb{R}$ now.

Knowing that the well-posedness of the problem (1.123) is equivalent to

$$\int_{\partial\Omega} \frac{\partial h}{\partial \nu}\, ds = 0,$$

so let $\Omega_r = \Omega \cap B(O, r)$, then by the Green's formula we get

$$\int_{\partial\Omega} \frac{\partial h}{\partial \nu}\, ds = \int_{\Omega_r} \triangle h\, dx - \int_{\partial D(r)} \frac{\partial h}{\partial \nu}\, ds = -2\pi b_1.$$

Therefore $b_1 = 0$ is sufficient and necessary in order that (1.123) admits a unique solution in $H^{1,*}(\Omega)/\mathbb{R}$.

Once $b_1 = 0$, there exist $f, g \in H^{1,*}(\Omega)/\mathbb{R}$, such that

$$\psi = a_0 f(x) - b_0 g(x),$$

where f, g are the functions to (1.123) with $h(x) = x_2, x_1$ respectively. Consequently

$$\phi(x) = a_0(x_2 + f(x)) + b_0(x_1 + g(x)).$$

On the other hand it is easy to see that the function u given in the Lemma for all a, b are in V_0. $\qquad\square$

Let $V = \{v \in H_{0c}(\Omega); < v \cdot \nu, 1 >_{\partial\Omega} = 0\}$, and let the closure of the quotient space V/V_0 with respect to the norm $\|\cdot\|_*$ be W. Besides, let $\mathcal{Q} = \{p \in L^2(\Omega); \operatorname{supp} p \subset\subset \Omega\}$, equipped with norm

$$\|p\|_\sharp = \left(\|p\|_0^2 + |\int_\Omega p\, dx|^2 \right)^{\frac{1}{2}}.$$

Then we take closure to obtain a Hilbert space Q. If $p \in Q$ and $\lim p_n = p$, $p_n \in \mathcal{Q}$, then we define $\int_\Omega p\, dx = \lim \int_\Omega p_n\, dx$. Notice this can be seen as a generalized integral. The subspace of Q, $\{p \in Q; \int_\Omega p\, dx = 0\}$ is denoted by Q_0.

We define bilinear forms on $W \times W$ and $W \times Q_0$,

$$d(u, v) = \int_\Omega (\nabla \times u) \cdot (\nabla \times v)\, dx + \int_\Omega (\nabla \cdot u)(\nabla \cdot v)\, dx$$

$$+ < u \cdot \nu, 1 >_{\partial\Omega} < v \cdot \nu, 1 >_{\partial\Omega}, \quad u, v \in W,$$

$$b(u, p) = \int_\Omega (\nabla \cdot u) p\, dx, \quad u \in W, p \in Q_0.$$

Now the variational formulation for the problem will be: Find $u \in W$, $p \in Q_0$, such that

$$d(u, v) + b(v, p) = \int_\Omega B \cdot (\nabla \times v)\, dx, \qquad \forall v \in W, \qquad (1.124)$$

$$b(u, q) = 0, \qquad \forall q \in Q_0, \qquad (1.125)$$

In order to prove that there is a unique solution to the problem (1.124),(1.125), we introduce one result on the inf-sup condition, which appears in Section 1.7, and will be used later on in some different problems. [Girault, V. and Raviart, P.A. (1988)]

Let W, Q be two Hilbert spaces, and $b(u, p)$ be a bounded linear form defined on $W \times Q$. Define

$$Z = \{w \in W; b(w, q) = 0, \forall q \in Q\},$$

$$Z^0 = \{f \in W'; < f, w >= 0, \forall w \in Z\},$$

$$Z^\perp = \{w \in W; (w, v)_W = 0, \forall v \in Z\},$$

where $(\cdot, \cdot)_W$ is the inner product in W. Define two operators $B : W \to Q'$ and $B' : Q \to W'$, satisfying

$$< Bw, q >= b(w, q), \qquad \forall w \in W, q \in Q,$$

$$< B'q, w >= b(w, q), \qquad \forall w \in W, q \in Q,$$

then ker $B = Z$.

Theorem 39. *The three following properties are equivalent:*
(a) there exists a constant $\beta > 0$, such that

$$\inf_{\substack{q \in Q \\ q \neq 0}} \sup_{\substack{w \in W \\ w \neq 0}} \frac{b(w, q)}{\|w\|_W \|q\|_Q} \geq \beta; \qquad (1.126)$$

(b) the operator B' is an isomorphism from Q onto Z^0 and

$$\|B'q\|_{W'} \geq \beta \|q\|_Q, \qquad \forall q \in Q; \qquad (1.127)$$

(c) the operator B is an isomorphism from $Z^\perp$ onto Q' and

$$\|Bw\|_{Q'} \geq \beta \|w\|_W, \qquad \forall w \in W. \qquad (1.128)$$

Proof. (i) Let us show that (a)$\Longleftrightarrow$(b). The statement (a) is equivalent to

$$\sup_{\substack{w \in W \\ w \neq 0}} \frac{<B'q, w>}{\|w\|_W} \geq \beta\|q\|_Q, \qquad q \in Q,$$

that is, (1.126) is equivalent to (1.127). It remains to prove that B' is an isomorphism. (1.127) implies that B' is a one-to-one operator from Q onto its range $R(B')$. Moreover, it also implies that the inverse of B' is continuous. Hence B' is an isomorphism from Q onto $R(B')$. Therefore $R(B')$ is a closed subspace of W', since B' is an isomorphism. We can apply the closed range theorem of Banach spaces (see [Yosida, K. (1974)]) which says that $R(B') = (\ker(B))^0 = Z^0$.

(ii) (b)$\Longleftrightarrow$(c).

We observe that Z^0 can be identified with $(Z^\perp)'$. Indeed, for $w \in W$ let $w^\perp$ denote the orthogonal projection of w on $Z^\perp$. Then with each $f \in (Z^\perp)'$ we associate the element $\tilde{f}$ of W' defined by

$$<\tilde{f}, w>=<f, w^\perp>, \qquad \forall w \in W.$$

Obviously $\tilde{f} \in Z^0$ and it is easy to check that the correspondence $f \to \tilde{f}$ maps isometrically $(Z^\perp)'$ onto Z^0. This permits to identity $(Z^\perp)'$ and Z^0. As a consequence, statements (b) and (c) are equivalent. $\square$

Besides, we need an auxiliary lemma.

Lemma 34. *For a given $q \in Q_0$ there is a $(v_q + V_0) \in W$ such that*

$$\nabla \cdot v_q = q, \quad \nabla \times v_q = 0, \quad v_q \times \nu|_{\partial\Omega} = 0,$$

consequently

$$\|v_q\|_*^2 = \|q\|_0^2 = \|q\|_\sharp^2.$$

Proof. For a given $\varepsilon > 0$ we have $q_\varepsilon \in \mathcal{Q}$ so that $\|q - q_\varepsilon\|_\sharp < \varepsilon$. Then consider the following problem: Find $\phi \in H_0^{1,*}(\Omega)$, such that

$$-(\nabla\phi, \nabla\psi) = (q_\varepsilon, \psi), \qquad \forall\psi \in H_0^{1,*}(\Omega).$$

By Lax-Milgram theorem there is a unique solution, and $\phi \in H^{1,*}(\Omega) \cap H_{loc}^2(\Omega)$.

Take a cut-off function $\zeta \in C_0^\infty(\mathbb{R}^2)$ such that $\zeta \equiv 1$ for $|x| < 1$, $\zeta \equiv 0$ for $|x| > 2$, and $0 \leq \zeta \leq 1$. Define $\zeta_a(x) = \zeta(x/a)$ and $\phi_a = \phi\zeta_a$. Let $q_{\varepsilon a} =$

$\Delta\phi_a$, then $q_{\varepsilon a} \in Q$. When a is large enough, we have $\zeta_a \Delta\phi = \Delta\phi = q_\varepsilon$, therefore for sufficiently large a it holds that

$$
\begin{aligned}
\|q_{\varepsilon a} - q_\varepsilon\|_\sharp &= \|\Delta\phi_a - \Delta\phi\|_\sharp \\
&= \|\Delta\zeta_a\phi + 2\nabla\zeta_a \cdot \nabla\phi + \zeta_a\Delta\phi - \Delta\phi\|_\sharp \\
&= \|\Delta\zeta_a\phi + 2\nabla\zeta_a \cdot \nabla\phi\|_\sharp \\
&= \left(\|\Delta\zeta_a\phi + 2\nabla\zeta_a \cdot \nabla\phi\|_0^2 + |\int_\Omega (\Delta\zeta_a\phi + 2\nabla\zeta_a \cdot \nabla\phi)\, dx|^2 \right)^{\frac{1}{2}}.
\end{aligned}
$$

$$(1.129)$$

Let us estimate the right hand side of (1.129). Since $|x|^2 \ln^2 |x| \le |x|^3 \le 8a^3$ in the domain $D = \{x; a \le |x| \le 2a\}$, we have

$$
\begin{aligned}
\|\Delta\zeta_a\phi\|_0^2 &\le \frac{1}{a^4}\|\Delta\zeta\|_{0,\infty}^2 \int_D |\phi(x)|^2\, dx \le \frac{8}{a}\|\Delta\zeta\|_{0,\infty}^2 \int_D \frac{|\phi(x)|^2}{|x|^2 \ln^2 |x|}\, dx \\
&\le \frac{8C}{a}\|\Delta\zeta\|_{0,\infty}^2 \|\nabla\phi\|_{0,\Omega}^2 \to 0, \quad (a \to \infty).
\end{aligned}
$$

Analogously we have

$$
\|2\nabla\zeta_a \cdot \nabla\phi\|_0^2 \le \frac{4}{a^2}\|\nabla\zeta\|_{0,\infty}^2 \|\nabla\phi\|_{0,\Omega}^2 \to 0.
$$

For the second term of (1.129) we notice that

$$
\int_\Omega \Delta\zeta_a\phi\, dx = -\int_\Omega \nabla\zeta_a\nabla\phi\, dx,
$$

therefore

$$
\begin{aligned}
&|\int_\Omega (\Delta\zeta_a\phi + 2\nabla\zeta_a \cdot \nabla\phi)\, dx| \\
&\le \int_D |\nabla\zeta_a \cdot \nabla\phi|\, dx \le \frac{1}{2}\|\nabla\zeta\|_{0,\Omega}\|\nabla\phi\|_{0,D} \le C\|\nabla\phi\|_{0,D} \to 0.
\end{aligned}
$$

To conclude there is a a_0 such that $\|q_{\varepsilon a} - q_\varepsilon\|_\sharp < \varepsilon$ for $a \ge a_0$. We fix such an a_ε.

Let $v_{q\varepsilon} = \nabla\phi_{a_\varepsilon}$ then $v_{q\varepsilon} \in H_{0c}(\Omega)$, and $\nabla \cdot v_{q\varepsilon} = q_{\varepsilon a_\varepsilon}$, $\nabla \times v_{q\varepsilon} = 0$, $< v_{q\varepsilon} \cdot \nu, 1 >= \int_\Omega q_{\varepsilon a_\varepsilon}\, dx$. Consequently

$$
\begin{aligned}
&\left(\|\nabla \cdot v_{q\varepsilon} - q\|_0^2 + | < v_{q\varepsilon} \cdot \nu, 1 >_{\partial\Omega} - \int_\Omega q\, dx|^2 \right)^{\frac{1}{2}} \\
&= \|q_{\varepsilon a_\varepsilon} - q\|_\sharp \le \|q_\varepsilon - q\|_\sharp + \|q_{\varepsilon a_\varepsilon} - q_\varepsilon\|_\sharp < 2\varepsilon.
\end{aligned}
$$

Therefore $v_{q\varepsilon}$ converges as $\varepsilon \to 0$. Let v_q be the limit, then $v_q + V_0$ is the element in W we are looking for. $\qquad\square$

Now we are ready to prove the well-posedness.

Theorem 40. *The problem (1.124),(1.125) admits a unique solution (u, p), and $p = 0$.*

Proof. Because

$$d(u, u) = \int_\Omega |\nabla \times u|^2 \, dx + \int_\Omega |\nabla \cdot u|^2 \, dx = \|u\|_*^2,$$

the bilinear form $d(\cdot, \cdot)$ is coercive. By Lemma 34

$$\sup_{\substack{v \in W \\ v \neq 0}} \frac{b(v, q)}{\|v\|_*} \geq \frac{b(v_q, q)}{\|v_q\|_*} = \frac{\|q\|_0^2}{\|q\|_\sharp} = \|q\|_\sharp. \tag{1.130}$$

Therefore the inf-sup condition with $\beta = 1$ holds.

We define $W_0 = \{u \in W; b(p, u) = 0, \forall p \in Q_0\}$, then being analogous to Theorem 26, the problem: find $u \in W_0$, such that

$$d(u, v) = \int_\Omega B \cdot (\nabla \times v) \, dx, \qquad \forall v \in W_0, \tag{1.131}$$

admits a unique solution.

To prove the existence and uniqueness of p, we note that the equation (1.124) defines a linear operator

$$F(v) = -d(u, v) + \int_\Omega B \cdot (\nabla \times v) \, dx$$

on W, and by the equation (1.131) it is in the space Z^0. Then by Theorem 39, p is determined by the inverse of B' uniquely.

Finally let us show $p = 0$. According to Lemma 34 there is v_p corresponding to the solution p. We take $v = v_p$ in (1.124), then we have

$$d(u, v_p) = \int_\Omega (\nabla \cdot u)(\nabla \cdot v_p) \, dx = \int_\Omega (\nabla \cdot u) p \, dx = b(u, p) = 0,$$

$$\int_\Omega B \cdot (\nabla \times v_p) \, dx = 0,$$

and

$$b(v_p, p) = (p, p).$$

Equation (1.124) implies $(p, p) = 0$, which gives $p = 0$. $\square$

Remark 1.

The solutions are not unique. They may differ from a function in V_0. It seems the most natural way to make the solution unique is to impose a boundary condition at the infinity, $u|_{|x|=\infty} = 0$, because a function in V_0 satisfying this condition is zero. Unfortunately we still can not prove the existence if this boundary condition is imposed.

Chapter 2

Boundary Element Method

The boundary element method is based on a reduction of a boundary value problem on a domain to an equivalent problem defined on the boundary. The dimension of the problem is thus deducted by one, and it does not care very much if the underlying domain is interior or exterior, so it is suitable to deal with exterior problems. The potential theory is classical. We will derive the boundary equations by using the potential theory. On the other hand we will introduce the approach by Feng and his students, using the theory of singular integrations. The advantage of this approach is that the energy norm of it is exactly the same as that of the original boundary value problem. For inhomogeneous equations a Newton potential is usually applied to reduce it to homogeneous ones, and a numerical scheme of integration can be applied to the Newton potential. Therefore we consider homogeneous equations only in this chapter.

2.1   Some typical domains

For some typical domains the solutions to exterior problems can be expressed explicitly. We work with complex-valued functions in this section.

2.1.1   *Harmonic equation*

Let us begin with the exterior domain Ω of the unit disk $B(O, 1)$ and the Dirichlet boundary value problem of the harmonic equation.

In polar coordinates the problem is

$$\frac{1}{r}\frac{\partial}{\partial r}\left(r\frac{\partial u}{\partial r}\right) + \frac{1}{r^2}\frac{\partial^2 u}{\partial \theta^2} = 0,$$

$$u|_{r=1} = g(\theta).$$

The solution can be expressed in terms of Fourier series,

$$u = \sum_{n=-\infty}^{\infty} \rho_n(r)e^{in\theta}.$$

Plug it into the equation and obtain

$$\sum_{n=-\infty}^{\infty} \left\{ \frac{1}{r}\frac{d}{dr}\left(r\frac{d\rho_n}{dr}\right) - \frac{n^2\rho_n}{r^2} \right\} e^{in\theta} = 0.$$

Therefore each term vanishes:

$$\frac{1}{r}\frac{d}{dr}\left(r\frac{d\rho_n}{dr}\right) - \frac{n^2\rho_n}{r^2} = 0.$$

The general solutions to these ordinary differential equations are:

$$\rho_0 = a_0 + b_0 \ln r, \quad \rho_n = a_n r^{-n} + b_n r^n, \ \forall n \neq 0.$$

Since $u \in H^{1,*}(\Omega)$, we get the solution

$$u = \sum_{n=-\infty}^{\infty} a_n r^{-|n|} e^{in\theta}.$$

By the boundary condition we can determine a_n:

$$a_n = \frac{1}{2\pi}\int_0^{2\pi} e^{-in\varphi}g(\varphi)\,d\varphi.$$

Therefore the solution is

$$u = \frac{1}{2\pi}\int_0^{2\pi}\left\{\sum_{n=-\infty}^{\infty} r^{-|n|}e^{in(\theta-\varphi)}\right\}g(\varphi)\,d\varphi. \tag{2.1}$$

The power series in (2.1) are equal to:

$$\sum_{n=0}^{\infty} r^{-n}e^{in(\theta-\varphi)} = \frac{1}{1-r^{-1}e^{i(\theta-\varphi)}}, \quad \sum_{n=-\infty}^{-1} r^{-|n|}e^{in(\theta-\varphi)} = \frac{r^{-1}e^{-i(\theta-\varphi)}}{1-r^{-1}e^{-i(\theta-\varphi)}}.$$

Finally we get

$$u = \frac{1}{2\pi}\int_0^{2\pi} \frac{r^2-1}{r^2+1-2r\cos(\theta-\varphi)}g(\varphi)\,d\varphi. \tag{2.2}$$

This is the Poisson formula for harmonic equations. If we replace the boundary condition by

$$u|_{r=R} = g(\theta),$$

and solve the problem exterior to $B(O, R)$, let $v(\frac{r}{R}, \theta) = u(r, \theta)$. Replacing r by $\frac{r}{R}$ in the Poisson formula we get for the general case,

$$u = \frac{1}{2\pi} \int_0^{2\pi} \frac{r^2 - R^2}{r^2 + R^2 - 2rR\cos(\theta - \varphi)} g(\varphi)\, d\varphi. \qquad (2.3)$$

Under a conformal mapping a harmonic function is transformed to a new harmonic function. Applying this property one can get some more solutions for other exterior domains. For example, if we define

$$w = az + \frac{b}{z}, \qquad a > b > 0,$$

where $z = x_1 + ix_2 = re^{i\theta}$, $w = \xi_1 + i\xi_2 = \rho e^{i\varphi}$, then the exterior of the unite circle $r = 1$ is transformed to Ω_1, the exterior of an ellipse, $\partial\Omega_1 = \{(\xi_1, \xi_2); \xi_1 = (a+b)\cos\theta, \xi_2 = (a-b)\sin\theta\}$. The function $v(\rho, \varphi) = u(r, \theta)$ can be expressed in terms of the Poisson formula,

$$v(\rho, \varphi) = \frac{1}{2\pi} \int_0^{2\pi} \frac{r^2 - 1}{r^2 + 1 - 2r\cos(\theta - \chi)} g(\chi)\, d\chi, \quad \rho e^{i\varphi} = are^{i\theta} + \frac{b}{re^{i\theta}},$$

and it is the solution to the problem,

$$\triangle v = 0, \qquad (\xi_1, \xi_2) \in \Omega_1,$$

$$v|_{\partial\Omega_1} = g(\theta).$$

There is another approach to derive the Poisson formula (2.2), which is the use of the potential theory. Let us investigate the integral equation (1.11). Since $\partial\Omega$ is the unit circle, we have

$$\frac{\partial}{\partial\nu_y} \ln\frac{1}{|x-y|} = \frac{\cos((x-y), \nu_y)}{|x-y|} = -\frac{1}{2},$$

so the equation is

$$\frac{1}{2}\sigma(x) - \frac{1}{4\pi}\int_{\partial\Omega} \sigma(y)\, ds_y = g(x).$$

We replace the boundary value $g(x)$ by $\tilde{g}(x) = g(x) - \frac{1}{2\pi} \int_{\partial\Omega} g(y) \, ds_y$, then the integral equation

$$\frac{1}{2}\sigma(x) - \frac{1}{4\pi} \int_{\partial\Omega} \sigma(y) \, ds_y = \tilde{g}(x),$$

admits a solution $\sigma(x) = 2g(x)$. The double layer potential

$$v(x) = \frac{1}{2\pi} \int_{\partial\Omega} \sigma(y) \frac{\cos((x-y), \nu_y)}{|x-y|} \, ds_y$$

satisfies (1.1),(1.4), and $v|_{\partial\Omega} = \tilde{g}$. Therefore

$$u(x) = \frac{1}{2\pi} \int_{\partial\Omega} \sigma(y) \frac{\cos((x-y), \nu_y)}{|x-y|} \, ds_y + \frac{1}{2\pi} \int_{\partial\Omega} g(y) \, ds_y$$

$$= \frac{1}{2\pi} \int_{\partial\Omega} \frac{2|x-y| \cos((x-y), \nu_y) + |x-y|^2}{|x-y|^2} g(y) \, ds_y.$$

Noting that $|x-y|^2 = r^2 + 1 - 2r\cos(\theta - \varphi)$ and $2|x-y|\cos((x-y), \nu_y) + |x-y|^2 = r^2 - 1$, we derive the Poisson formula (2.2) again.

The above approach can be applied to derive the three dimensional Poisson formula of the Dirichlet problem. Because the boundary $\partial\Omega$ is the unite sphere $S_3 = \partial B(O, 1)$, the integral equation (1.11) is

$$\frac{1}{2}\sigma(x) - \frac{1}{8\pi} \int_{S_3} \sigma(y) \frac{1}{|x-y|} \, ds_y = g(x).$$

We replace the boundary value $g(x)$ by

$$\tilde{g}(x) = g(x) - \frac{1}{4\pi} \int_{S_3} g(y) \frac{1}{|x-y|} \, ds_y,$$

then $\sigma(x) = 2g(x)$ is the solution. We notice that the difference between g and $\tilde{g}$ is nothing but a single layer potential, so the solution u is

$$u(x) = \frac{1}{4\pi} \int_{S_3} \sigma(y) \frac{\cos((x-y), \nu_y)}{|x-y|^2} \, ds_y + \frac{1}{4\pi} \int_{\partial\Omega} g(y) \frac{1}{|x-y|} \, ds_y$$

$$= \frac{1}{4\pi} \int_{S_3} \frac{2|x-y| \cos((x-y), \nu_y) + |x-y|^2}{|x-y|^3} g(y) \, ds_y.$$

Hence we get the Poisson formula

$$u(x) = \frac{1}{4\pi} \int_{S_3} \frac{r^2 - 1}{(r^2 + 1 - 2r\cos\varphi)^{\frac{3}{2}}} g(y) \, ds_y,$$

where $r = |x|$ and φ is the angle between the vectors x and $y \in S_3$. If we replace the boundary condition by

$$u|_{r=R} = g(x),$$

and solve the problem exterior to $B(O, R)$, let $v(\frac{r}{R}, \theta) = u(r, \theta)$. Replacing r by $\frac{r}{R}$ in the Poisson formula we get for the general case,

$$u(x) = \frac{1}{4\pi} \int_{S_3} \frac{r^2 - R^2}{(r^2 + R^2 - 2rR\cos\varphi)^{\frac{3}{2}}} g(y) R \, ds_y. \qquad (2.4)$$

2.1.2 Bi-harmonic equation

Let us solve the bi-harmonic equation next. The problem is:

$$\left(\frac{1}{r} \frac{\partial}{\partial r} \left(r \frac{\partial}{\partial r} \right) + \frac{1}{r^2} \frac{\partial^2}{\partial \theta^2} \right)^2 u = 0,$$

$$u|_{r=1} = g_1(\theta), \qquad \frac{\partial u}{\partial r} \bigg|_{r=1} = g_2(\theta).$$

The solution is expressed in terms of Fourier series,

$$u = \sum_{n=-\infty}^{\infty} \rho_n(r) e^{in\theta}.$$

Then we get

$$\left(\frac{1}{r} \frac{d}{dr} \left(r \frac{d}{dr} \right) - \frac{n^2}{r^2} \right)^2 \rho_n = 0.$$

Therefore we have

$$\frac{1}{r} \frac{d}{dr} \left(r \frac{d}{dr} \right) \rho_0 = a_0 + b_0 \ln r,$$

$$\left(\frac{1}{r} \frac{d}{dr} \left(r \frac{d}{dr} \right) - \frac{n^2}{r^2} \right) \rho_n = a_n r^{-n} + b_n r^n, \ \forall n \neq 0..$$

Solving these equations again and noting that $u \in H^{2,*}(\Omega)$, we get

$$\rho_0 = a_0 + b_0(1 + 2\ln r), \qquad \rho_n = a_n r^{-|n|} + b_n r^{-|n|+2}, \ \forall n \neq 0.$$

Taking the boundary conditions into account we solve the coefficients as follows:

$$g_1(\theta) = \sum_{n \neq 0} (a_n + b_n)e^{in\theta} + (a_0 + b_0),$$

$$g_2(\theta) = \sum_{n \neq 0} (-|n|a_n - (|n| - 2)b_n)e^{in\theta} + 2b_0.$$

Then

$$a_n + b_n = \frac{1}{2\pi} \int_0^{2\pi} e^{-in\varphi} g_1(\varphi) \, d\varphi,$$

$$-|n|a_n - (|n| - 2)b_n = \frac{1}{2\pi} \int_0^{2\pi} e^{-in\varphi} g_2(\varphi) \, d\varphi.$$

$$a_n = \frac{1}{2\pi} \int_0^{2\pi} e^{-in\varphi} \left(\left(1 - \frac{|n|}{2}\right) g_1 - \frac{1}{2}g_2 \right) d\varphi,$$

$$b_n = \frac{1}{2\pi} \int_0^{2\pi} e^{-in\varphi} \left(\frac{|n|}{2} g_1 + \frac{1}{2}g_2 \right) d\varphi.$$

Then the solution is

$$u = \frac{1}{2\pi} \int_0^{2\pi} e^{in(\theta - \varphi)}$$
$$\cdot \left\{ (g_1 - \frac{1}{2}g_2 + \frac{1}{2}g_2 r^2)r^{-|n|} - \frac{|n|}{2}g_1(1 - r^2)r^{-|n|} \right\} d\varphi$$
$$+ \frac{1}{2\pi} \int_0^{2\pi} (g_1 + g_2 \ln r) \, d\varphi.$$

After some calculation we finally get

$$u = \frac{1}{2\pi} \int_0^{2\pi} \frac{r^2 - 1}{r^2 + 1 - 2r\cos(\theta - \varphi)} \left(g_1(\varphi) - \frac{1}{2}g_2(\varphi) + \frac{1}{2}g_2(\varphi)r^2\right) d\varphi$$
$$+ \frac{1}{2\pi} \int_0^{2\pi} \frac{2r^2 - r(r^2 + 1)\cos(\theta - \varphi)}{(r^2 + 1 - 2r\cos(\theta - \varphi))^2} (1 - r^2)g_1(\varphi) \, d\varphi$$
$$+ \frac{1}{2\pi} \int_0^{2\pi} (\ln r + \frac{1}{2} - \frac{1}{2}r^2)g_2(\varphi) \, d\varphi.$$

If we replace the boundary condition by

$$u|_{r=R} = g_1(\theta), \qquad \frac{\partial u}{\partial r}\bigg|_{r=R} = g_2(\theta).$$

and solve the problem exterior to $B(O, R)$, let $v(\frac{r}{R}, \theta) = u(r, \theta)$. v satisfies the boundary condition,

$$v|_{r=1} = g_1(\theta), \qquad \frac{\partial v}{\partial r}\bigg|_{r=1} = Rg_2(\theta).$$

Then for this problem the solution is

$$
\begin{aligned}
u = &\frac{1}{2\pi} \int_0^{2\pi} \frac{r^2 - R^2}{r^2 + R^2 - 2rR\cos(\theta - \varphi)} \Big(g_1(\varphi) - \frac{R}{2}g_2(\varphi) + \frac{1}{2R}g_2(\varphi)r^2\Big) d\varphi \\
&+ \frac{1}{2\pi} \int_0^{2\pi} \frac{2R^2r^2 - Rr(r^2 + R^2)\cos(\theta - \varphi)}{(r^2 + R^2 - 2rR\cos(\theta - \varphi))^2} \Big(1 - \frac{r^2}{R^2}\Big)g_1(\varphi) d\varphi \\
&+ \frac{1}{2\pi} \int_0^{2\pi} (\ln\frac{r}{R} + \frac{1}{2} - \frac{1}{2}\frac{r^2}{R^2}) Rg_2(\varphi) d\varphi.
\end{aligned}
\tag{2.5}
$$

2.1.3 *Stokes equation*

For the exterior problem of the Stokes equation,

$$-\triangle u + \nabla p = 0, \tag{2.6}$$

$$\nabla \cdot u = 0, \tag{2.7}$$

$$u|_{r=R} = g(\theta), \tag{2.8}$$

we introduce the stream function

$$\psi(x) = \oint_{x_0}^x (-u_2\, dx_1 + u_1\, dx_2),$$

where r_0 is a fixed point. By the equation (2.7) the value of ψ is independent of the path of integral. And for different initial point x_0 the function ψ may differ from a constant only. The velocity u can be obtained from ψ:

$$u_1 = \frac{\partial \psi}{\partial x_2}, \qquad u_2 = -\frac{\partial \psi}{\partial x_1}.$$

Applying the operator

$$\left(\frac{\partial}{\partial x_2}, -\frac{\partial}{\partial x_1} \right)^T$$

to the equation (2.6) we get $\triangle^2 \psi = 0$, so the stream function ψ is the solution to the bi-harmonic equation.

We assume

$$\int_0^{2\pi} g \cdot \nu \, d\theta = 0$$

first, then in the polar coordinate $\psi(R, 2\pi) = \psi(R, 0)$. By this assumption ψ is single valued.

Let (u_r, u_θ) be the components of u in polar coordinates, that is

$$\begin{pmatrix} u_r \\ u_\theta \end{pmatrix} = \begin{pmatrix} \cos\theta & \sin\theta \\ -\sin\theta & \cos\theta \end{pmatrix} \begin{pmatrix} u_1 \\ u_2 \end{pmatrix}.$$

Under the polar coordinates it holds that

$$\psi(r, \theta) = \oint_{(r_0, \theta_0)}^{(r, \theta)} (-u_\theta \, dr + u_r r \, d\theta), \quad u_r = \frac{1}{r} \frac{\partial \psi}{\partial \theta}, \quad u_\theta = -\frac{\partial \psi}{\partial r}.$$

The boundary condition (2.8) becomes

$$u_r|_{r=R} = g_r(\theta), \qquad u_\theta|_{r=R} = g_\theta(\theta).$$

The boundary condition for ψ is

$$\psi|_{r=R} = R \int_0^\theta g_r(\eta) \, d\eta, \qquad \frac{\partial \psi}{\partial r}\bigg|_{r=R} = -g_\theta(\theta).$$

Applying the expression for the solutions to the bi-harmonic equation we obtain

$$\begin{aligned} \psi ={}& \frac{1}{2\pi} \int_0^{2\pi} \frac{r^2 - R^2}{r^2 + R^2 - 2rR\cos(\theta - \varphi)} \left(R \int_0^\varphi g_r(\eta) \, d\eta \right. \\ & + \frac{R}{2} g_\theta(\varphi) - \frac{1}{2R} g_\theta(\varphi) r^2 \bigg) \, d\varphi \\ & + \frac{1}{2\pi} \int_0^{2\pi} \frac{2R^2r^2 - Rr(r^2 + R^2)\cos(\theta - \varphi)}{(r^2 + R^2 - 2rR\cos(\theta - \varphi))^2} (1 - \frac{r^2}{R^2}) \\ & \cdot (R \int_0^\varphi g_r(\eta) \, d\eta) \, d\varphi - \frac{1}{2\pi} \int_0^{2\pi} (\ln\frac{r}{R} + \frac{1}{2} - \frac{1}{2}\frac{r^2}{R^2}) R g_\theta(\varphi) \, d\varphi. \end{aligned}$$

Then after some calculation we obtain the solution u:

$$
\begin{aligned}
u_r = & -\frac{1}{2\pi} \int_0^{2\pi} \frac{R^2(-r^2 + 3R^2)\sin(\theta - \varphi)}{(r^2 + R^2 - 2rR\cos(\theta - \varphi))^2} \int_0^\varphi g_r(\eta) \, d\eta \, d\varphi \\
& -\frac{1}{2\pi} \int_0^{2\pi} \frac{(r^2 - R^2)^2 \sin(\theta - \varphi)}{(r^2 + R^2 - 2rR\cos(\theta - \varphi))^2} g_\theta(\varphi) \, d\varphi \\
& +\frac{1}{2\pi} \int_0^{2\pi} \frac{4R^2(2R^2r^2 - Rr(r^2 + R^2)\cos(\theta - \varphi))\sin(\theta - \varphi)}{(r^2 + R^2 - 2rR\cos(\theta - \varphi))^3} \\
& \cdot (1 - \frac{r^2}{R^2}) \int_0^\varphi g_r(\eta) \, d\eta \, d\varphi,
\end{aligned}
\tag{2.9}
$$

and

$$
\begin{aligned}
u_\theta = & \frac{1}{2\pi} \int_0^{2\pi} \frac{2(r^2 - R^2)(r - R\cos(\theta - \varphi))}{(r^2 + R^2 - 2rR\cos(\theta - \varphi))^2} (R \int_0^\varphi g_r(\eta) \, d\eta \\
& + \frac{R}{2} g_\theta(\varphi) - \frac{1}{2R} g_\theta(\varphi) r^2) \, d\varphi \\
& + \frac{1}{2\pi} \int_0^{2\pi} \frac{2r}{r^2 + R^2 - 2rR\cos(\theta - \varphi)} (R \int_0^\varphi g_r(\eta) \, d\eta \\
& + \frac{R}{2} g_\theta(\varphi) - \frac{1}{2R} g_\theta(\varphi) r^2) \, d\varphi \\
& + \frac{1}{2\pi} \int_0^{2\pi} \frac{r^2 - R^2}{r^2 + R^2 - 2rR\cos(\theta - \varphi)} \frac{r}{R} g_\theta(\varphi) \, d\varphi \\
& - \frac{1}{2\pi} \int_0^{2\pi} \frac{(2R^2r^2 - Rr(r^2 + R^2)\cos(\theta - \varphi))(r - R\cos(\theta - \varphi))}{(r^2 + R^2 - 2rR\cos(\theta - \varphi))^3} \\
& \cdot (1 - \frac{r^2}{R^2}) R \int_0^\varphi g_r(\eta) \, d\eta \, d\varphi \\
& + \frac{1}{2\pi} \int_0^{2\pi} \frac{4R^3r - 4Rr^3 + (5r^4 - R^4)\cos(\theta - \varphi)}{(r^2 + R^2 - 2rR\cos(\theta - \varphi))^2} \int_0^\varphi g_r(\eta) \, d\eta \, d\varphi \\
& - \frac{1}{2\pi} \int_0^{2\pi} (\frac{R}{r} - \frac{r}{R}) g_\theta(\varphi) \, d\varphi.
\end{aligned}
\tag{2.10}
$$

If

$$
\frac{1}{2\pi} \int_0^{2\pi} g_r(\theta) \, d\theta = \bar{g} \neq 0,
$$

we define a particular solution

$$u_0 = \begin{pmatrix} u_{0r} \\ u_{0\theta} \end{pmatrix} = \begin{pmatrix} \bar{g}/r \\ 0 \end{pmatrix}, \quad p_0 = 0$$

to the equations (2.6),(2.7). Then

$$\int_0^{2\pi} (u - u_0)|_{r=R} \cdot \nu \, d\theta = 0.$$

$u - u_0$ can be expressed in term of the above expressions.

2.1.4   *Plane elasticity*

For the plane elasticity problem,

$$-\mu \bigtriangleup u - (\lambda + \mu)\text{grad div}u = 0, \quad x \in \Omega.$$

$$u_1|_{|x|=1} = g_1, u_2|_{|x|=1} = g_2,$$

we introduce the complex expression for the displacement $u = (u_1, u_2)^T$ first.

Let $f(z)$ be an analytic function, $f = p + iq$, $z = x + iy$. Let $L = -\mu \bigtriangleup -(\lambda + \mu)\text{grad div}$. By the Cauchy-Riemann equation, we have

$$\text{div} \begin{pmatrix} p \\ -q \end{pmatrix} = 0, \quad \text{div} \begin{pmatrix} q \\ p \end{pmatrix} = 0,$$

and $\bigtriangleup p = \bigtriangleup q = 0$, therefore

$$L \begin{pmatrix} p \\ -q \end{pmatrix} = 0, \quad L \begin{pmatrix} p \\ q \end{pmatrix} = -2(\lambda + \mu) \begin{pmatrix} \frac{\partial^2 p}{\partial x^2} \\ \frac{\partial^2 p}{\partial x \partial y} \end{pmatrix}.$$

For the function $z\overline{f(z)} = (xp + yq) + i(yp - xq)$, we have

$$\text{div} \begin{pmatrix} xp + yq \\ yp - xq \end{pmatrix} = 2p, \quad \text{grad div} \begin{pmatrix} xp + yq \\ yp - xq \end{pmatrix} = 2 \begin{pmatrix} \frac{\partial p}{\partial x} \\ \frac{\partial p}{\partial y} \end{pmatrix},$$

and

$$\bigtriangleup \begin{pmatrix} xp + yq \\ yp - xq \end{pmatrix} = \begin{pmatrix} \frac{\partial p}{\partial x} + \frac{\partial q}{\partial y} \\ \frac{\partial p}{\partial y} - \frac{\partial q}{\partial x} \end{pmatrix}.$$

Therefore

$$L \begin{pmatrix} xp + yq \\ yp - xq \end{pmatrix} = -2(\lambda + 2\mu) \begin{pmatrix} \frac{\partial p}{\partial x} \\ \frac{\partial p}{\partial y} \end{pmatrix}.$$

Let $af(z) - z\overline{f'(z)} = U + iV$, then

$$L \begin{pmatrix} U \\ V \end{pmatrix} = -2a(\lambda + \mu) \begin{pmatrix} \frac{\partial^2 p}{\partial x^2} \\ \frac{\partial^2 p}{\partial x \partial y} \end{pmatrix} + 2(\lambda + 2\mu) \begin{pmatrix} \frac{\partial^2 p}{\partial x^2} \\ \frac{\partial^2 p}{\partial x \partial y} \end{pmatrix}.$$

We set $a = \frac{\lambda + 2\mu}{\lambda + \mu}$, then $L \begin{pmatrix} U \\ V \end{pmatrix} = 0$.

Let us return to the Lamé equation(1.35). If $u_1 + iu_2 = a\varphi(z) - z\overline{\varphi'(z)} - \overline{\psi(z)}$, where φ, ψ are two analytic functions, then $u = (u_1, u_2)^T$ is the solution to the equation. Since $u_1, u_2 \in H^{1,*}(\Omega)$, the functions φ, ψ are expressed in terms of Laurent series

$$\varphi(z) = \sum_{n=0}^{\infty} a_n r^{-n} e^{-in\theta}, \qquad \psi(z) = \sum_{n=0}^{\infty} b_n r^{-n} e^{-in\theta}.$$

Thus we get the expression,

$$u_1 + iu_2 = a \sum_{n=0}^{\infty} a_n r^{-n} e^{-in\theta} + \sum_{n=0}^{\infty} n\overline{a_n} r^{-n} e^{i(n+2)\theta} - \sum_{n=0}^{\infty} \overline{b_n} r^{-n} e^{in\theta}.$$

Applying the boundary conditions we get

$$g = g_1 + ig_2 = a \sum_{n=0}^{\infty} a_n e^{-in\theta} + \sum_{n=0}^{\infty} n\overline{a_n} e^{i(n+2)\theta} - \sum_{n=0}^{\infty} \overline{b_n} e^{in\theta}.$$

The coefficients can be evaluated with $b_0 = 0$,

$$a_n = \frac{1}{2\pi a} \int_0^{2\pi} e^{in\phi} g(\phi) \, d\phi, \quad n = 0, 1, \cdots,$$

$$\overline{b_n} = -\frac{1}{2\pi} \int_0^{2\pi} e^{-in\phi} g(\phi) \, d\phi, \quad n = 1, 2,$$

$$\overline{b_n} = -\frac{1}{2\pi} \int_0^{2\pi} e^{-in\phi} g(\phi) \, d\phi + \frac{n-2}{2\pi a} \int_0^{2\pi} e^{-i(n-2)\phi} \overline{g}(\phi) \, d\phi, \quad n = 3, \cdots.$$

Finally we get

$$
\begin{aligned}
u_1 + iu_2 =& \frac{1}{2\pi} \int_0^{2\pi} \left\{ \sum_{n=0}^{\infty} r^{-n} e^{-in(\theta-\phi)} g(\phi) + \sum_{n=0}^{\infty} \frac{nr^{-n}}{a} e^{2i\theta} e^{in(\theta-\phi)} \bar{g}(\phi) \right. \\
& \left. + \sum_{n=1}^{\infty} r^{-n} e^{in(\theta-\phi)} g(\phi) - \sum_{n=3}^{\infty} \frac{(n-2)r^{-n}}{a} e^{2i\theta} e^{i(n-2)(\theta-\phi)} \bar{g}(\phi) \right\} d\phi \\
=& \frac{1}{2\pi} \int_0^{2\pi} \left\{ \frac{g(\phi)}{1 - r^{-1} e^{-i(\theta-\varphi)}} + \frac{r^{-1} e^{i(\theta-\varphi)} g(\phi)}{1 - r^{-1} e^{i(\theta-\varphi)}} \right. \\
& \left. - \frac{e^{2i\theta}}{a} \frac{r^{-1} e^{i(\theta-\varphi)} \bar{g}(\phi)}{(1 - r^{-1} e^{i(\theta-\varphi)})^2} (1 - r^{-2}) \right\} d\phi \\
=& \frac{1}{2\pi} \int_0^{2\pi} \left\{ \frac{r^2 - 1}{r^2 + 1 - 2r\cos(\theta-\varphi)} g(\varphi) \right. \\
& \left. - \frac{e^{2i\theta}}{a} \frac{e^{i(\theta-\varphi)}(r^2-1)}{r(r - e^{i(\theta-\varphi)})^2} \bar{g}(\phi) \right\} d\phi.
\end{aligned}
$$

If we replace the boundary condition by

$$
u_1|_{|x|=R} = g_1, u_2|_{|x|=R} = g_2,
$$

then

$$
\begin{aligned}
u_1 + iu_2 =& \frac{1}{2\pi} \int_0^{2\pi} \left\{ \frac{r^2 - R^2}{r^2 + R^2 - 2rR\cos(\theta-\varphi)} g(\varphi) \right. \\
& \left. - \frac{e^{2i\theta}}{a} \frac{e^{i(\theta-\varphi)} R(r^2 - R^2)}{r(r - Re^{i(\theta-\varphi)})^2} \bar{g}(\phi) \right\} d\phi.
\end{aligned}
\tag{2.11}
$$

2.1.5 *Helmholtz equation*

In order to solve the exterior problem of the Helmholtz equation let us introduce the definition and some basic properties of the Bessel functions. The Bessel equation is

$$
\frac{d^2 y}{dx^2} + \frac{1}{x} \frac{dy}{dx} + \left(1 - \frac{s^2}{x^2} \right) = 0.
$$

One solution of it is the Bessel function of the first kind of order s, which can be developed into a series,

$$
J_s(x) = \sum_{k=0}^{\infty} (-1)^k \frac{1}{\Gamma(s+k+1)} \left(\frac{x}{2} \right)^{2k+s},
$$

where Γ is the Gamma function. The asymptotic behavior of it as $x \to \infty$ is

$$J_s(x) = \sqrt{\frac{2}{\pi x}} \cos\left(x - \frac{s\pi}{2} - \frac{\pi}{4}\right) + O(x^{-3/2}).$$

Another linear independent solution of the Bessel equation is usually denoted by $Y_s(x)$, the Bessel function of the second kind of order s, which possesses a singular point $x = 0$. We omit its expression, but give the asymptotic expression of it here:

$$Y_s(x) = \sqrt{\frac{2}{\pi x}} \sin\left(x - \frac{s\pi}{2} - \frac{\pi}{4}\right) + O(x^{-3/2}).$$

Moreover the Bessel functions of the third kind of order s are called the Hankel functions of the first kind of order s and the second kind of order s. The definitions of them are:

$$H_s^{(1)}(x) = J_s(x) + iY_s(x),$$

$$H_s^{(2)}(x) = J_s(x) - iY_s(x).$$

Therefore the asymptotic behavior of them is the following:

$$H_s^{(1)}(x) = \sqrt{\frac{2}{\pi x}} e^{i\left(x - \frac{s\pi}{2} - \frac{\pi}{4}\right)} + O(x^{-3/2}), \tag{2.12}$$

and

$$H_s^{(2)}(x) = \sqrt{\frac{2}{\pi x}} e^{-i\left(x - \frac{s\pi}{2} - \frac{\pi}{4}\right)} + O(x^{-3/2}). \tag{2.13}$$

Multiplying them by a factor $e^{-i\omega t}$, we find $H_s^{(1)}(x)e^{-i\omega t}$ represents an outgoing plane wave and $H_s^{(2)}(x)e^{-i\omega t}$ represents an incoming plane wave.

For the Dirichlet boundary value problem of the homogeneous Helmholtz equation

$$(\triangle + \omega^2)u = 0, \qquad x \in \Omega, \tag{2.14}$$

$$u = g, \qquad x \in \partial\Omega, \tag{2.15}$$

we consider the case of two dimension first and let $\partial\Omega = \{x; r = R\}$. The solution can be expressed in terms of Fourier series,

$$u(r, \theta) = \sum_{n=-\infty}^{\infty} a_n H_n^{(1)}(\omega r)e^{in\theta}. \qquad (2.16)$$

Taking into account of the boundary condition we get

$$g(\theta) = \sum_{n=-\infty}^{\infty} a_n H_n^{(1)}(\omega R)e^{in\theta}, \qquad (2.17)$$

therefore

$$a_n = \frac{1}{2\pi} \int_0^{2\pi} g(\theta)e^{-in\theta}\, d\theta / H_n^{(1)}(\omega R).$$

Lemma 35. *If $g \in H^{\frac{3}{2}}(\partial\Omega)$, the series (2.16) converges and solves the problem (2.14),(2.15).*

Proof.   By (2.17) the boundary condition (2.15) is satisfied, provided the series converges. Making use of separation of variables it is easy to see that each term of (2.16) is a solution to the equation (2.14). It represents an outgoing sphere wave, so the Sommerfield radiation condition is satisfied.

It remains to prove convergence. We recall the estimate

$$\|u\|_V < C\|f\|_{0,\sigma}, \qquad (2.18)$$

where f is the right hand side of the equation (1.25) and the support of f is bounded. For the problem (2.14),(2.15) we apply the inverse trace theorem of Sobolev spaces to take a function u_0 in $H^2(\Omega)$ with a compact support, so that $u_0|_{\partial\Omega} = g$, and $\|u_0\|_2 \leq C\|g\|_{\frac{3}{2}}$. Let $u_1 = u - u_0$, then it is the solution to

$$(\triangle + \omega^2)u_1 = -(\triangle + \omega^2)u_0.$$

By the estimate (2.18) we have

$$\|u\|_V \leq \|u_0\|_V + \|u_1\|_V \leq C\|g\|_{\frac{3}{2}}.$$

Therefore the convergence of the series (2.17) in $H^{\frac{3}{2}}(\partial\Omega)$ implies the convergence of (2.16) in V.   □

The treatment for three dimensional case is analogous. The solution can be represented as a series of spherical functions:

$$u(r,\theta) = \sum_{n=0}^{\infty} \sum_{m=-n}^{n} a_{nm} \zeta_{n+\frac{1}{2}}^{(2)}(\omega r) Y_{nm}(\theta, \varphi), \qquad (2.19)$$

where

$$\zeta_{n+\frac{1}{2}}^{(2)}(x) = \sqrt{\frac{\pi}{2x}} H_{n+\frac{1}{2}}^{(1)}(x), \quad Y_{nm}(\theta, \varphi) = P_n^{(m)}(\cos\theta)e^{im\varphi},$$

$$P_n^{(m)}(x) = \frac{(1-x^2)^{\frac{m}{2}}}{2^n n} \frac{d^{n+m}}{dx^{n+m}}(x^2-1)^n.$$

2.2   General domains

Let us consider the harmonic equation first. The integral equations (1.11) and (1.13) can be regarded as a boundary reduction, so the corresponding boundary element method is based on the approximation to these integral equations. However for Dirichlet boundary value problems we have shown that the integral equations are not in fact equivalent to the original problems, which causes some inconvenience. Another choice is representing the solutions of Dirichlet problems in terms of single layer potentials. We start from the two dimensional case and consider the problem (1.1),(1.2),(1.4). Let the solution u be expressed as

$$u(x) = \frac{1}{2\pi} \int_{\partial\Omega} \omega(y) \ln\frac{1}{|x-y|} ds_y + C, \qquad (2.20)$$

then the boundary condition (1.2) leads to a Fredholm integral equation of the first type:

$$g(x) = \frac{1}{2\pi} \int_{\partial\Omega} \omega(y) \ln\frac{1}{|x-y|} ds_y + C, \quad x \in \partial\Omega, \qquad (2.21)$$

where the density ω and the constant C are to be determined.

Theorem 41. *If $g \in C^2(\partial\Omega)$ then the equation (2.21) admits a unique solution (ω, C) satisfying*

$$\int_{\partial\Omega} \omega(y) ds_y = 0. \qquad (2.22)$$

Proof. Owing to the theory of elliptic equations the solution u to the problem (1.1),(1.2) satisfies $u \in C^1(\overline{\Omega})$ under the hypothesis of this theorem. Therefore $\frac{\partial u}{\partial \nu} \in C(\partial\Omega)$. By Theorem 7 u can be expressed in terms of a double layer potential plus a constant. It can be derived from (1.7) that $\nabla u = O(|x|^{-2})$ for large $|x|$, then we notice $\int_\Omega \Delta u \, dx = 0$ and make use of the Green's formula to obtain $\int_{\partial\Omega} \frac{\partial u}{\partial \nu} \, ds = 0$. Let us regard $\frac{\partial u}{\partial \nu}$ as a boundary value, then by Theorem 8 u can be expressed in terms of a single layer potential plus one constant and the density satisfies $\int_{\partial\Omega} \omega(y) \, ds_y = 0$, which proves the existence.

To prove uniqueness, we assume that $g = 0$ in (2.21), then define u by (2.20). By (2.21) u is a constant on the boundary $\partial\Omega$. By uniqueness of the Dirichlet problem u is identity to a constant on the entire plane. Then by Lemma 3 $\omega = 0$. Then the equation (2.21) implies $C = 0$. □

To consider more general $g's$, we need to study weak solutions and the variational formulation. We will see that the variational formulation is convenient to implementation by using the finite element method. Letting $\chi \in C(\partial\Omega)$ and $\int_{\partial\Omega} \chi(x) \, ds_x = 0$, we multiply the equation (2.21) by χ and integrate the equation on $\partial\Omega$ to obtain

$$b(\omega,\chi) \cong \frac{1}{2\pi} \int_{\partial\Omega} \int_{\partial\Omega} \omega(y)\chi(x) \ln \frac{1}{|x-y|} \, ds_y \, ds_x = \int_{\partial\Omega} g(x)\chi(x) \, ds_x.$$
(2.23)

To study the equation (2.23), we take $v \in C_0^\infty(\mathbb{R}^2)$, and consider the bilinear form

$$a(u,v) = \int_{\mathbb{R}^2} \nabla u \cdot \nabla v \, dx = \int_{\mathbb{R}^2} \nabla \left(\frac{1}{2\pi} \int_{\partial\Omega} \omega(y) \ln \frac{1}{|x-y|} \, ds_y + C \right) \cdot \nabla v \, dx.$$

By exchanging the order of integrations we get

$$a(u,v) = \int_{\partial\Omega} \omega(y) \, ds_y \int_{\mathbb{R}^2} \nabla_x \frac{1}{2\pi} \ln \frac{1}{|x-y|} \cdot \nabla_x v \, dx.$$

Let $f = -\Delta v$, then by the property of Newton potential (see Section 1.2)

$$v(y) = \int_{\mathbb{R}^2} \frac{1}{2\pi} \ln \frac{1}{|x-y|} f(x) \, dx = \int_{\mathbb{R}^2} \nabla_x \frac{1}{2\pi} \ln \frac{1}{|x-y|} \nabla v(x) \, dx.$$

Therefore

$$a(u,v) = \int_{\partial\Omega} \omega(y)v(y) \, ds_y.$$
(2.24)

Now we set $v \in H^{1,*}(\mathbb{R}^2)$ and take a series $\{v_n\} \subset C_0^\infty(\mathbb{R}^2)$ tending to v. Passing to the limit in (2.24) we find that (2.24) holds for all $v \in H^{1,*}(\mathbb{R}^2)$. Letting it be an equation for u, we will prove it is well-posed. By the trace theorem of Sobolev spaces, $v \in H^{1/2}$ on the boundary $\partial\Omega$, so we may assume that $\omega \in H^{-1/2}(\partial\Omega)$, the dual space of $H^{1/2}(\partial\Omega)$. The equation (2.24) is thus written in the form of distributions: Find $u \in H^{1,*}(\mathbb{R}^2)$, such that

$$a(u,v) = <\omega, v>_{\partial\Omega}, \qquad \forall v \in H^{1,*}(\mathbb{R}^2). \qquad (2.25)$$

We define $W = \{\omega \in H^{-1/2}(\partial\Omega); <\omega, 1>_{\partial\Omega} = 0\}$, then we have

Lemma 36. *The problem (2.25) admits a solution for all $\omega \in W$, and the difference of any two solutions is a constant.*

Proof. We define $V = \{u \in H^{1,*}(\mathbb{R}^2); \int_{B(O,R_0)} u\, dx = 0\}$ with $R_0 > 0$, and consider the problem: Find $u \in V$, such that

$$a(u,v) = <\omega, v>_{\partial\Omega}, \qquad \forall v \in V.$$

Since $|\cdot|_1$ is a norm of V, it admits a unique solution u. For any $v \in H^{1,*}(\mathbb{R}^2)$ we can divide it into two parts: $v = v_1 + C$, where $v_1 \in V$ and C is a constant. Noting that $<\omega, 1>_{\partial\Omega} = 0$, we find u is also a solution to the problem (2.25).

If u is a solution to the homogeneous problem, then because $a(u,u) = 0$, u is a constant. $\qquad\square$

We have shown that a classical solution u can be expressed in terms of a single layer potential (2.20), and satisfies (2.25). Now we set $\omega \in W$, then solve (2.25) to get $u \in H^{1,*}(\mathbb{R}^2)$. The mapping from ω to u is continuous. Therefore we can take the closure in (2.20) and define a generalized single layer potential $u(x) = <\frac{1}{2\pi} \ln \frac{1}{|x-\cdot|}, \omega>_{\partial\Omega}$, which defines a continuous operator from W to $H^{1,*}(\mathbb{R}^2)$.

Let us return to the problem (2.23). $b(\omega, \chi)$ is now ready to be defined on $W \times W$. The problem is: For a given $g \in H^{1/2}(\partial\Omega)$, find $\omega \in W$, such that

$$b(\omega, \chi) = <g, \chi>_{\partial\Omega}, \qquad \forall \chi \in W \qquad (2.26)$$

Theorem 42. *The problem (2.26) admits a unique solution.*

Proof. We consider a single layer potential $v(x) = <\frac{1}{2\pi} \ln \frac{1}{|x-\cdot|}, \chi>_{\partial\Omega}$, then $b(\omega, \chi) = <\omega, v>_{\partial\Omega}$. By (2.25) we get $b(\omega, \chi) = a(u,v)$. For a given $v \in H^{1/2}(\partial\Omega)$, we consider the solutions to the interior and exterior problems with Dirichlet boundary value v, still denoted by v, then $v \in H^{1,*}(\mathbb{R}^2)$, and $\|v\|_{1,*,\mathbb{R}^2} \leq M\|v\|_{1/2,\partial\Omega}$ with a constant M. Hence for all $\omega \in W$ it holds that

$$\|\omega\|_{-1/2} = \sup_{v \in H^{1/2}} \frac{<\omega, v>_{\partial\Omega}}{\|v\|_{1/2,\partial\Omega}} \leq M \sup_{v \in H^{1/2}} \frac{<\omega, v>_{\partial\Omega}}{\|v\|_{1,*,\Omega}}$$

$$\leq M \sup_{v \in H^{1/2}} \frac{<\omega, v>_{\partial\Omega}}{|v|_1} = M \sup_{v \in H^{1/2}} \frac{a(u, v)}{|v|_1} \leq M|u|_1.$$

We get $b(\omega, \omega) \geq \frac{1}{M^2}\|\omega\|^2_{-1/2}$. Owing to the Lax-Milgram theorem uniqueness and existence follows. □

Using the boundary element method to solve (1.1),(1.2),(1.4), we first solve (2.26) for a given g, then evaluate $u(x) =< \frac{1}{2\pi} \ln \frac{1}{|x-\cdot|}, \omega >_{\partial\Omega} +C$, where the constant C is determined by $g(x) =< \frac{1}{2\pi} \ln \frac{1}{|x-\cdot|}, \omega >_{\partial\Omega} +C$, $x \in \partial\Omega$. It is easy to verify that u is the desired solution.

For three dimensional problems the situation is similar. The solution to (1.1),(1.2),(1.4) can be expressed in terms of a single layer potential,

$$u(x) = \frac{1}{4\pi} \int_{\partial\Omega} \omega(y) \frac{1}{|x - y|} ds_y. \tag{2.27}$$

The kernel ω satisfies the problem: Find $\omega \in H^{-1/2}(\partial\Omega)$, such that

$$b(\omega, \chi) =< g, \chi >_{\partial\Omega}, \qquad \forall \chi \in H^{-1/2}(\partial\Omega), \tag{2.28}$$

where

$$b(\omega, \chi) = \frac{1}{4\pi} \int_{\partial\Omega} \int_{\partial\Omega} \omega(y)\chi(x) \frac{1}{|x - y|} ds_y \, ds_x.$$

Because $\omega, \chi \in H^{-1/2}(\partial\Omega)$, the bilinear form $b(\cdot, \cdot)$ should be understood in the sense of distributions. Being the same, (2.27) is also in the sense of distributions. For a given $g \in H^{1/2}(\partial\Omega)$ the problem (2.28) admits a unique solution ω. It is easy to verify that u defined by (2.27) is the desired solution.

For the Neŭmann problem (1.1),(1.3),(1.4) if we express the solution in terms of a double layer potential, then

$$u(x) = \frac{1}{2\pi} \int_{\partial\Omega} \sigma(y) \frac{\partial}{\partial\nu_y} \ln \frac{1}{|x - y|} ds_y,$$

for two dimensional case. Then the kernel σ satisfies

$$g(x) = \frac{1}{2\pi} \int_{\partial\Omega} \sigma(y) \frac{\partial}{\partial\nu_x} \frac{\partial}{\partial\nu_y} \ln \frac{1}{|x - y|} ds_y, \qquad x \in \partial\Omega.$$

The kernel is hyper-singular. We will discuss this kind of integral equations later on.

Next let us consider the Dirichlet boundary value problem of the bi-harmonic equation, (1.53). The fundamental solution $K(r) = \frac{1}{8\pi} r^2 \ln r$, $r = |x|$, satisfies

$$\triangle^2 u = \delta(x),$$

and K depends on r only. The solution u to the problem (1.53) is expressed in terms of a potential form,

$$u(x) = \int_{\partial\Omega} \omega(y) K(x-y)\, ds_y + \int_{\partial\Omega} \sigma(y) \frac{\partial}{\partial\nu_y} K(x-y)\, ds_y + p(x), \quad p \in P_1(\Omega).$$
$$(2.29)$$

Being analogous to (2.22) we require that the densities ω, σ satisfy the condition,

$$\int_{\partial\Omega} \left(\omega(y)v(y) + \sigma(y)\frac{\partial v(y)}{\partial\nu_y} \right) ds_y = 0, \qquad \forall v \in P_1. \qquad (2.30)$$

We verify that under the condition (2.30) the function u determined by (2.29) is in the space $H^{2,*}(\Omega)$. Then we derive the variational formulations for u and (ω, σ). Finally we verify that u is the solution to (1.53).

Lemma 37. *If ω and σ are continuous functions and satisfy (2.30), then $u \in H^{2,*}(\Omega)$.*

Proof. Let $x_0 \in \Omega^c$, then by the condition (2.30) we have

$$\int_{\partial\Omega} (\omega(y)(K(x - x_0) + \nabla K(x - x_0) \cdot (x_0 - y)) + \sigma(y)\frac{\partial}{\partial\nu_y}(K(x - x_0)$$
$$+ \nabla K(x - x_0) \cdot (x_0 - y)))\, ds_y = 0.$$

Then we get

$$u(x) = \int_{\partial\Omega} \omega(y)\{K(x - y) - K(x - x_0) - \nabla K(x - x_0) \cdot (x_0 - y)\}\, ds_y$$
$$+ \int_{\partial\Omega} \sigma(y)\frac{\partial}{\partial\nu_y}\{K(x - y) - K(x - x_0) - \nabla K(x - x_0) \cdot (x_0 - y)\}\, ds_y + p(x).$$

Applying Taylor's expansion we find

$$K(x - y) - K(x - x_0) - \nabla K(x - x_0) \cdot (x_0 - y) = O(|D^2 K| \cdot |x_0 - y|^2),$$

as $x \to \infty$, with x_0 being a fixed point and y bounded. Then by the expression of K we easily find

$$\int_\Omega \frac{u^2}{|x|^4 \ln^4 |x/R_0|} \, dx < \infty.$$

The other two terms in (1.52) can be verified in the same way. $\qquad \square$

We substitute (2.29) into (1.47) with $v \in C_0^\infty(\mathbb{R}^2)$. Let us evaluate the first term as the following:

$$\int_{\mathbb{R}^2} \tilde{\nu} \triangle \left\{ \int_{\partial\Omega} \omega(y) K(x-y) \, ds_y \right.$$

$$+ \int_{\partial\Omega} \sigma(y) \frac{\partial}{\partial\nu_y} K(x-y) \, ds_y + p(x) \biggr\} \cdot \triangle v(x) \, dx$$

$$= \int_{\partial\Omega} \omega(y) \, ds_y \int_{\mathbb{R}^2} \tilde{\nu} \triangle_x K(x-y) \triangle v(x) \, dx$$

$$+ \int_{\partial\Omega} \sigma(y) \frac{\partial}{\partial\nu_y} \int_{\mathbb{R}^2} \tilde{\nu} \triangle_x K(x-y) \cdot \triangle v(x) \, dx \, ds_y$$

$$= \int_{\partial\Omega} \omega(y) \, ds_y \int_{\mathbb{R}^2} \tilde{\nu} K(x-y) \triangle^2 v(x) \, dx$$

$$+ \int_{\partial\Omega} \sigma(y) \frac{\partial}{\partial\nu_y} \int_{\mathbb{R}^2} \tilde{\nu} K(x-y) \cdot \triangle^2 v(x) \, dx \, ds_y.$$

By the property of the Newton potential $\int_{\mathbb{R}^2} K(x-y) \triangle^2 v(x) \, dx = v(y)$, so the first term of (1.47) is

$$\tilde{\nu} \int_{\partial\Omega} \left(\omega(y) v(y) + \sigma(y) \frac{\partial v(y)}{\partial\nu_y} \right) ds_y.$$

The second term can be evaluated in the same way, and we get

$$a(u, v) = \int_{\partial\Omega} \left(\omega(y) v(y) + \sigma(y) \frac{\partial v(y)}{\partial\nu_y} \right) ds_y. \qquad (2.31)$$

Lemma 38. *Let $\omega \in H^{-3/2}(\partial\Omega)$, $\sigma \in H^{-1/2}(\partial\Omega)$, and (2.30) be satisfied, then the solution $u \in H^{2,*}(\mathbb{R}^2)$ to the equation (2.31) for all $v \in H^{2,*}(\mathbb{R}^2)$ exists, and the difference of any two solutions belongs to P_1.*

Proof. We define $V = \{u \in H^{2,*}(\mathbb{R}^2); \int_{\mathbb{R}^2} up \, dx = 0, \forall p \in P_1\}$, then follow the same line as the proof of Lemma 36 to obtain the result. $\qquad \square$

We remark that some integrals are understood in the sense of distributions. Since there is no misunderstanding we will not distinguish them with normal integrals.

From the boundary conditions of (1.53) and the expression (2.29) we get

$$g_1(x) = \int_{\partial\Omega} \omega(y)K(x-y)\, ds_y + \int_{\partial\Omega} \sigma(y)\frac{\partial}{\partial\nu_y}K(x-y)\, ds_y + p(x),$$

$$x \in \partial\Omega,$$

and

$$g_2(x) = \frac{\partial}{\partial\nu_x}\left\{ \int_{\partial\Omega} \omega(y)K(x-y)\, ds_y \right.$$

$$\left. + \int_{\partial\Omega} \sigma(y)\frac{\partial}{\partial\nu_y}K(x-y)\, ds_y + p(x) \right\}, \qquad x \in \partial\Omega.$$

We define

$$v(y) = \int_{\partial\Omega} \chi(x)K(x-y)\, ds_x - \int_{\partial\Omega} \tau(x)\frac{\partial}{\partial\nu_x}K(x-y)\, ds_x,$$

with

$$\int_{\partial\Omega} \left(\chi(x)q(x) + \tau(x)\frac{\partial q(x)}{\partial\nu_x} \right)\, ds_x = 0, \qquad \forall q \in P_1,$$

then we consider the dual products

$$<g_1,\chi> + <g_2,\tau> = \int_{\partial\Omega}\left\{ \int_{\partial\Omega} \omega(y)K(x-y)\, ds_y \right.$$

$$+ \int_{\partial\Omega} \sigma(y)\frac{\partial}{\partial\nu_y}K(x-y)\, ds_y + p(x) \Big\}\chi(x)\, ds_x$$

$$+ \int_{\partial\Omega} \frac{\partial}{\partial\nu_x}\left\{ \int_{\partial\Omega} \omega(y)K(x-y)\, ds_y \right. \tag{2.32}$$

$$+ \int_{\partial\Omega} \sigma(y)\frac{\partial}{\partial\nu_y}K(x-y)\, ds_y + p(x) \Big\}\tau(x)\, ds_x.$$

By the expression of v we have

$$<g_1,\chi> + <g_2,\tau> = <\omega,v> + <\sigma,v>. \tag{2.33}$$

With regard to (2.32) we define a bilinear form on $(H^{-3/2}(\partial\Omega) \times H^{-1/2}(\partial\Omega)) \times (H^{-3/2}(\partial\Omega) \times H^{-1/2}(\partial\Omega))$ as follows:

$$
b((\omega,\sigma),(\chi,\tau))
$$

$$
= \int_{\partial\Omega} \left\{ \int_{\partial\Omega} \omega(y) K(x-y)\, ds_y + \int_{\partial\Omega} \sigma(y) \frac{\partial}{\partial\nu_y} K(x-y)\, ds_y \right\} \chi(x)\, ds_x
$$

$$
+ \int_{\partial\Omega} \frac{\partial}{\partial\nu_x} \left\{ \int_{\partial\Omega} \omega(y) K(x-y)\, ds_y + \int_{\partial\Omega} \sigma(y) \frac{\partial}{\partial\nu_y} K(x-y)\, ds_y \right\} \tau(x)\, ds_x.
$$

We define a space

$$
W = \{(\omega,\sigma) \in (H^{-3/2}(\partial\Omega) \times H^{-1/2}(\partial\Omega));
$$

$$
<\omega, v>_{\partial\Omega} + <\sigma, \frac{\partial v}{\partial\nu}>_{\partial\Omega} = 0, \forall v \in P_1\}.
$$

Theorem 43. *The problem: find* $(\omega,\sigma) \in W$, *such that*

$$
b((\omega,\sigma),(\chi,\tau)) = <g_1,\chi> + <g_2,\tau>, \quad \forall (\chi,\tau) \in W, \tag{2.34}
$$

admits a unique solution.

Proof.   By (2.31) we see that $b((\omega,\sigma),(\chi,\tau)) = a(u,v)$, then following the same lines as the proof of Theorem 42 we get the result.   □

With the above argument the problem (1.53) is reduced to a boundary equation (2.34). Then we use (ω,σ) and (2.29) to evaluate u. The polynomial p is determined by (2.32) in the following way: We define a set $S = \{(\chi,\tau); \chi = p, \tau = \frac{\partial p}{\partial\nu}, p \in P_1\}$ on $\partial\Omega$, which is three dimensional. Then by the definition of W the dual products of the solutions to (2.34), $(\omega,\sigma) \in W$, applied to S is zero. We set $(\chi,\tau) \in S$ in (2.32), then we get three equations to determine the three coefficients of p.

To see (2.29) is the solution to (1.53) we only need to verify the boundary conditions, that is, (2.32) should be satisfied for all χ and τ. Let $(\chi,\tau) \in (L^2(\partial\Omega))^2$, then we can divide it into two parts, one is in W and the other one is in S, then it is straight forward to see that (2.32) holds for all $(\chi,\tau) \in (L^2(\partial\Omega))^2$, so u is the solution to (1.53).

2.3 Subdivision of the domain

If the domain Ω is in a complicated shape, the simplest way to deal with exterior problems is to truncate the domain. Let us consider the problem

$$\triangle u = 0, \tag{2.35}$$

$$u = g, \qquad x \in \partial\Omega, \tag{2.36}$$

in three dimension first. Let Ω_0 be a domain with a simple shape, and $\Omega_0 \supset \overline{\Omega^c}$. We define $\Omega_1 = \Omega_0 \cap \Omega$. The domain Ω is replaced by Ω_1, and some artificial boundary condition is imposed on the artificial boundary $\partial\Omega_0$.

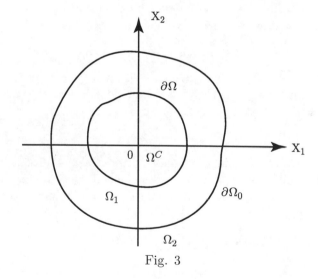

Fig. 3

The most self evident attempt is to impose

$$u = 0, \qquad x \in \partial\Omega_0. \tag{2.37}$$

Let u_h be the solution to (2.35),(2.36),(2.37) on Ω_1. It is indeed an approximation, because we have the following estimate.

Lemma 39. *We suppose that dim* $(\Omega) = 3$. *Let* $R = \min\{|x|; x \in \partial\Omega_0\}$, *then*

$$|u - u_h| = O\left(\frac{1}{R}\right).$$

Proof. By the expression in Section 1.1 we get $u = O(1/R)$ on $\partial\Omega_0$. Then applying the maximum principle for $u - u_h$ the conclusion follows.□

For two dimensional problems the situation is worse, because the solution u does not tend to zero at the infinity. However we still have the following:

Lemma 40. *We suppose that* $dim(\Omega) = 2$. *Let* $g \in H^{1/2}(\partial\Omega)$. *Then the approximate solution* u_h *converges to* u *on any bounded domains weakly in* H^1 *and strongly in* L^2, *as* $R \to \infty$.

Proof. Owing to the inverse trace theorem of Sobolev spaces, we can take a function $u_0 \in H^1(\Omega)$ with a compact support, such that $u_0|_{\partial\Omega} = g$ and $\|u_0\|_1 \leq C\|g\|_{\frac{1}{2}}$. Let R be large enough, so that $u_0 = 0$ on $\partial\Omega_0$. Let $u_1 = u_h - u_0$, then by the weak formulation (1.20) it holds that

$$a(u_1, v) = -a(u_0, v), \qquad \forall v \in H_0^1(\Omega_1).$$

Let $v = u_1$, then $|u_1|_1^2 \leq |u_0|_1|u_1|_1$, which gives $|u_1|_1 \leq |u_0|_1 \leq C\|g\|_{\frac{1}{2}}$, where the constant C is independent of the domain Ω_1. We extend u_h by zero on $\Omega_2 = \mathbb{R}^2 \setminus \overline{\Omega}$, then $u_h \in H^{1,*}(\Omega)$. u_h satisfies

$$\|u_h\|_{1,*} \leq C\|g\|_{\frac{1}{2}}.$$

For a sequence of R tending to infinity, we extract a subsequence of u_h converging weakly. Owing to the compact embedding of Sobolev spaces, the subsequence strongly converges in L^2-norm on any bounded sub-domain. Let u be the limit of u_h in $H^{1,*}(\Omega)$. We take the limit in

$$a(u_h, v) = 0, \qquad \forall v \in H_0^1(\Omega_1),$$

to obtain

$$a(u, v) = 0, \qquad \forall v \in H_0^1(\Omega_1).$$

Since $R \to \infty$, u is the solution to the problem (2.35),(2.36). Because u is unique, u_h converges to u as $R \to \infty$. □

Although the above strategy generates convergent approximate solutions, the rate is very slow. The artificial boundary condition (2.37) should be improved. Using the theory in Section 1.1 and Section 2.2 the solution

u on the domain Ω_2 can be expressed in terms of single or double layer potentials. For definiteness, let us consider two dimensional domains. By (2.20)

$$u(x) = \frac{1}{2\pi} \int_{\partial\Omega_2} \omega(y) \ln \frac{1}{|x-y|} \, ds_y + C, \quad \int_{\partial\Omega_2} \omega(y) \, ds_y = 0, \qquad (2.38)$$

so we can impose a boundary condition on $\partial\Omega_0$ instead of (2.37),

$$u(x) = \frac{1}{2\pi} \int_{\partial\Omega_2} \omega(y) \ln \frac{1}{|x-y|} \, ds_y + C, x \in \partial\Omega_0, \quad \int_{\partial\Omega_0} \omega(y) \, ds_y = 0.$$
$$(2.39)$$

Here ω and C are unknown, so we need another boundary condition. By (1.9)

$$\frac{\partial u(x_0)}{\partial\nu^+} = -\frac{1}{2}\omega(x_0) + \frac{1}{2\pi} \int_{\partial\Omega_0} \omega(y) \frac{\partial}{\partial\nu_x} \ln \frac{1}{|x_0-y|} \, ds_y,$$

for $x_0 \in \partial\Omega_0$. Therefore we get another boundary condition,

$$\frac{\partial u}{\partial\nu^+} = -\frac{1}{2}\omega(x) + \frac{1}{2\pi} \int_{\partial\Omega_0} \omega(y) \frac{\partial}{\partial\nu_x} \ln \frac{1}{|x-y|} \, ds_y, \quad x \in \partial\Omega_0. \qquad (2.40)$$

Theorem 44. *Let $g \in H^{1/2}(\partial\Omega)$. The problem (2.35),(2.36),(2.39), (2.40) with unknowns u, ω, C on the domain Ω_1 is equivalent to the problem (2.35),(2.36) on the domain Ω.*

Proof. Let u be the solution to (2.35),(2.36), then u is a harmonic function, so it is analytic. By Theorem 42 u can be expressed by (2.38) on Ω_2, so u is the solution to (2.35),(2.36),(2.39),(2.40).

Conversely, if u is a solution to (2.35),(2.36),(2.39),(2.40), then we can define u on Ω_2 by (2.38). Then $u \in H^{1,*}(\Omega)$, u is a harmonic function on Ω_1 and Ω_2 respectively, and u satisfies the boundary condition (2.36). We only need to show that u is a harmonic function on the entire domain Ω.

The weak formulations are:

$$\int_{\Omega_1} \nabla u \cdot \nabla v \, dx = < -\frac{1}{2}\omega + \frac{1}{2\pi} \int_{\partial\Omega_0} \omega(y) \frac{\partial}{\partial\nu} \ln \frac{1}{|\cdot - y|} \, ds_y, v >_{\partial\Omega_0},$$
$$\forall v \in H^1(\Omega_1), v|_{\partial\Omega} = 0$$

and

$$\int_{\Omega_2} \nabla u \cdot \nabla v \, dx = < \frac{1}{2}\omega - \frac{1}{2\pi} \int_{\partial\Omega_0} \omega(y) \frac{\partial}{\partial\nu} \ln \frac{1}{|\cdot - y|} \, ds_y, v >_{\partial\Omega_0},$$
$$\forall v \in H^{1,*}(\Omega_2).$$

Now we take $v \in H_0^{1,*}(\Omega)$ and add them together, then we get

$$a(u,v) = \int_{\Omega_1} \nabla u \cdot \nabla v \, dx + \int_{\Omega_2} \nabla u \cdot \nabla v \, dx = 0.$$

Therefore u is a harmonic function on Ω.   □

The artificial boundary conditions (2.39)(2.40) are exact, so they yield better results.

2.4   Dirichlet to Neŭmann operator

The above mapping $u|_{\partial \Omega_0} \to (\omega, C) \to \frac{\partial u}{\partial \nu}|_{\partial \Omega_0}$ defines an operator K : $H^{1/2}(\partial \Omega_0) \to H^{-1/2}(\partial \Omega_0)$, which is called a "Dirichlet to Neŭmann operator". Using the operator an exact artificial boundary condition is designed to get a problem equivalent to (2.35)(2.36):

$$\triangle u = 0, \qquad x \in \Omega_1,$$

$$u = g, \qquad x \in \partial \Omega,$$

$$\frac{\partial u}{\partial \nu} = Ku, \qquad x \in \partial \Omega_0.$$

The operator K depends only on the the equation and the artificial boundary $\partial \Omega_0$. To obtain K, one needs to solve an integral equation for ω. However the choosing of Ω_0 is quite flexible. If on some occasions $\Omega_0 = B(O,R)$, then using the results in Section 2.1 K can be expressed explicitly. This idea is due to Feng and named after "natural boundary reduction". See [Feng, K. (1982)][Feng, K. (1983)][Feng, K. (1984)] [Feng, K. (1994)][Feng, K. and Yu, D. (1983)], also see [Han, H. and Ying, L. (1980)].

Let us consider the exterior problem of harmonic functions first. Differential of (2.3) gives

$$\frac{\partial u}{\partial r} = \frac{1}{2\pi} \int_0^{2\pi} \frac{4rR^2 - 2R(r^2 + R^2)\cos(\theta - \varphi)}{(r^2 + R^2 - 2rR\cos(\theta - \varphi))^2} g(\varphi) \, d\varphi.$$

Let $r \to R$, then we get the following formula formally,

$$\frac{\partial u}{\partial r}\bigg|_{r=R} = \frac{1}{4\pi R} \int_0^{2\pi} \frac{1}{\sin^2 \frac{\theta - \varphi}{2}} g(\varphi) \, d\varphi. \tag{2.41}$$

Let ν denote the outward unit normal vector along the boundary, then we get an explicit expression for K:

$$\frac{\partial u}{\partial \nu} = -\frac{1}{4\pi R} \int_0^{2\pi} \frac{1}{\sin^2 \frac{\theta - \varphi}{2}} g(\varphi) \, d\varphi.$$

For the Helmholtz equation differential of (2.16) gives

$$u_r(r, \theta) = \sum_{n=-\infty}^{\infty} a_n \omega H_n^{(1)'}(\omega r) e^{in\theta}, \qquad (2.42)$$

then formally we get

$$u_r(R, \theta) = \sum_{n=-\infty}^{\infty} a_n \omega H_n^{(1)'}(\omega R) e^{in\theta}. \qquad (2.43)$$

Let ν be directed to the exterior of the domain Ω^c, we define $tr(u)(\theta) = u(R, \theta) = g(\theta)$, $tr(u_\nu)(\theta) = u_\nu(R, \theta)$, then (2.16) and (2.43) together defines an operator K, such that

$$tr(u_\nu) = K(tr(u)),$$

or explicitly

$$-u_r(R, \theta) = K(\theta) * u(R, \theta) = \int_0^{2\pi} K(\theta - \theta') u(R, \theta') \, d\theta', \qquad (2.44)$$

where the function K is

$$K(\theta) = \frac{1}{2\pi} \sum_{n=-\infty}^{\infty} \left(-\omega \frac{H_n^{(2)'}(\omega R)}{H_n^{(1)}(\omega R)} \right) e^{i\omega\theta}.$$

The integral in (2.44) is hyper-singular. K is in fact a pseudo-differential operator of order 1 on the boundary manifold $\partial\Omega$, and defines a linear continuous map

$$K : H^s(\partial\Omega) \to H^{s-1}(\partial\Omega).$$

Parallel to the association of the operator $\triangle + \omega^2$ in the domain Ω with the bilinear functional

$$a(u, v) = \int_\Omega (\nabla u \cdot \nabla \bar{v} + \omega^2 u \bar{v}) \, dx,$$

there is also an association of the operator K on the boundary $\partial\Omega$ with the bilinear functional

$$b(\varphi, \psi) = R \int_0^{2\pi} \bar{\psi}(\theta) \, d\theta \int_0^{2\pi} K(\theta - \theta')\varphi(\theta') \, d\theta',$$

which is inherently related to $a(\cdot, \cdot)$ by the equality,

$$a(u, v) = b(tr(u), tr(v)).$$

Being the same, for three dimensional case differential of (2.19) gives

$$u_r(r, \theta) = \sum_{n=0}^{\infty} \sum_{m=-n}^{n} \omega a_{nm} \zeta_{n+\frac{1}{2}}^{(2)}(\omega r) Y_{nm}(\theta, \varphi), \qquad (2.45)$$

then (2.19) and (2.45) together define an operator K.

We will provide a rigorous mathematical setting of the operator K in the next section.

2.5   Finite part of divergent integrals

Sometimes some integrals in the expressions of some DtN operators are divergent in the sense of Riemann integral. In fact they are singular integrals in the sense of Cauchy or Hadamard. To provide a rigorous meaning of these integrals, let us introduce some basic definitions. [Feng, K. (1982)]

Let

$$x_+^m = \begin{cases} x^m, & x > 0, \\ 0, & x < 0, \end{cases}$$

where m is a complex number. Letting $\varphi \in C_0^{\infty}(\mathbb{R})$, we consider the integral

$$\int_{-\infty}^{\infty} x_+^m \varphi(x) \, dx = \int_0^{\infty} x_+^m \varphi(x) \, dx, \qquad (2.46)$$

which converges in the sense of Riemann if $\mathrm{Re}(m) > -1$. Therefore x_+^m is a distribution, denoted by $[x_+^m]$. It is easy to see that

$$\frac{d^p}{dx^p}\left\{ \frac{1}{\Gamma(m+p+1)}[x_+^{m+p}] \right\} = \frac{1}{\Gamma(m+1)}[x_+^m],$$

for positive p.

Let us define distributions for $\mathrm{Re}(m) \le -1$. The dual product $([x_+^m], \varphi)$ defines an analytic function for $\mathrm{Re}(m) > -1$. The analytic continuation can be defined as the following: For $m \neq$ negative integers, and a non-negative integer p, satisfying $\mathrm{Re}(m) + p > -1$, define a distribution,

$$Pfx_+^m = \frac{1}{(m+p)\cdots(m+1)}\frac{d^p}{dx^p}[x_+^{m+p}] = \frac{\Gamma(m+1)}{\Gamma(m+p+1)}\frac{d^p}{dx^p}[x_+^{m+p}].$$

The above definition is independent of the choosing of the parameter p, and $Pfx_=^m = [x_+^m]$ provided $\mathrm{Re}(m) > -1$.

By the definition of distributions one gets a finite part of (2.46),

$$(Pfx_+^m, \varphi) = (-1)^p \frac{\Gamma(m+1)}{\Gamma(m+p+1)}\int_0^\infty x^{m+p}\varphi^{(p)}(x)\,dx, \qquad (2.47)$$

where $\varphi^{(p)} = \frac{d^p\varphi}{dx^p}$. It is easy to check that (Pfx_+^m, φ) is an analytic function with first order poles $m =$ negative integers.

One explicit expression of (2.47) is derived as follows. For $\varepsilon > 0$, we consider $\int_\varepsilon^\infty x^m\varphi(x)\,dx$. Let $a > \varepsilon$ be a fixed number, and

$$\varphi_p = \sum_{k=0}^p \frac{\varphi^{(k)}(0)}{k!}x^k,$$

then

$$\int_\varepsilon^\infty x^m\varphi(x)\,dx = \int_\varepsilon^a x^m(\varphi(x)-\varphi_p(x))\,dx + \int_\varepsilon^a x^m\varphi_p(x)\,dx + \int_a^\infty x^m\varphi(x)\,dx.$$

If m is not a negative integer, then

$$\int_\varepsilon^a x^m\varphi_p(x)\,dx = \sum_{k=0}^p \frac{\varphi^{(k)}(0)}{k!}\frac{a^{m+k+1}}{m+k+1} - \sum_{k=0}^p \frac{\varphi^{(k)}(0)}{k!}\frac{\varepsilon^{m+k+1}}{m+k+1}.$$

If m is a negative integer, then

$$\int_\varepsilon^a x^m\varphi_p(x)\,dx = \sum_{0<k<p, k\neq -m-1} \frac{\varphi^{(k)}(0)}{k!}\frac{a^{m+k+1}}{m+k+1} + \frac{\varphi^{(-m-1)}(0)}{(-m-1)!}\ln a$$

$$- \sum_{0\leq k\leq p, k\neq -m-1} \frac{\varphi^{(k)}(0)}{k!}\frac{\varepsilon^{m+k+1}}{m+k+1} - \frac{\varphi^{(-m-1)}(0)}{(-m-1)!}\ln\varepsilon.$$

If $k \neq -m-1$, the limit of $\frac{\varepsilon^{m+k+1}}{m+k+1}$ is either zero or infinity as $\varepsilon \to 0$. Naturally the finite part of the divergent integral is defined by: If m is not a negative integer, then

$$Pf \int_0^\infty x^m \varphi(x)\, dx = \int_0^a x^m (\varphi(x) - \varphi_p(x))\, dx_p + \int_a^\infty x^m \varphi(x)\, dx$$
$$+ \sum_{k=0}^{p} \frac{\varphi^{(k)}(0)}{k!} \frac{a^{m+k+1}}{m+k+1}. \tag{2.48}$$

If m is a negative integer, then

$$Pf \int_0^\infty x^m \varphi(x)\, dx = \int_0^a x^m (\varphi(x) - \varphi_p(x))\, dx_p + \int_a^\infty x^m \varphi(x)\, dx$$
$$+ \sum_{0 \le k \le p, k \ne -m-1} \frac{\varphi^{(k)}(0)}{k!} \frac{a^{m+k+1}}{m+k+1} + \frac{\varphi^{(-m-1)}(0)}{(-m-1)!} \ln a. \tag{2.49}$$

Lemma 41. *If m is not a negative integer, then*

$$Pf \int_0^\infty x^m \varphi(x)\, dx = (Pf x_+^m, \varphi). \tag{2.50}$$

Proof.   It is easy to see that

$$\int_0^\infty x^m \varphi(x)\, dx = ([x_+^m], \varphi)$$

if $\mathrm{Re}(m) > -1$. However both sides are analytic functions with respect to m, and both sides of (2.50) are analytic continuations, therefore by the uniqueness they are equal.   □

It remains to prove that the definition of (2.48)(2.49) is independent of the choosing of parameters p and a. If m is not a negative integer, the assertion is already contained in the previous lemma. Here we give another proof which is valid for all complex numbers m.

Lemma 42. *The definition of (2.48),(2.49) is independent of p and a.*

Proof.   Let k_* be the smallest non-negative integer p satisfying $\mathrm{Re}(m) + p + 1 > 0$. If $k > k_*$ and $\varepsilon \to 0$, then

$$\frac{\varepsilon^{m+k+1}}{m+k+1} \to 0.$$

By the definition, if m is not a negative integer,

$$Pf \int_0^\infty x^m \varphi(x)\,dx = \lim_{\varepsilon \to 0} \left[\int_\varepsilon^\infty x^m \varphi(x)\,dx + \sum_{k=0}^p \frac{\varphi^{(k)}(0)}{k!} \frac{\varepsilon^{m+k+1}}{m+k+1} \right]$$

$$= \lim_{\varepsilon \to 0} \left[\int_\varepsilon^\infty x^m \varphi(x)\,dx + \sum_{k=0}^{k_*} \frac{\varphi^{(k)}(0)}{k!} \frac{\varepsilon^{m+k+1}}{m+k+1} \right].$$

The right hand side is independent of p and a.

Noting that if m is a negative integer, $k_* = -m - 1$, then we get

$$Pf \int_0^\infty x^m \varphi(x)\,dx$$

$$= \lim_{\varepsilon \to 0} \left[\int_\varepsilon^\infty x^m \varphi(x)\,dx + \sum_{k=0}^{k_*-1} \frac{\varphi^{(k)}(0)}{k!} \frac{\varepsilon^{m+k+1}}{m+k+1} + \frac{\varphi^{(k_*)}(0)}{k_*!} \ln \varepsilon \right],$$

which is also independent of p and a. $\qquad\qquad\qquad\qquad\square$

We have derived (2.41) formally in the previous section. Now we are ready to prove it rigorously. Let us assume that $g \in C^\infty(\partial\Omega)$ first. By (2.1) we get

$$\frac{\partial u}{\partial r} = -\frac{1}{2\pi} \int_0^{2\pi} \left\{ \sum_{n=-\infty}^\infty |n| r^{-|n|-1} e^{in(\theta-\varphi)} \right\} g(\varphi)\,d\varphi,$$

where for simplicity we take $R = 1$. We define a function

$$K_r(\theta) = \frac{1}{2\pi} \sum_{n=-\infty}^\infty |n| r^{-|n|-1} e^{in\theta}, \qquad (2.51)$$

then for a fixed $r \in (1, \infty)$ the right hand side can be expressed in terms of a convolution product, or a dual product on the space $C^\infty(\partial\Omega)$:

$$\frac{\partial u}{\partial r} = -K_r * g = - < K_r(\theta - \cdot), g > .$$

For this boundary condition the solution u to the exterior problem belongs to $C^\infty(\bar\Omega)$, Therefore $\frac{\partial u}{\partial r}$ converges as $r \to 1$. The functions K_r is thus convergent weakly with respect to r in the dual space $C^\infty(\bar\Omega)'$. Let the limit be $K \in C^\infty(\bar\Omega)'$, which is a distribution by definition. We obtain the derivative at $r = 1$,

$$\frac{\partial u}{\partial r} = - < K(\theta - \cdot), g > . \qquad (2.52)$$

Now let us derive the expression of K. Let us define

$$K_{1r}(\theta) = \frac{1}{2\pi i} \sum_{\substack{n=-\infty \\ n \neq 0}}^{\infty} \frac{|n|}{n} r^{-|n|-1} e^{in\theta}, \tag{2.53}$$

and

$$K_{2r}(\theta) = -\frac{1}{2\pi} \sum_{\substack{n=-\infty \\ n \neq 0}}^{\infty} \frac{1}{|n|} r^{-|n|-1} e^{in\theta}, \tag{2.54}$$

then

$$\frac{\partial K_{1r}}{\partial \theta} = K_r, \qquad \frac{\partial K_{2r}}{\partial \theta} = K_{1r},$$

for $r \in (1, \infty)$. Consequently

$$< K_r(\theta - \cdot), g > = < K_{1r}(\theta - \cdot), g' > = < K_{2r}(\theta - \cdot), g'' > .$$

Letting $r \to 1$, by (2.51) K_{2r} converges in $L^2(\partial\Omega)$ to

$$\lim_{r \to 1} K_{2r}(\theta) = -\frac{1}{2\pi} \sum_{n=-\infty}^{\infty} \frac{1}{|n|} e^{in\theta} = -\frac{1}{\pi} \ln \left| 2 \sin \frac{\theta}{2} \right|,$$

which is a function in the classical sense. Therefore

$$\left. \frac{\partial u}{\partial r} \right|_{r=1} = - < K(\theta - \cdot), g > = \frac{1}{\pi} \int_0^{2\pi} \ln \left| 2 \sin \frac{\theta - \varphi}{2} \right| g''(\varphi) \, d\varphi. \tag{2.55}$$

This is the expression of the distribution $-K$. Taking classical derivatives on $\theta \in (0, 2\pi)$ we get

$$\frac{d}{d\theta} \left\{ \frac{1}{\pi} \ln \left| 2 \sin \frac{\theta}{2} \right| \right\} = \frac{1}{2\pi} \cot \frac{\theta}{2}, \quad \frac{d}{d\theta} \left\{ \frac{1}{2\pi} \cot \frac{\theta}{2} \right\} = -\frac{1}{4\pi \sin^2 \frac{\theta}{2}}.$$

Noting (2.53),(2.54) we can define some distributions as the limit of some divergent series,

$$K = Pf \left\{ -\frac{1}{4\pi \sin^2 \frac{\theta}{2}} \right\} = Pf \left\{ \frac{1}{2\pi} \sum_{n=-\infty}^{\infty} |n| e^{in\theta} \right\},$$

and

$$K_1 = Pf\left\{\frac{1}{2\pi}\cot\frac{\theta}{2}\right\} = Pf\left\{\frac{1}{2\pi i}\sum_{\substack{n=-\infty \\ n\neq 0}}^{\infty}\frac{|n|}{n}e^{in\theta}\right\}.$$

The derivative of u is the finite part of a hyper-singular integral,

$$\left.\frac{\partial u}{\partial r}\right|_{r=1} = Pf\left\{\frac{1}{4\pi}\int_0^{2\pi}\frac{1}{\sin^2\frac{\theta-\varphi}{2}}g(\varphi)\,d\varphi\right\}. \qquad (2.56)$$

For general R, the formulas are,

$$K = Pf\left\{\frac{1}{2\pi R}\sum_{n=-\infty}^{\infty}|n|e^{in\theta}\right\},$$

and

$$\left.\frac{\partial u}{\partial r}\right|_{r=R} = Pf\left\{\frac{1}{4\pi R}\int_0^{2\pi}\frac{1}{\sin^2\frac{\theta-\varphi}{2}}g(\varphi)\,d\varphi\right\}.$$

2.6 Numerical approximation

We investigate the implementation of the boundary element method in this section. Some meshes on the boundaries are required in the numerical computation of the above integral equations, and if the exterior domains are truncated, some meshes on some bounded domains are required as well. First of all let us consider the integral equation (1.11) for two dimensional problems as an example.

Let us fix a point on $\partial\Omega$ and let $s = 0$ at this point. $s \in [0, L]$ is defined by the arc length along anticlockwise direction. The curve $\partial\Omega$ is divided into some arc segments by some nodes $s_1 = 0, s_2, \cdots, s_N$, and the elements, e_i, $i = 1, 2, \cdot, N$, are the arc segments $s_1\breve{s}_2, s_2\breve{s}_3, \cdots, s_N\breve{s}_1$. Let the approximate solution be $\sigma_h(s)$, which is piecewise smooth and continuous on $\partial\Omega$. For example we can use linear interpolation on each element, that is, we define

$$\sigma_h(s) = \frac{s - s_i}{h_i}\sigma_h(s_{i+1}) + \frac{s_{i+1} - s}{h_i}\sigma_h(s_i)$$

on $s_i \tilde{s}_{i+1}$, where h_i is the arc length of this element. Usually there are two methods to solve σ_h:

1. Collocation method

Let us write the function K by $K(s,t)$, $s \in [0, L], t \in [0, L]$. We assume that the equation (1.11) is satisfied at nodes,

$$\frac{1}{2}\sigma_h(s_i) + \int_0^L \sigma_h(t)K(s_i, t)\, dt = g(s_i), \quad i = 1, 2, \cdots, N, \qquad (2.57)$$

which is an algebraic system with N equations and N unknowns $\sigma_h(s_i)$. We can define a finite dimensional space

$$S(\partial\Omega) = \{\sigma_h \in C([o, L]); \sigma_h(0) = \sigma_h(L), \sigma_h|_{e_i} \in P_1(e_i), i = 1, 2, \cdots, N\},$$

where P_1 is the set of linear functions. Then we can define a set of basis functions in S_h:

$$L_i \in S(\partial\Omega), \qquad L_i(s_i) = 1, L_i(s_j) = 0, j \neq i, \quad i = 1, 2, \cdots, N.$$

The function σ_h can be developed in terms of the basis functions:

$$\sigma_h(s) = \sum_{i=1}^N \sigma_h(s_i)L_i(s). \qquad (2.58)$$

Then the equation (2.57) can be written as

$$\frac{1}{2}\sigma_h(s_i) + \sum_{j=1}^N \left(\int_0^L L_j(t)K(s_i, t)\, dt\right)\sigma_h(s_j) = g(s_i), \quad i = 1, 2, \cdots, N.$$

2. Galerkin method

Multiply the equation by $\tau \in C(\partial\Omega)$ and integrate the equation over $\partial\Omega$, then we have

$$\frac{1}{2}\int_{\partial\Omega} \sigma(x)\tau(x)\, dx + \int_{\partial\Omega}\int_{\partial\Omega} \sigma(y)K(x, y)\tau(x)\, ds_y\, ds_x = \int_{\partial\Omega} g(x)\tau(x)\, ds_x.$$

The Galerkin scheme is: Find $\sigma_h \in S(\partial\Omega)$, such that

$$\frac{1}{2}\int_{\partial\Omega} \sigma_h(x)\tau_h(x)\, dx + \int_{\partial\Omega}\int_{\partial\Omega} \sigma_h(y)K(x, y)\tau_h(x)\, ds_y\, ds_x$$
$$= \int_{\partial\Omega} g(x)\tau_h(x)\, ds_x \quad \forall \tau_h \in S(\partial\Omega). \qquad (2.59)$$

Since $S(\partial\Omega)$ is a N-dimensional linear space, the equation (2.59) is a $N \times N$ algebraic system. Let the unknown function σ_h be developed in terms of the basis functions as (2.58), and

$$\tau_h(s) = \sum_{i=1}^{N} \tau_h(s_i) L_i(s),$$

then the equation (2.59) can be written as

$$\sum_{j=1}^{N} k_{ij}\sigma_h(s_j) = g_i, \qquad i = 1, 2, \cdots, N,$$

where

$$k_{ij} = \frac{1}{2}\int_0^L L_j(s)L_i(s)\,ds + \int_0^L \int_0^L L_j(t)K(s,t)L_i(s)\,dt\,ds,$$

and

$$g_i = \int_0^L g(s)L_i(s)\,ds.$$

The strategy for solving the integral equation (1.11) for three dimensional problems is similar. However, it is difficult to construct a continuous and piecewise polynomial function on a closed smooth surface $\partial\Omega$, so usually the boundary $\partial\Omega$ is approximated by an surface $\partial\Omega_h$, which causes an additional error.

Next let us consider the problem (2.26). Let us define a subspace of $S(\partial\Omega)$:

$$S_0(\partial\Omega) = \left\{ \omega \in S(\partial\Omega); \int_0^L \omega(s)\,ds = 0 \right\},$$

then the approximation to (2.26) is: Find $\omega_h \in S_0(\partial\Omega)$, such that

$$b(\omega_h, \chi_h) = <g, \chi_h>_{\partial\Omega}, \qquad \forall \chi \in S_0(\partial\Omega). \qquad (2.60)$$

The bilinear form $b(\cdot, \cdot)$ defined in (2.23),

$$b(\omega, \chi) = \frac{1}{2\pi}\int_{\partial\Omega}\int_{\partial\Omega} \omega(y)\chi(x)\ln\frac{1}{|x-y|}\,ds_y\,ds_x,$$

is symmetric, so the equation (2.60) is equivalent to

$$\frac{d}{d\omega_h}\left\{ \frac{1}{2}b(\omega_h, \omega_h) - <g, \omega_h>_{\partial\Omega} \right\} = 0.$$

With respect to the constraint $\int_0^L \omega_h(s)\,ds = 0$, we introduce a Lagrangian multiplier λ and define a function

$$\frac{1}{2}b(\omega_h, \omega_h) - < g, \omega_h >_{\partial\Omega} + \lambda \int_0^L \omega_h(s)\,ds.$$

The problem (2.60) is equivalent to: Find $(\omega_h, \lambda) \in S(\partial\Omega) \times \mathbb{R}$, such that

$$b(\omega_h, \chi_h) + \lambda \int_0^L \chi_h(s)\,ds = \int_0^L g(s)\chi_h(s)\,ds, \qquad \forall \chi \in S(\partial\Omega),$$

and

$$\int_0^L \omega_h(s)\,ds = 0.$$

This is a $(N+1) \times (N+1)$ algebraic system.

Finally let us consider the exterior problem of the Poisson equation,

$$-\triangle u = f, \qquad x \in \Omega, \tag{2.61}$$

$$u = 0, \qquad x \in \partial\Omega \tag{2.62}$$

in two dimensional space, where $f \in C_0(\bar{\Omega})$. For simplicity of the domain partition, let us assume that Ω^c is a polygon. Let Ω_0 be a disk $B(O, R)$ with R large enough, so that supp $f \subset B(O, R)$. Let $\Omega_1 = \Omega_0 \cap \Omega$, then by the argument in Section 2.4 the problem is equivalent to

$$-\triangle u = f, \qquad x \in \Omega_1, \tag{2.63}$$

$$u = 0, \qquad x \in \partial\Omega, \tag{2.64}$$

$$\frac{\partial u}{\partial \nu} = Ku, \qquad x \in \partial\Omega_0, \tag{2.65}$$

where K is the Dirichlet to Neŭmann operator. Let

$$a(u, v) = \int_{\Omega_1} \nabla u \cdot \nabla v \, dx,$$

then the variational formulation of the problem is: Find $u \in H^1(\Omega_1)$, such that $u|_{\partial\Omega} = 0$ and

$$a(u, v) - (Ku, v)_{\partial\Omega_0} = (f, v)_{\Omega_1}, \qquad \forall v \in H^1(\Omega_1), v|_{\partial\Omega} = 0.$$

Applying the finite element method to solve this variational problem, we set a mesh. For example, the domain Ω_1 is divided into triangular elements $\{e_i\}$, and linear interpolation is designed on each element with respect to the vertices. Since $\partial\Omega_0$ is a circle, each element neighboring the circle possesses an arc edge. The finite element space is defined by,

$$S(\Omega_1) = \{u \in H^1(\Omega_1); u|_{e_i} \in P_1(e_i), \forall i, u|_{\partial\Omega} = 0\}.$$

The finite element formulation is: Find $u_h \in S(\Omega_1)$, such that

$$a(u_h, v) - (Ku_h, v)_{\partial\Omega_0} = (f, v)_{\Omega_1}, \qquad \forall v \in S(\Omega_1). \tag{2.66}$$

The evaluation of $a(u_h, v)$ and (f, v) is routine. However the evaluation of $(Ku_h, v)_{\partial\Omega_0}$ requires some special consideration. There are two kinds of approaches:

1. Truncation

$$(Ku_h, v)_{\partial\Omega_0} = \int_0^{2\pi} Pf\left\{-\frac{1}{2\pi}\int_0^{2\pi}\sum_{n=-\infty}^{\infty}|n|e^{in(\theta-\varphi)}u_h(R, \varphi)\,d\varphi\right\}v(R, \theta)\,d\theta.$$

Let N be a finite number, then approximately

$$(Ku_h, v)_{\partial\Omega_0} \approx \int_0^{2\pi}\left\{-\frac{1}{2\pi}\int_0^{2\pi}\sum_{n=-N}^{N}|n|e^{in(\theta-\varphi)}u_h(R, \varphi)\,d\varphi\right\}v(R, \theta)\,d\theta.$$

$$\tag{2.67}$$

2. Integration by parts
   If u_h and v are infinitely differentiable, we denote

$$g(\theta) = u_h(R, \theta), \gamma(\theta) = v(R, \theta),$$

and

$$K_2(\theta) = -\frac{1}{\pi R}\ln\left|2\sin\frac{\theta}{2}\right|,$$

then by (2.55) it holds that

$$Ku_h(\theta) = -<K_2(\theta - \cdot), g''>.$$

Consequently we have

$$(Ku_h, v)_{\partial\Omega_0} = -\int_0^{2\pi} < K_2(\theta - \cdot), g'' > \gamma(\theta) R\, d\theta$$

$$= \int_0^{2\pi} < K_2'(\theta - \cdot), g' > \gamma(\theta) R\, d\theta$$

$$= -\int_0^{2\pi} < K_2(\theta - \cdot), g' > \gamma'(\theta) R\, d\theta,$$

that is

$$(Ku_h, v)_{\partial\Omega_0} = \frac{1}{\pi} \int_0^{2\pi} \int_0^{2\pi} \ln\left| 2\sin\frac{\theta - \varphi}{2} \right| g'(\varphi)\gamma'(\theta)\, d\varphi\, d\theta. \qquad (2.68)$$

The definition of $S(\Omega_1)$ implies that $u|_{\partial\Omega_0} \in H^1(\partial\Omega_0)$, so applying the fact that $C^\infty(\partial\Omega_0)$ is dense in $H^1(\partial\Omega_0)$ we can show that (2.68) holds for $u_h, v \in S(\Omega_1)$. The integrals in (2.68) are convergent, and can be evaluated using a numerical scheme of integration.

2.7 Error estimates

Let us prove convergence and estimate errors to some of the above schemes. The first example is (2.60). We assume that the exact solution $\omega \in H^2(\partial\Omega)$, and let Π denote the interpolation operator. Let h be the maximum length of all elements. Then owing to the error estimates of interpolation: [Ciarlet, P.G. (1978)]

$$\|\Pi\omega - \omega\|_{-1/2} \le C\|\Pi\omega - \omega\|_0 \le Ch^2|\omega|_2.$$

Let $e_h = < \Pi\omega, 1 > / \int_{\partial\Omega} ds$, then

$$|e_h| = \frac{|< \Pi\omega - \omega, 1 >|}{\int_{\partial\Omega} ds} \le C\frac{\|\Pi\omega - \omega\|_{-1/2}\|1\|_{1/2}}{\int_{\partial\Omega} ds} \le Ch^2|\omega|_2.$$

Let $\omega_I = \Pi\omega - e_h$, then $\omega_I \in S_0(\partial\Omega)$, and

$$\|\omega_I - \omega\|_{-1/2} \le Ch^2|\omega|_2.$$

The following error estimate is routine:

$$\frac{\alpha}{M}\|\omega - \omega_h\|_{-1/2}^2 \le b(\omega - \omega_h, \omega - \omega_h) = b(\omega - \omega_h, \omega - \omega_I)$$

$$\le \|\omega - \omega_h\|_{-1/2}\|\omega - \omega_I\|_{-1/2}.$$

Consequently we have the following:

Theorem 45. *It holds that*

$$\|\omega - \omega_h\|_{-1/2} \le Ch^2|\omega|_2.$$

The second example is the scheme (2.66). We assume that the partition is regular, that is the interior angles of all elements possess a common lower bound $\theta_0 > 0$.

First of all let us show the existence and uniqueness of the solution to the problem (2.66). We extend a function $u_h \in S(\Omega_1)$ to Ω_2, the exterior of $\partial\Omega_0$, by solving a boundary value problem of the harmonic equation on Ω_2 with the boundary value $u_h|_{\partial\Omega_0}$, then $u_h \in H^{1,*}(\Omega)$, and the application of Green's formula leads to

$$a(u_h, v) - (Ku_h, v)_{\partial\Omega_0} = \int_\Omega \nabla u_h \cdot \nabla v \, dx, \qquad \forall v \in H^{1,*}(\Omega).$$

Owing to the estimate for exterior problems of the harmonic equation and the trace theorem of Sobolev spaces, we obtain

$$|u_h|_{1,\Omega_2} \le C\|u_h\|_{1/2,\partial\Omega_0} \le C\|u_h\|_{1,\Omega_1}$$

Consequently it holds that

$$|a(u_h, v) - (Ku_h, v)_{\partial\Omega_0}| \le C\|u_h\|_{1,\Omega_1}\|v\|_{1,\Omega_1}, \quad \forall u_h, v \in S(\Omega_1),$$

and

$$|a(u_h, u_h) - (Ku_h, u_h)_{\partial\Omega_0}| \ge \alpha\|u_h\|_{1,\Omega_1}^2, \quad \forall u_h \in S(\Omega_1),$$

with $\alpha > 0$. Owing to the Lax-Milgram theorem there is a unique solution. Let u be the exact solution to (2.61),(2.62), then a standard argument leads to the following: (see [Ciarlet, P.G. (1978)])

Theorem 46. *It holds that*

$$\|u - u_h\|_{1,\Omega_1} \le Ch|u|_{2,\Omega_1}.$$

Finally let us consider the error of the truncation (2.67) Let u be the solution to (2.63)-(2.65), and u_N be the solution to

$$-\triangle u_N = f, \qquad x \in \Omega_1,$$

$$u_N = 0, \qquad x \in \partial\Omega,$$

$$\frac{\partial u_N}{\partial \nu} = K_N u_N, \qquad x \in \partial\Omega_0,$$

where

$$K_N u_N = -\frac{1}{2\pi} \int_0^{2\pi} \sum_{n=-N}^{N} |n| e^{in(\theta-\varphi)} u_N(R,\varphi) \, d\varphi.$$

The variational formulation is: Find $u_N \in H^1(\Omega_1)$, such that $u_N|_{\partial\Omega} = 0$, and

$$b_N(u_N, v) = (f, v), \qquad \forall v \in H^1(\Omega_1), v|_{\partial\Omega} = 0, \qquad (2.69)$$

where

$$b_N(u_N, v) = a(u_N, v) - \int_0^{2\pi} K_N u_N(\theta)\overline{v(R,\theta)} R \, d\theta,$$

and

$$a(u_N, v) = \int_{\Omega_1} \nabla u_N \cdot \overline{\nabla v} \, dx,$$

where we notice that although the problem and functions are real, the Fourier series are expressed in terms of complex functions.

Define

$$b(u, v) = a(u, v) - (Ku, v)_{\partial\Omega_0},$$

then the variational formulation of the problem (2.63)-(2.65) is: Find $u \in H^1(\Omega_1)$ such that $u|_{\partial\Omega} = 0$, and

$$b(u, v) = (f, v), \qquad \forall v \in H^1(\Omega_1), v|_{\partial\Omega} = 0. \qquad (2.70)$$

Let us define a norm:

$$\||u_N\|| = b_N(u_N, u_N)^{\frac{1}{2}}$$

$$= \left\{ |u_N|_{1,\Omega_1}^2 + \left| \frac{1}{\sqrt{2\pi}} \int_0^{2\pi} \sum_{n=-N}^{N} \sqrt{|n|} e^{-in\varphi} u_N(R,\varphi) \, d\varphi \right|^2 \right\}^{\frac{1}{2}},$$

then

$$|b_N(u_N, v)| \leq \||u_N\|| \cdot \||v\||.$$

Owing to the Lax-Milgram theorem the problem (2.69) admits a unique solution.

We turn now to prove the convergence of (2.69). We have

$$b_N(u - u_N, u - u_N)$$
$$= b_N(u, u - u_N) - b(u, u - u_N) + b(u, u - u_N) - b_N(u_N, u - u_N)$$
$$= b_N(u, u - u_N) - b(u, u - u_N).$$

By the definition of b and b_N,

$$b_N(u, u - u_N) - b(u, u - u_N) = (Ku - K_N u, u - u_N)_{\partial \Omega_0}.$$

We apply the Schwarz inequality to get

$$|b_N(u, u - u_N) - b(u, u - u_N)| \leq \|Ku - K_N u\|_{0, \partial \Omega_0} \|u - u_N\|_{0, \partial \Omega_0}. \quad (2.71)$$

Let us estimate $\|u - u_N\|_{0, \partial \Omega_0}$. Let R_0 be the maximum positive constant, such that $B(O, R_0) \subset \Omega^c$, R_1 be the minimum constant, such that supp $(f) \subset B(O, R_1)$, and let the constant R satisfy $R > R_1$. Extend the function $u - u_N$ by zero on Ω^c, then we get

$$(u - u_N)(R, \theta) = \int_{R_0}^{R} \frac{\partial(u - u_N)}{\partial r} \, dr.$$

By the Schwarz inequality we obtain

$$((u - u_N)(R, \theta))^2 \leq R \int_{R_0}^{R} \left| \frac{\partial(u - u_N)}{\partial r} \right|^2 dr.$$

Hence we have

$$\int_0^{2\pi} ((u - u_N)(R, \theta))^2 \, d\theta \leq \frac{R}{R_0} \int_{R_0}^{R} r \, dr \int_0^{2\pi} \left| \frac{\partial(u - u_N)}{\partial r} \right|^2 d\theta \leq \frac{R}{R_0} |u - u_N|_{1, \Omega_1}^2,$$

that is

$$\|u - u_N\|_{0, \partial \Omega_0}^2 \leq \frac{R^2}{R_0} |u - u_N|_{1, \Omega_1}^2 \leq \frac{R^2}{R_0} \||u - u_N\||^2. \quad (2.72)$$

Next let us estimate $\|Ku - K_N u\|_{0, \partial \Omega_0}$. The solution u can be developed into a series

$$u = \sum_{n=-\infty}^{\infty} a_n r^{-|n|} e^{in\theta},$$

with

$$a_n = \frac{R_1^{|n|}}{2\pi} \int_0^{2\pi} e^{-in\varphi} u(R_1, \varphi) \, d\varphi,$$

for $r > R_1$. Therefore

$$Ku = \left.\frac{\partial u}{\partial r}\right|_{r=R} = -\sum_{n=-\infty}^{\infty} |n|a_n R^{-|n|-1}e^{in\theta},$$

and

$$Ku - K_N u = -\sum_{|n|>N} |n|a_n R^{-|n|-1}e^{in\theta}.$$

The norm is equal to

$$\|Ku - K_N u\|_{0,\partial\Omega_0}^2 = \int_{\partial\Omega_0} |Ku - K_N u|^2\,ds = \int_0^{2\pi} |Ku - K_N u|^2 R\,d\theta$$
$$= 2\pi R \sum_{|n|>N} |n|^2 |a_n|^2 R^{-2|n|-2}$$

(2.73)

Integrating by parts gives

$$a_n = \frac{R_1^{|n|}}{2\pi n i} \int_0^{2\pi} e^{-in\varphi} \frac{\partial u(R_1,\varphi)}{\partial\varphi}\,d\varphi,$$

(2.74)

so

$$|n|^2 |a_n|^2 = R_1^{2|n|} \left| \frac{1}{2\pi} \int_0^{2\pi} e^{-in\varphi} \frac{\partial u(R_1,\varphi)}{\partial\varphi}\,d\varphi \right|^2.$$

Plugging it into (2.73) gives

$$\|Ku - K_N u\|_{0,\partial\Omega_0}^2 = 2\pi R \sum_{|n|>N} R^{-2|n|-2} R_1^{2|n|} \left| \frac{1}{2\pi} \int_0^{2\pi} e^{-in\varphi} \frac{\partial u(R_1,\varphi)}{\partial\varphi}\,d\varphi \right|^2$$

$$\leq \frac{1}{2\pi} R^{-1} \left(\frac{R}{R_1}\right)^{-2(N+1)} \sum_{|n|>N} \left| \int_0^{2\pi} e^{-in\varphi} \frac{\partial u(R_1,\varphi)}{\partial\varphi}\,d\varphi \right|^2$$

$$\leq \frac{1}{2\pi} R^{-1} \left(\frac{R}{R_1}\right)^{-2(N+1)} \sum_{n=-\infty}^{\infty} \left| \int_0^{2\pi} e^{-in\varphi} \frac{\partial u(R_1,\varphi)}{\partial\varphi}\,d\varphi \right|^2$$

$$\leq R^{-1} \left(\frac{R}{R_1}\right)^{-2(N+1)} \int_0^{2\pi} \left(\frac{\partial u(R_1,\varphi)}{\partial\varphi}\right)^2 d\varphi$$

$$= \left(\frac{R}{R_1}\right)^{-2N-3} \int_{\partial B(O,R_1)} \left(\frac{\partial u(R_1,\varphi)}{\partial s}\right)^2 ds.$$

Therefore

$$\|Ku - K_N u\|_{0,\partial\Omega_0} \leq \left(\frac{R}{R_1}\right)^{-N-\frac{3}{2}} |u|_{1,\partial B(O,R_1)}. \tag{2.75}$$

The combination of (2.71),(2.72),(2.75) gives the following:

Theorem 47. *It holds that*

$$\||u - u_N\|| \leq \frac{R_1}{R_0^{\frac{1}{2}}} \left(\frac{R_1}{R}\right)^{N+\frac{1}{2}} |u|_{1,\partial B(O,R_1)}.$$

Moreover, the solution u is infinitely differentiable on Ω. If we replace (2.74) by integrating by parts for k times, then the result is generalized to

Theorem 48. *It holds that*

$$\||u - u_N\|| \leq \frac{R_1}{R_0^{\frac{1}{2}}} \left(\frac{R_1}{N+1}\right)^{k-1} \left(\frac{R_1}{R}\right)^{N+\frac{1}{2}} |u|_{k,\partial B(O,R_1)}, \quad \forall k = 1, 2, \cdots.$$

2.8 Domain decomposition

The stiffness matrices of the finite element method are usually sparse, while the matrices of the natural boundary element method are usually circulant. It is known that effective numerical methods to solve these two kinds of matrices are different. For the coupling of these two methods one strategy to separate these two matrices is using an iterative procedure. For example the problem (2.66) can be solved by

$$a(u_h^n, v) - (K\lambda^n, v)_{\partial\Omega_0} = (f, v)_{\Omega_1}, \qquad \forall v \in S(\Omega_1),$$

$$\lambda^{n+1} = \theta_n u_h^n + (1 - \theta_n)\lambda^n,$$

where $n = 1, 2, \cdots$, and $\theta_n \in (0,1)$ is the relaxation factor. This is the conventional Dirichlet-Neŭmann alternating scheme in the non-overlapping domain decomposition method.

Another approach is the overlapping domain decomposition method. Again for the problem (2.66) let $R > R_1$, $B(O, R_1) \supset$ supp (f), and define $\Omega_1 = \Omega \cap B(O, R)$, $\Omega_2 = \{x; r > R_1\}$. The scheme reads: $u_{1h}^n \in S(\Omega_1)$, $u_{1h}|_{r=R} = \Pi u_{2h}^{n-1}|_{r=R}$, and

$$a(u_{1h}^n, v) = (f, v)_{\Omega_1}, \qquad \forall v \in S(\Omega_1), v|_{r=R} = 0,$$

and u_{2h} is given by the Poisson formula on Ω_2:

$$u_{2h}^n = \frac{1}{2\pi} \int_0^{2\pi} \frac{r^2 - R_1^2}{r^2 + R_1^2 - 2rR_1 \cos(\theta - \varphi)} u_{1h}^n(R_1, \varphi)\, d\varphi,$$

where $n = 1, 2, \cdots$, and Π is the interpolation operator to the finite element space.

Convergence proof is essentially standard. For details, see [Yu, D. (2002)].

2.9   Boundary perturbation

For two dimensional case if the boundary is a small perturbation of a circle: $\partial\Omega = \{(r, \theta); r = R_0 + \varepsilon\gamma(\theta)\}$, then the solution to the exterior problem can be developed into a power series in terms of the parameter ε. Following the argument by [Bruno, O.P. and Reitich, F. (1996)] let us introduce this approach for the exterior boundary value problem of the Helmholtz equation:

$$(\triangle + \omega^2)u = 0, \qquad x \in \Omega, \tag{2.76}$$

with the boundary conditions,

$$u = g, \qquad x \in \partial\Omega, \tag{2.77}$$

$$\frac{\partial u}{\partial r} - i\omega u = O(\frac{1}{r^2}). \tag{2.78}$$

Let $B(O, R) \supset \overline{\Omega^c}$, and $\Omega_1 = \{(r, \theta); R_0 < r < R\}$. The solution u can be developed as

$$u = \sum_{n=0}^{\infty} u_n(r, \theta)\varepsilon^n. \tag{2.79}$$

The functions u_n satisfy

$$(\triangle + \omega^2)u_n = 0, \qquad x \in \Omega_1,$$

$$u_n = g_n, \qquad r = R_0,$$

$$\frac{\partial u_n}{\partial \nu} = K u_n, \qquad r = R,$$

where K is the Dirichlet to Neŭmann operator given by (2.44), and g_n are equal to

$$g_n(\theta) = \delta_{n0}g(\theta) - \sum_{l=0}^{n-1} \frac{\gamma^{n-l}(\theta)}{(n-l)!} \frac{\partial^{n-l} u_l(R_0, \theta)}{\partial r^{n-l}}.$$

It is observed that the above algorithm suffers from ill conditioning due to significant cancelations [Nicholis, D.P. and Reitich, F. (2004a)][Nicholis, D.P. and Reitich, F. (2004b)]. There is another approach. First of all a mapping is defined to transfer the domain $\Omega \bigcap B(O, R)$ to Ω_1:

$$r' = \frac{(R - R_0)r - R\varepsilon\gamma(\theta)}{(R - R_0) - \varepsilon\gamma(\theta)}, \quad \theta' = \theta.$$

The system is modified, upon dropping primes, to:

$$r\frac{\partial}{\partial r}\left(r\frac{\partial u}{\partial r}\right) + \frac{\partial^2 u}{\partial \theta^2} + r^2\omega^2 u = F(r, \theta, u), \quad R_0 < r < R,$$

$$u = g, \qquad r = R_0,$$

$$\frac{\partial u}{\partial \nu} = Ku + J(\theta), \qquad , r = R,$$

where F and J are some known functions. Then following the same steps u is developed into a series (2.79), and u_n are solved by recursion. A spectral-Galerkin method is given to solve this problem. For details, see [Nicholis, D.P. and Shen, J.].

Chapter 3

Infinite Element Method

Using an infinite number of degree of freedom, one can make a direct discretization of the original exterior problem. The problem is the implementation of the approximate problem which contains infinite unknowns. The introducing of a structured mesh can solve the problem. It is possible to analyze the discrete problem and eliminate infinite number of unknowns in advance. Then only a very small amount of unknowns are needed to be taken into account in the real computation to result in a solution with infinite unknowns exactly. We will present this approach and study the related mathematical problems in this chapter.

3.1 Harmonic equation-two dimensional problems

3.1.1 *Infinite element formulation*

We start with the problem (1.1),(1.2) and (1.4). We assume that Ω^c is a polygon in a plane, and it is in star-shape with respect to the point O, that is, the line segment linking each point on $\partial\Omega$ with the point O is included entirely in the interior of $\partial\Omega$. Applying the approach in Section 1.3, we have the variational formulation: For a given $g \in H^{1/2}(\partial\Omega)$ find $u \in H^{1,*}(\Omega)$ such that $u|_{\partial\Omega} = g$, and

$$a(u, v) = 0, \qquad \forall v \in H_0^{1,*}(\Omega), \tag{3.1}$$

where

$$a(u, v) = \int_\Omega \nabla u \cdot \nabla v \, dx.$$

To solve (3.1) Ω is divided into an infinite number of triangular elements. Let us denote $\partial\Omega$ by Γ_0. Taking a constant $\xi > 1$, we draw the similar curves of Γ_0 with center O and the constants of proportionality $\xi, \xi^2, \cdots, \xi^k, \cdots$, which are denoted by $\Gamma_1, \Gamma_2, \cdots, \Gamma_k, \cdots$ respectively. The domain between two polygons is named a "layer". Afterwards each layer is further divided into elements. It may be proceeded in such a way: some points on Γ_0 are selected as nodes, where the vertices of Γ_0 must be nodes and some other nodes on the line segments can be properly selected according to our requirement, then rays are drawn from the origin to the nodes, consequently every layer is divided into some quadrilaterals which are similar to each other among the layers, finally each quadrilateral is divided into two triangles such that the manner of partition for each layer is the same. The triangular elements are denoted by e_i, $i = 1, 2, \cdots$. We denote $\xi^k\Omega$ the exterior of Γ_k and $\Omega_k = (\xi^{k-1}\Omega) \setminus \overline{(\xi^k\Omega)}$.

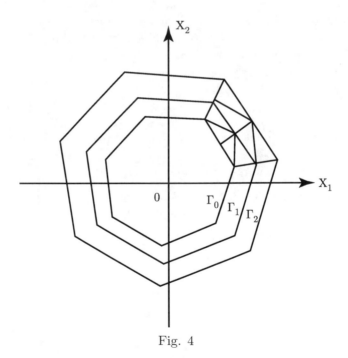

Fig. 4

Sometimes the domain Ω^c may not be in star-shape, and the curve Γ_0 may be a polygon in very complicated shape. Then we may decompose the exterior domain Ω to an unbounded domain with regular shape and a bounded domain with complicated shape like Section 2.3. We still denote by Ω the unbounded domain, which can be divided in the above way, while the bounded domain can be divided into finite elements in conventional way. We need the two styles of partition to be conform to each other, that is, the nodes and line segments of the elements are coincide to each other along the interior boundary. We still first analyze the domain Ω for this situation. We will state an algorithm for calculating the combined stiffness matrix of Ω, which is a matrix with finite order, and which is applied either to solve a boundary value problem directly, or to solve a boundary value problem for a domain with complicated shape by assembling it with the stiffness matrices of other elements. For the later case Ω is treated as one element.

we consider the infinite element subspaces of $H^{1,*}(\Omega)$. Let the nodes be the vertices of all triangular elements, and linear interpolation is applied in all elements. Then let the interpolating functions belong to the space $H^{1,*}(\Omega)$. The infinite element spaces is defined as the following:

$$S(\Omega) = \left\{ u \in H^{1,*}(\Omega); u\big|_{e_i} \in P_1(e_i), i = 1, 2, \cdots \right\},$$

$$S_0(\Omega) = \left\{ u \in S(\Omega); u\big|_{\partial\Omega} = 0 \right\},$$

$$S(\Gamma_k) = \left\{ u\big|_{\Gamma_k}; u \in S(\Omega) \right\}.$$

Let $g_I \in S(\Gamma_0)$ be an approximation of g. The formulation of the infinite element scheme is: find $u_h \in S(\Omega)$ such that $u|_{\partial\Omega} = g_I$, and

$$a(u_h, v) = 0, \qquad \forall v \in S_0(\Omega). \tag{3.2}$$

Lemma 43. *The problem (3.2) admits a unique solution.*

Proof. We define a function $u_0 \in S(\Omega)$, such that $u_0|_{\partial\Omega} = g_I$ and $u_0 \equiv 0$ for all $x \in \xi\Omega$, which can be obtained by interpolation with respect to all nodes. Let $w_h = u_h - u_0$, then w_h is the solution to: find $w_h \in S_0(\Omega)$, such that

$$a(w_h, v) = -a(u_0, v), \qquad \forall v \in S_0(\Omega).$$

Owing to the Lax-Milgram theorem, existence and uniqueness follows. $\square$

3.1.2   *Tranfer matrix*

We now turn to the implementation of the above scheme. In order to carry out the real computation, we deduce an equivalent and finite form of it. We consider the polygon Γ_k, and starting from one node, we arrange the nodes of Γ_k in an order according to the anticlockwise direction. Let the number of nodes be n. The nodal values of the solution of (3.2) are also arranged in an order, which are denoted by $y_k^{(1)}, y_k^{(2)}, \cdots, y_k^{(n)}$, and they form a n-dimensional column vecter $y_k = (y_k^{(1)}, y_k^{(2)}, \cdots, y_k^{(n)})^T$. Let us denote $y_k = B_k u_h$, where B_k is the trace operator. We will identify $y_k \in \mathbb{R}^n$ with $u|_{\Gamma_k}$ rather than distinguish them.

We consider the k-th layer Ω_k. The element stiffness matrix of each triangular element can be evaluated by conventional way (for instance, see [Ciarlet, P.G. (1978)]). Upon assembling the element stiffness matrices by the nodes in the k-th layer, we obtain a stiffness matrix of one layer, which is denoted by

$$\begin{pmatrix} K_0 & -A^T \\ -A & K_0' \end{pmatrix}, \tag{3.3}$$

that is,

$$\int_{\Omega_k} \nabla u_h \cdot \nabla v \, dx = \begin{pmatrix} y_{k-1}^T & y_k^T \end{pmatrix} \begin{pmatrix} K_0 & -A^T \\ -A & K_0' \end{pmatrix} \begin{pmatrix} z_{k-1} \\ z_k \end{pmatrix},$$

where $z_{k-1} = B_{k-1} v$ and $z_k = B_k v$, and K_0, K_0', and A are $n \times n$ matrices. It is known by the fundamental theory of the finite element method that (3.3) is a symmetric matrix, therefore A^T and A are the transpose of each other, and K_0, K_0' are symmetric.

By the equation (1.1) and the similarity of the partition it is easy to see that the stiffness matrices for all layers are the same. Upon assembling the layer stiffness matrices by the nodes, we obtain an infinite by infinite total stiffness matrix. Noting the equation and the boundary condition, we get an algebraic system of equations of infinite order. Let $K = K_0 + K_0'$, then it is

$$-Ay_0 + Ky_1 - A^T y_2 = 0,$$

$$\cdots\cdots,$$

$$-Ay_{k-1} + Ky_k - A^T y_{k+1} = 0, \tag{3.4}$$

$$\cdots\cdots,$$

which takes the form of a block tri-diagonal Toeplitz matrix. If the boundary condition is the Neŭmann boundary condition (1.3), there is another equation for y_0,

$$K_0 y_0 - A^T y_1 = f_0, \tag{3.5}$$

where f_0 is the "load vecter", which can be evaluated from the boundary value g by conventional finite element method. Now our aim is giving the algorithm of the combined stiffness matrix on Ω, thus we ignore the difference between these two boundary conditions at this occasion. In order to reduce the algebraic system of equations of infinite order, we need a careful study of the properties of the solutions.

Lemma 44. *If $u \in S(\Omega)$, $y_k = B_k u$, then there is a constant C independent of u, such that*

$$\frac{1}{C} \sum_{k=1}^{\infty} |y_k - y_{k-1}|^2 \le \int_{\Omega} |\nabla u|^2 \, dx \le C \sum_{k=0}^{\infty} |y_k|^2. \tag{3.6}$$

Proof. Let $e_j \subset \Omega$ be an arbitrary triangular element. Let the length of the side pointing to the origin O be L, and the values of u at the both ends of it be $y_k^{(i)}$ and $y_{k-1}^{(i)}$. Let mease_j denote the area of the triangle e_j, and ρ the least height of e_j. Then

$$\frac{|y_k^{(i)} - y_{k-1}^{(i)}|}{L} \le |\nabla u| \le \frac{|y_k| + |y_{k-1}|}{\rho}.$$

Therefore it holds that

$$\frac{\text{mease}_j}{L^2} |y_k^{(i)} - y_{k-1}^{(i)}|^2 \le \int_{e_j} |\nabla u|^2 \, dx \le \frac{\text{mease}_j}{\rho^2} (|y_k| + |y_{k-1}|)^2.$$

By similarity $\frac{\text{mease}_j}{L^2}$ and $\frac{\text{mease}_j}{\rho^2}$ depend only on the position of e_j in Ω_k and are independent of k. Summing up the above inequalities with respect to all triangles e_j we obtain (3.6). $\qquad \square$

Lemma 45. *K_0 and K_0' are positive definite matrices.*

Proof. We take an arbitrary $y_0 \ne 0$ and set $y_1 = 0$. The interpolating function u of them is not a constant, hence

$$a(u, u) = \int_{\Omega_1} |\nabla u|^2 \, dx > 0.$$

On the other hand

$$a(u, u) == (y_0^T \ y_1^T) \begin{pmatrix} K_0 & -A^T \\ -A & K_0' \end{pmatrix} \begin{pmatrix} y_0^T \\ y_1^T \end{pmatrix}.$$

Thus

$$y_0^T K_0 y_0 > 0, \qquad \forall y_0 \neq 0,$$

so K_0 is a positive definite matrix. The argument for K_0' is similar. □

By Lemma 43 the problem (3.2) admits a unique solution u_h for any $y_0 \in \mathbb{R}^n$ such that $B_0 u_h = y_0$. Let $y_k = B_k u_h$, then y_1 is determined by y_0 uniquely. It is easy to verify that the mapping $y_0 \to y_1$ is linear. Therefore there is a real matrix X such that

$$y_1 = X y_0.$$

Let $w(x) = u_h(\xi^k x)$, then $w \in S(\Omega), B_0 w = y_k$ and w satisfies the problem (3.2), hence

$$y_{k+1} = X y_k.$$

We obtain by induction that

$$y_k = X^k y_0, \quad k = 1, 2, \cdots . \tag{3.7}$$

The matrix X is the transfer matrix for the problem.

We will make use of the complex function solutions to the Laplace equation in the remaining part of this section. The above argument can be applied for this case without any modification except we regard that $y_1, y_2, \ldots, y_k, \ldots$ are vectors over the field of complex numbers. We observe that the transfer matrix X is real as we have proved.

Substituting (3.7) into (3.4) we obtain

$$(-A + KX - A^T X^2) y_0 = 0.$$

y_0 is an arbitrary vector, hence X satisfies the equation

$$A^T X^2 - KX + A = 0. \tag{3.8}$$

The solutions to (3.8) are certainly not unique. We need the following lemmas to identify a unique solution.

Lemma 46. *If λ is an eigenvalue of the transfer matrix X and $|\lambda| \geq 1$, then $\lambda = 1$.*

Proof. Let g be the eigenvector associated with λ. Taking $y_0 = g$ as the boundary value of problem (3.2), we get $y_1, y_2, \ldots, y_k, \ldots$, and by (3.7)

$$y_k = X^k g = \lambda^k g.$$

Hence

$$\sum_{k=1}^{\infty} |y_k - y_{k-1}|^2 = \sum_{k=1}^{\infty} |\lambda^k - \lambda^{k-1}|^2 |g|^2 = \sum_{k=1}^{\infty} |\lambda|^{2k-2} |\lambda - 1|^2 |g|^2.$$

By Lemma 44, the series on the right hand side converges. We have $|g| \neq 0$, hence $\lambda = 1$. $\square$

Lemma 47. *The elementary divisor corresponding to an eigenvalue $\lambda = 1$ of the transfer matrix X is linear.*

Proof. Suppose it were not the case, then there would be nonvanishing vectors g and g_0 such that

$$(X - I)g_0 = 0, \quad (X - I)g = g_0,$$

where I is a unit matrix, i.e.

$$Xg = g_0 + g.$$

By induction we would get

$$X^k g = k g_0 + g.$$

Taking $y_0 = g$ as the boundary value of the problem (3.2), we obtain y_1, $y_2, \ldots, y_k, \ldots$, then by

$$y_k = k g_0 + g$$

we would get

$$\sum_{k=1}^{\infty} |y_k - y_{k-1}|^2 = \sum_{k=1}^{\infty} |g_0|^2 = +\infty,$$

which contradicts $u \in H^{1,*}(\Omega)$ according to Lemma 44. $\square$

Lemma 48. *There is an eigenvalue $\lambda = 1$ of the transfer matrix X, and the associated eigenvector is $g_1 = (1, 1, \ldots, 1)^T$.*

Proof. We take $y_0 = g_1$ as the boundary value of the problem (3.2) and get a constant solution $u_h \equiv 1$. Let $B_k u_h = y_k$, then $y_k = g_1$. By (3.7) we have

$$X g_1 = g_1. \qquad \square$$

Lemma 49. *As $k \to \infty$, $X^k y_0 \to y_\infty$ for any complex vector y_0, where*

$$y_\infty = \alpha g_1,$$

and α is a constant.

Proof. On the basis of Lemma 46 and Lemma 47, we conclude that the limit of X^k exists as $k \to \infty$. Denote by y_∞ the limit of $X^k y_0$, which is $(y_\infty^{(1)}, y_\infty^{(2)}, \ldots, y_\infty^{(n)})^T$ in component form.

It will suffice to prove $y_\infty^{(1)} = y_\infty^{(2)} = \cdots = y_\infty^{(n)}$. If it were not true, we might assume that $y_\infty^{(i)} \neq y_\infty^{(i+1)}$, which correspond to the i-th node and the $(i+1)$-th node respectively. There must be one element e_j in Ω_k such that the connecting line segment of the above two nodes is one side of e_j. Let the length of this side be L, then we have

$$\int_{e_j} |\nabla u|^2 \, dx \geq \frac{\text{mease}_j}{L^2} |y_k^{(i)} - y_k^{(i+1)}|^2.$$

By similarity $\frac{\text{mease}_j}{L^2}$ is independent of k. Summing up the above inequality with respect to k, we obtain

$$\int_\Omega |\nabla u|^2 \, dx \geq \frac{\text{mease}_j}{L^2} \sum_{k=1}^{\infty} |y_k^{(i)} - y_k^{(i+1)}|^2.$$

The general term of the series on the right hand side would take $|y_\infty^{(i)} - y_\infty^{(i+1)}|^2 \neq 0$ as its limit as $k \to \infty$, hence

$$\int_\Omega |\nabla u|^2 \, dx = \infty,$$

which leads to a contradiction. $\qquad \square$

Lemma 50. *There is a unique eigenvalue $\lambda = 1$ of the transfer matrix X, and the absolute values of all other eigenvalues are less than 1.*

Proof. If it were not true, then by Lemma 46, Lemma 47 and Lemma 48, there would be another eigenvector g' associated with the eigenvalue $\lambda = 1$. According to Lemma 49 we have $X^k g' \to \alpha g_1$ $(k \to \infty)$, but $X^k g' = g'$, hence $g' = \alpha g_1$, which contradicts the fact that g_1 and g' are linearly independent. $\qquad \square$

Let

$$R_1 = \begin{pmatrix} K & -A \\ I & 0 \end{pmatrix}, \quad R_2 = \begin{pmatrix} A^T & 0 \\ 0 & I \end{pmatrix}, \tag{3.9}$$

where I is the unit matrix. The equation (3.4) can be written as

$$R_1 \begin{pmatrix} y_k \\ y_{k-1} \end{pmatrix} = R_2 \begin{pmatrix} y_{k+1} \\ y_k \end{pmatrix}. \tag{3.10}$$

Let λ, g be a couple of eigenvalue and eigenvector, then

$$Xg = \lambda g. \tag{3.11}$$

Taking $k = 1$ and setting $y_0 = g$ in (3.10), and noting (3.7),(3.11) we get

$$R_1 \begin{pmatrix} \lambda g \\ g \end{pmatrix} = \lambda R_2 \begin{pmatrix} \lambda g \\ g \end{pmatrix}. \tag{3.12}$$

Therefore λ and $\begin{pmatrix} \lambda g \\ g \end{pmatrix}$ are the generalized eigenvalue and generalized eigen-vector of the matrices bundle $R_1 - \lambda R_2$. To find matrix X, we first solve the above generalized eigenvalue problem to get all $2n$ eigenvalues and eigen-vectors, then pick out the n eigenvalues and eigenvectors of X.

We will always denote by det the determinant of a matrix. Now we evaluate

$$\det(R_1 - \lambda R_2) = \det \begin{pmatrix} K - \lambda A^T & -A \\ I & -\lambda I \end{pmatrix}.$$

This matrix is split into blocks. We add the second column by the product of the first column and λ, then obtain

$$\det(R_1 - \lambda R_2) = \det \begin{pmatrix} K - \lambda A^T & -A + \lambda K - \lambda^2 A^T \\ I & 0 \end{pmatrix}$$

$$= (-1)^n \det(-A + \lambda K - \lambda^2 A^T).$$

K is symmetric, so this expression is a symmetric polynomial with the independent variable λ. Therefore if $\lambda \neq 0$ is an eigenvalue, then so is $1/\lambda$. Therefore we only need to pick out the eigenvalues, which satisfy $|\lambda| < 1$, of the matrices bundle $R_1 - \lambda R_2$, then supply a $\lambda = 1$. Let the eigenvalues be $\lambda_1, \lambda_2, \cdots, \lambda_n$, and the corresponding eigenvectors be $g_1, g_2, \cdots, g_n$, and we set $T = (g_1, g_2, \cdots, g_n), \Lambda = \mathrm{diag}(\lambda_1, \lambda_2, \cdots, \lambda_n)$, where diag means a diagonal matrix or a block diagonal matrix. By (3.11) we have

$$XT = T\Lambda,$$

that is

$$X = T\Lambda T^{-1}. \tag{3.13}$$

(3.13) is the formula for evaluating the transfer matrix X. Once X is obtained, then (3.7) gives the solution on the entire domain Ω.

If A is invertible, then the above generalized eigenvalue problem can be reduced to a conventional eigenvalue problem. Multiplying R_2^{-1} on the left of (3.12), we get

$$\begin{pmatrix} (A^T)^{-1}K & -(A^T)^{-1}A \\ I & 0 \end{pmatrix} \begin{pmatrix} \lambda g \\ g \end{pmatrix} = \lambda \begin{pmatrix} \lambda g \\ g \end{pmatrix},$$

which is a conventional eigenvalue problem.

Eigenvalue λ may be a complex number, thus the above calculation may encounter complex numbers. To prevent from this inconvenience, we can vary (3.13) slightly. Eigenvalues and eigenvectors must appear in conjugate pairs. Let $\lambda = \alpha \pm i\beta, g = p \pm iq$ be a pair of eigenvalues and eigenvectors, then by (3.11) we have

$$X(p \pm iq) = (\alpha \pm i\beta)(p \pm iq).$$

Separating the real and imaginary parts we obtain

$$Xp = \alpha p - \beta q, \quad Xq = \beta p + \alpha q,$$

that is

$$X \begin{pmatrix} p & q \end{pmatrix} = \begin{pmatrix} p & q \end{pmatrix} \begin{pmatrix} \alpha & \beta \\ -\beta & \alpha \end{pmatrix}.$$

We substitute the corresponding two columns in matrices T and Λ, and let

$$T_1 = (\cdots, p, q, \cdots),$$

$$\Lambda_1 = \text{diag}\left(\cdots, \begin{pmatrix} \alpha & \beta \\ -\beta & \alpha \end{pmatrix}, \cdots\right).$$

It is easy to see

$$X = T_1\Lambda_1 T_1^{-1}.$$

After substituting with respect to all complex eigenvalues, the calculation of the transfer matrix X becomes real.

3.1.3 *Further discussion for the transfer matrix*

We need to prove that the above conditions for X are not only necessary but sufficient. The argument is the following. Let $\varphi_1, \varphi_2, \ldots \varphi_i, \cdots \in S_0(\Omega)$ and φ_i be equal to 1 at a certain node of $\bigcup_{k=1}^{\infty} \Gamma_k$ and vanish at other nodes, which can be named a "shape function". The set $\tilde{S}_0(\Omega)$ consists of all finite linear combinations

$$\varphi = c_1\varphi_1 + c_2\varphi_2 + \cdots + c_i\varphi_i$$

of $\varphi_i's$, then $\tilde{S}_0(\Omega) \subset S_0(\Omega)$. Each function in $\tilde{S}_0(\Omega)$ has a bounded support.

Theorem 49. $\tilde{S}_0(\Omega)$ *is dense in* $S_0(\Omega)$.

Proof. We take an arbitrary $u \in S_0(\Omega)$. Since $S_0(\Omega) \subset H_0^{1,*}(\Omega)$, there exists $u_1 \in C_0^{\infty}(\Omega)$, such that $\|u - u_1\|_{1,*} < \varepsilon$ for any $\varepsilon > 0$. We consider the following problem: Find $u_2 \in S_0(\Omega)$, such that

$$a(u_2, v) = a(u_1, v), \qquad \forall v \in S_0(\Omega). \tag{3.14}$$

We can prove that the problem (3.14) admits a unique solution as we did for the conclusion of Lemma 43. By the positive definiteness of the following quadratic functional we obtain

$$\frac{1}{2}a(u_2, u_2) \quad a(u_1, u_2) = \min_{v \in S_0(\Omega)} \left\{ \frac{1}{2}a(v, v) - a(u_1, v) \right\},$$

hence

$$\frac{1}{2}a(u_2, u_2) - a(u_1, u_2) \leq \frac{1}{2}a(u, u) - a(u_1, u).$$

By adding $\frac{1}{2}a(u_1, u_1)$ to the both sides of the above inequality we obtain

$$a(u_2 - u_1, u_2 - u_1) \leq a(u - u_1, u - u_1),$$

thus

$$a(u_2 - u_1, u_2 - u_1) < \varepsilon.$$

Applying the inequality (1.19) we get

$$\|u_2 - u_1\|_{1,*} < 5\varepsilon.$$

But $u_1 \in C_0^\infty(\Omega)$, so there exists a natural number k_0 such that $u_1 \equiv 0$ on the domain $\xi^{k_0}\Omega$. The equation (3.14) imply that

$$a(u_2, v) = 0, \qquad \forall v \in S_0(\xi^{k_0}\Omega),$$

where we accept that every function in $S_0(\xi^{k_0}\Omega)$ is extended to zero on $\Omega \setminus \xi^{k_0}\Omega$. According to the conclusion of Lemma 49, u_2 tends to a constant α as $|x| \to \infty$, therefore u_2 is a bounded function.

We construct truncated functions $u_{2,k} \in S_0(\Omega), k = 1, 2, \ldots$ such that

$$u_{2,k} = \begin{cases} u_2, \text{ for } x \in \Omega \setminus \xi^k\Omega, \\ 0, \text{ for } x \in \xi^{k+1}\Omega. \end{cases}$$

Let $y_k = B_k u_2$, then by Lemma 44 we have

$$\int_\Omega |\nabla u_{2,k}|^2 \, dx \leq \int_\Omega |\nabla u_2|^2 \, dx + \frac{1}{C}|y_k|^2.$$

Therefore $\{u_{2,k}\}$ is bounded in $S_0(\Omega)$. Taking any $v \in S_0(\Omega)$ and making use of the Schwarz inequality we obtain

$$\left| \int_\Omega \nabla v \cdot \nabla(u_2 - u_{2,k}) \, dx \right|$$

$$= \left| \int_{\Omega_{k+1}} \nabla v \cdot \nabla(u_2 - u_{2,k}) \, dx + \int_{\xi^{k+1}\Omega} \nabla v \cdot \nabla u_2 \, dx \right|$$

$$\leq \left(\int_{\Omega_{k+1}} |\nabla v|^2 \, dx \right)^{\frac{1}{2}} \left(\int_{\Omega_{k+1}} |\nabla(u_2 - u_{2,k})|^2 \, dx \right)^{\frac{1}{2}} + \left| \int_{\xi^{k+1}\Omega} \nabla v \cdot \nabla u_2 \, dx \right|$$

Because both u_2 and $u_{2,k}$ are bounded functions, $\int_{\Omega_{k+1}} |\nabla(u_2 - u_{2,k})|^2 \, dx$ are uniformly bounded with respect to k. And on account of $v, u_2 \in S_0(\Omega)$,

$$\int_{\Omega_{k+1}} |\nabla v|^2 \, dx \to 0, \qquad \left| \int_{\xi^{k+1}\Omega} \nabla v \cdot \nabla u_2 \, dx \right| \to 0$$

as $k \to \infty$, therefore

$$\int_\Omega \nabla v \cdot \nabla(u_2 - u_{2,k}) \, dx \to 0$$

as $k \to \infty$, which means $u_{2,k}$ weakly converges to u_2. By the Mazur Theorem [Yosida, K. (1974)], there is a linear combination,

$$u_3 = c_1 u_{2,1} + c_2 u_{2,2} + \cdots + c_i u_{2,i},$$

such that $\|u_3 - u_2\|_{1,*} < \varepsilon$. Thus $\|u - u_3\|_{1,*} < 7\varepsilon$. But $u_3 \in \tilde{S}_0(\Omega)$, which leads to the conclusion. $\qquad\square$

Now we continue to discuss the Laplace equation. We give another formulation for the infinite element method: Find $u \in S(\Omega)$, such that $u_h\big|_{\partial\Omega} = g_I$ and

$$a(u_h, \varphi_i) = 0, \qquad i = 1, 2, \cdots . \tag{3.15}$$

Theorem 50. *The formulation (3.15) is equivalent to (3.2).*

Proof. It will suffice to prove that (3.15) implies (3.2). Let u_h be the solution to (3.15), then u_h satisfies

$$a(u_h, v) = 0, \qquad \forall v \in \tilde{S}_0(\Omega),$$

which yields (3.2 by noting Theorem 49. $\qquad\square$

Expressing (3.15) in terms of matrices, we get the system of equations (3.4), and the equation (3.8) for the transfer matrix X. We should bear in mind that the solutions to (3.8) are not unique.

Theorem 51. *The necessary and sufficient conditions to determine the transfer matrix X are*
(a) *X satisfies (3.8);*
(b) *X admits an eigenvalue $\lambda = 1$ which corresponds to a linear elementary divisor and the eigenvector g_1;*
(c) *The absolute values of all other eigenvalues of X are less than 1.*

Proof. The lemmas from Lemma 46 to Lemma 50 and the equation (3.8) show that those conditions are necessary. It remains to prove they are sufficient. If a matrix X satisfies (a),(b),(c), let us prove it is the transfer matrix. We take an arbitrary $g \in \mathbb{R}^n$, and set $y_k = X^k g$, $k = 0, 1, \ldots$, then y_k satisfy (3.4). Let u_h be the interpolating function. If we can prove $u_h \in S(\Omega)$, then owing to Theorem 50, u_h is the solution to the problem (3.2), which means X is the desired one.

By (b) and (c), the limit of $X^k y_0$ exists as $k \to \infty$. Let it be y_∞. We have

$$X y_\infty = X \lim_{k\to\infty} X^k y_0 = \lim_{k\to\infty} X^{k+1} y_0 = y_\infty,$$

therefore $y_\infty = \alpha g_1$. Let $y_k' = y_k - y_\infty$, then the interpolating function of y_k' is just $u' = u_h - \alpha$. We only need to show $u' \in S(\Omega)$.

We express X in the Jordan canonical form,

$$X = TJT^{-1}, \tag{3.16}$$

where

$$J = \begin{pmatrix} 1 & \\ & J_1 \end{pmatrix},$$

where J_1 consists of those Jordan blocks, the absolute values of the eigenvalues corresponding to which are less than 1. By the definition of y_k,

$$Xy'_{k-1} = X(y_{k-1} - y_\infty) = y_k - y_\infty = y'_k. \tag{3.17}$$

Let

$$z_k = T^{-1}y'_k,$$

then (3.16) and (3.17) yield

$$z_k = Jz_{k-1}, \qquad k = 1, 2, \cdots.$$

By induction we obtain

$$z_k = J^k z_0,$$

that is

$$z_k = \begin{pmatrix} 1 & \\ & J_1^k \end{pmatrix} z_0. \tag{3.18}$$

Because $y'_k \to 0$ as $k \to \infty$, we have $z_k \to 0$. (3.18) implies that the first component of z_0 is zero, otherwise the limit of z_k would not be zero. Thus

$$z_k = \begin{pmatrix} 0 & \\ & J_1^k \end{pmatrix} z_0.$$

Let $\|\cdot\|$ be the spectral norm of a matrix. Denote by $\rho_1 = \rho(J_1)$ the spectral radius of the matrix J_1, and p the highest order among the Jordan blocks of J_1 whose spectral radius are equal to ρ_1, then [Varga, R.S. (1962)]

$$\|J_1^k\| \leq c(p)C_k^{p-1}\rho_1^{k-p+1}, \tag{3.19}$$

where $k \geq p - 1$, $c(p)$ is a constant depending only on p, and C_k^{p-1} is the combinational number. We have

$$\sum_{k=p-1}^{\infty} |z_k|^2 \leq \sum_{k=p-1}^{\infty} \|J_1^k\|^2 |z_0|^2 \leq \sum_{k=p-1}^{\infty} c^2(p) \left\{ C_k^{p-1} \rho_1^{k-p+1} \right\}^2 |z_0|^2.$$

It is known that $\rho_1 < 1$, hence the series on the right hand side converges. Therefore

$$\sum_{k=0}^{\infty} |y_k'|^2 = \sum_{k=0}^{\infty} |Tz_k|^2 \leq \|T\|^2 \sum_{k=0}^{\infty} |z_k|^2$$

converges too. On account of Lemma 44

$$\int_{\Omega} |\nabla u'|^2 \, dx < +\infty.$$

Applying Lemma 7 we have $u' \in H^{1,*}(\Omega)$. □

We remark that using the Jordan canonical form (3.16) and (3.7) we can get the limit of the solution at the infinity,

$$y_\infty = X^\infty y_0 = T \begin{pmatrix} 1 & \\ & 0 \end{pmatrix} T^{-1}.$$

3.1.4 Combined stiffness matrix

The combined stiffness matrix is the Dirichlet to Neŭmann operator in discrete form. Substituting (3.7) into (3.5) we obtain

$$(K_0 - A^T X)y_0 = f_0.$$

We define $K_z = K_0 - A^T X$. The following theorem indicates the important role of this matrix.

Theorem 52. *Let u_h be the solution to the problem (3.2), then for an arbitrary $v \in S(\Omega)$ it holds that*

$$a(u_h, v) = z_0^T K_z y_0, \qquad y_0 = B_0 u_h, z_0 = B_0 v. \tag{3.20}$$

Proof. We decompose $v = v_1 + v_2$ in $S(\Omega)$, where $B_0 v_1 = B_0 v = z_0$, $B_k v_1 = 0$, $k = 1, 2, \cdots$, and $B_0 v_2 = 0$, $B_k v_2 = B_k v$, $k = 1, 2, \cdots$. Then $v_2 \in S_0(\Omega)$, and by (3.2) we get $a(u_h, v_2) = 0$. By (3.3) we have

$$a(u_h, v) = a(u_h, v_1) = \int_{\Omega_1} \nabla u_h \cdot \nabla v_1 \, dx = \begin{pmatrix} y_0^T & y_1^T \end{pmatrix} \begin{pmatrix} K_0 & -A^T \\ -A & K_0' \end{pmatrix} \begin{pmatrix} z_0 \\ z_1 \end{pmatrix}.$$

Since the stiffness matrix is symmetric, we have

$$a(u_h, v) == \begin{pmatrix} z_0^T & 0 \end{pmatrix} \begin{pmatrix} K_0 & -A^T \\ -A & K_0' \end{pmatrix} \begin{pmatrix} y_0 \\ X y_0 \end{pmatrix},$$

which gives (3.20). □

Therefore K_z is defined as a combined stiffness matrix. We should note that the combined stiffness matrix is different from the total stiffness matrix. At this circumstance, the total stiffness matrix is an infinite by infinite matrix, by which we can get the strain energy arising from any nodal values, but the combined stiffness matrix is only a $n \times n$ matrix, by which we can only get the strain energy when the nodal values $y_0, y_1, \cdots$ satisfy the equations (3.4). The procedure from the the total stiffness matrix to the combined stiffness matrix can be viewed as a procedure of elimination. By this procedure, $y_1, y_2, \cdots$ are all eliminated.

Lemma 51. *The eigenvectors of the transfer matrix X associated with the eigenvalue $\lambda = 1$ are necessarily the null eigenvectors of the combined stiffness matrix K_z.*

Proof. Let g be such an eigenvector. Since it corresponds to a constant solution u_h, we have

$$a(u_h, u_h) = \begin{pmatrix} g^T & g^T \end{pmatrix} \begin{pmatrix} K_0 & -A^T \\ -A & K_0' \end{pmatrix} \begin{pmatrix} g \\ g \end{pmatrix} = 0.$$

Here $\begin{pmatrix} K_0 & -A^T \\ -A & K_0' \end{pmatrix}$ is a symmetric and semi-positive definite matrix, therefore $\begin{pmatrix} g \\ g \end{pmatrix}$ is its null eigenvector, that is

$$\begin{pmatrix} K_0 & -A^T \\ -A & K_0' \end{pmatrix} \begin{pmatrix} g \\ g \end{pmatrix} = 0.$$

We obtain from the first row that

$$(K_0 - A^T)g = 0,$$

then by substituting $g = Xg$ into it we get

$$K_z g = (K_0 - A^T X)g = 0.$$ $\square$

Theorem 53. *K_z is a symmetric and semi-positive definite matrix.*

Proof. $a(\cdot, \cdot)$ is symmetric, hence by (3.20) we have $z_0^T K_z y_0 = y_0^T K_z z_0$ provided v is a solution to the problem (3.2). y_0 and z_0 are arbitrary, so K_z is symmetric. Because $a(u, u) \geq 0$ for any $u \in S(\Omega)$, K_z is semi-positive definite. $\square$

If the domain Ω^c is not in star-shape, or the curve Γ_0 is in complicated shape, we can decompose the exterior domain Ω to an unbounded domain with regular shape and a bounded domain with complicated shape, as we have stated previously. Then we can find the combined stiffness matrix for the unbounded domain, which can be regarded as a single element, owing to the conclusion of Theorem 52. Upon assembling the combined stiffness matrix with the stiffness matrices of the other elements, we get an algebraic system of equations. Then the problem can be solved.

3.2 General elements

The pattern of partition in Section 3.1 can be generalized. On the one hand the elements are not restricted to be linear triangular elements, on the other hand each layer may not be divided by rays. We should notice that we have not made any other assumption to the stiffness matrices in the preceding argument except that they are the same for all layers. Therefore the previous argument is applicable for more general partitions. The triangular elements can be of higher order, and quadrilateral elements can also be applied. There are only two restrictions, namely, one to one correspondence of the nodes between Γ_{k-1} and Γ_k and the geometry of the layer mesh is similar to each other.

It should be noticed that there may be some nodes which do not belong to Γ_k or Γ_{k-1}. It is needed to eliminate those nodal values from the systems and obtain the combined stiffness matrix for one layer. Let $y_{k-\frac{1}{2}}$ be a column vector consisting of the interior nodal values, then we get the matrix expression for the general cases:

$$a(u,v) = \sum_{k=1}^{\infty} \left(y_{k-1}^T \ y_{k-\frac{1}{2}}^T \ y_k^T \right) \begin{pmatrix} K_{11} & K_{12} & K_{13} \\ K_{21} & K_{22} & K_{23} \\ K_{31} & K_{32} & K_{33} \end{pmatrix} \begin{pmatrix} z_{k-1} \\ z_{k-\frac{1}{2}} \\ z_k \end{pmatrix} = 0, \quad (3.21)$$

in terms of the total stiffness matrix. v is arbitrary in $S_0(\Omega)$. Let it vanish on all sub-domains except one Ω_k for a fixed k. Then

$$\left(y_{k-1}^T \ y_{k-\frac{1}{2}}^T \ y_k^T \right) \begin{pmatrix} K_{12} \\ K_{22} \\ K_{32} \end{pmatrix} z_{k-\frac{1}{2}} = 0.$$

Besides, $z_{k-\frac{1}{2}}$ is arbitrary. Therefore

$$y_{k-1}^T K_{12} + y_{k-\frac{1}{2}}^T K_{22} + y_k^T K_{32} = 0.$$

$y_{k-\frac{1}{2}}$ can be solved explicitly:

$$y_{k-\frac{1}{2}} = -K_{22}^{-1}(K_{12}^T y_{k-1} + K_{32}^T y_k).$$

Upon substituting it into (3.21), and rearranging the terms, we can get

$$a(u, v) = \sum_{k=1}^{\infty} \begin{pmatrix} y_{k-1}^T & y_k^T \end{pmatrix} \begin{pmatrix} K_0 & -A^T \\ -A & K_0' \end{pmatrix} \begin{pmatrix} z_{k-1} \\ z_k \end{pmatrix}.$$

This expression is the same as that of the linear triangular elements, except the stiffness matrices may be different. Then all above algorithm can be applied here without any change. .

3.3   Harmonic equation-three dimensional problems

The discussion for the Laplace equation in three space dimension is similar to that for the two dimensional case. Let Γ_0 be a closed convex surface in a three dimensional space, it is stitched up with space quadrilaterals, and the point O is in the interior of Γ_0. Taking $\xi > 1$, we draw the similar surfaces to Γ_0 with center O and the constants of proportionality $\xi, \xi^2, \cdots, \xi^k, \cdots$, which are denoted by $\Gamma_1, \Gamma_2, \cdots, \Gamma_k, \cdots$ respectively. The domain between Γ_{k-1} and Γ_k may be divided into many sorts of elements, for instance eight nodes hexahedron isoparametric elements. Let

$$\begin{pmatrix} K_0 & -A^T \\ -A & K_0' \end{pmatrix}$$

be the stiffness matrix of the layer between Γ_0 and Γ_1, where the degree of freedom with respect to $y_{\frac{1}{2}}$ has already been eliminated, as we have done in Section 3.2. Then the stiffness matrices of the other layers are

$$\xi^{k-1} \begin{pmatrix} K_0 & -A^T \\ -A & K_0' \end{pmatrix}, \quad k = 2, 3, \cdots.$$

Let $K = \xi^{\frac{1}{2}} K_0 + \xi^{-\frac{1}{2}} K_0'$, then the equation corresponding to (3.4) is

$$-A y_{k-1} + \xi^{\frac{1}{2}} K y_k - \xi A^T y_{k+1} = 0.$$

Hence the transfer matrix X satisfies the equation

$$\xi A^T X^2 - \xi^{\frac{1}{2}} KX + A = 0.$$

If we set $\xi^{\frac{1}{2}} X = Y$, then Y satisfies the equation

$$A^T Y^2 - KY + A = 0,$$

the form of which is the same as (3.8). The solutions are not unique, so it is necessary to study the properties of X.

Lemma 52. *If* $u \in S(\Omega)$, $y_k = B_k u$, *then there is a constant* C *independent of* u, *such that*

$$\frac{1}{C} \sum_{k=1}^{\infty} \xi^k |y_k - y_{k-1}|^2 \le \int_{\Omega} |\nabla u|^2 \, dx \le C \sum_{k=0}^{\infty} \xi^k |y_k|^2.$$

The proof of it is analogous to that of Lemma 44, thus omitted.

Being analogous to two dimensional cases, K_0 and K_0' are symmetric and positive definite matrices. Moreover we have

Lemma 53. *Each eigenvalue* λ *of the transfer matrix* X *satisfies* $|\lambda| < \xi^{-\frac{1}{2}}$.

Proof. Let g be the eigenvector associated with λ. We take $y_0 = g$ as the boundary value of the problem (3.2) (three dimensional case), and get $y_1, y_2, \ldots, y_k, \ldots$, then $y_k = \lambda^k g$. Hence

$$\sum_{k=1}^{\infty} \xi^k |y_k - y_{k-1}|^2 = \sum_{k=1}^{\infty} \xi^k |\lambda|^{2k-2} |\lambda - 1|^2 |g|^2.$$

By Lemma 52 the above series converges. If $\lambda \ne 1$, then $\xi^k |\lambda|^{2k} < 1$, that is, $|\lambda| < \xi^{-\frac{1}{2}}$. If $\lambda = 1$, then $y_0 = y_1 = \cdots = y_k = \cdots = g$. We have

$$\int_{\Omega_k} \frac{u^2(x)}{|x|^2} \, dx \ge C\xi^k,$$

thus $|u|_{1,*} = \infty$, which is unsuited to present needs. Therefore $\lambda = 1$ is not an eigenvalue. ⊔

By this result all eigenvalues λ of the matrix Y satisfy $|\lambda| < 1$. We define matrices R_1, R_2 by (3.9), then we can evaluate the general eigenvalues and eigenvectors of the matrices bundle $R_1 - \lambda R_2$. Finally we get the matrix Y by the approach in Section 3.1.

Theorem 54. *The necessary and sufficient conditions to determine the transfer matrix X are*

(a) X *satisfies the equation*

$$\xi A^T X^2 - \xi^{\frac{1}{2}} KX + A = 0;$$

(b) *The absolute values of all eigenvalues of X are less than $\xi^{-\frac{1}{2}}$.*

3.4   Inhomogeneous equations

We study the infinite element method for the Poisson equation

$$-\triangle u = f, \qquad x \in \Omega, \tag{3.22}$$

in two space dimension in this section. We assume that f is compactly supported, otherwise the computing procedure can not be terminated in finite steps. The variational formulation for the Dirichlet problem is: For given $f \in L^2(\Omega)$ and $g \in H^{1/2}(\partial\Omega)$ find $u \in H^{1,*}(\Omega)$ such that $u|_{\partial\Omega} = g$, and

$$a(u, v) = (f, v), \qquad \forall v \in H_0^{1,*}(\Omega). \tag{3.23}$$

For simplicity we still use the linear triangular elements and the same partition as that in Section 3.1. The infinite element formulation is: find $u_h \in S(\Omega)$ such that $u_h|_{\partial\Omega} = g_I$, and

$$a(u_h, v) = (f, v), \qquad \forall v \in S_0(\Omega). \tag{3.24}$$

Being analogous to (3.3) there are vectors f_k, g_k such that

$$\int_{\Omega_k} \nabla u_h \cdot \nabla v \, dx - \int_{\Omega_k} f \cdot v \, dx$$
$$= \left(y_{k-1}^T \ y_k^T \right) \begin{pmatrix} K_0 & -A^T \\ -A & K_0' \end{pmatrix} \begin{pmatrix} z_{k-1} \\ z_k \end{pmatrix} + \left(z_{k-1}^T \ z_k^T \right) \begin{pmatrix} f_k \\ g_k \end{pmatrix}. \tag{3.25}$$

Let $\{y_k^*\}$ be a solution of the infinite element scheme to the equation, but it does not necessary satisfy the boundary conditions. Let $\tilde{y}_k = y_k - y_k^*$, then $\tilde{y}_{k+1} = X\tilde{y}_k$. We have

$$y_0 = \tilde{y}_0 + y_0^*,$$

$$y_1 = \tilde{y}_1 + y_1^* = X\tilde{y}_0 + y_1^* = X(y_0 - y_0^*) + y_1^* \equiv Xy_0 + q_1,$$

. . . . . .

$$y_k = X y_{k-1} + q_k,$$

. . . . . .

Let

$$W_k = \sum_{l=k+1}^{\infty} \left\{ \frac{1}{2} \int_{\Omega_l} |\nabla u_h|^2 \, dx - \int_{\Omega_l} f \cdot u_h \, dx \right\} = \frac{1}{2} y_k^T K_z y_k + y_k^T h_k. \quad (3.26)$$

On the other hand

$$W_k = \min_{y_{k+1}} \left\{ \frac{1}{2} \begin{pmatrix} y_k^T & y_{k+1}^T \end{pmatrix} \begin{pmatrix} K_0 & -A^T \\ -A & K_0' \end{pmatrix} \begin{pmatrix} y_k \\ y_{k+1} \end{pmatrix} \right.$$
$$\left. + \begin{pmatrix} y_k^T & y_{k+1}^T \end{pmatrix} \begin{pmatrix} f_{k+1} \\ g_{k+1} \end{pmatrix} + \frac{1}{2} y_{k+1}^T K_z y_{k+1} + y_{k+1}^T h_{k+1} \right\}. \quad (3.27)$$

We take the partial derivatives with respect to y_{k+1}, then let it be zero to obtain

$$(K_0' + K_z) y_{k+1} - A y_k + g_{k+1} + h_{k+1} = 0.$$

We solve y_{k+1} from this equation then substitute it into (3.27) and compare (3.27) with (3.26). It is deduced that

$$h_k = A^T (K_0' + K_z)^{-1} (g_{k+1} + h_{k+1}) + f_{k+1}, \quad (3.28)$$

and we have

$$q_k = -(K_0' + K_z)^{-1} (g_k + h_k). \quad (3.29)$$

Thus the procedure to solve inhomogeneous equations is:
 (1) evaluate h_k from the recurrence formula (3.28);
 (2) evaluate q_k from the formula (3.29);
 (3) then get y_k.
 The recurrence is backward in the first step. Since f is compactly supported, $h_{k+1} = 0$ for k large enough, then it is the initial point.

3.5 Plane elasticity

We consider plane stress or plane strain here. The Lamé equation for the displacement $u = (u_1, u_2)^T$ is

$$-\mu \triangle u - (\lambda + \mu)\text{grad div} u = f, \quad x \in \Omega. \tag{3.30}$$

The only difference for these two cases lies in the constants. We consider the Dirichlet boundary condition,

$$u = 0, \quad x \in \partial\Omega, \tag{3.31}$$

as an example, and the approach for the Neŭmann problem is similar. Inhomogeneous problems can be dealt with applying the approach in Section 3.4, so we consider homogeneous problems only in this section. We will work in the spaces $V = \left(H^{1,*}(\Omega)\right)^2$, and $V_0 = \{u \in V; u|_{\partial\Omega} = 0\}$, and the corresponding bilinear form is

$$a(u, v) = \int_\Omega \sum_{i,j=1}^2 \sigma_{ij}(u)\epsilon_{ij}(u) \, dx$$

$$= \int_\Omega \left\{ \lambda \, \text{div} \, u \, \text{div} \, v + 2\mu \sum_{i,j=1}^2 \epsilon_{ij}(u)\epsilon_{ij}(v) \right\} dx,$$

where the strains are

$$\epsilon_{ij}(u) = \epsilon_{ji} = \frac{1}{2}(\frac{\partial u_i}{\partial x_j} + \frac{\partial u_j}{\partial x_i}), \quad 1 \leq i, j \leq 2,$$

and the stresses are

$$\sigma_{ij}(u) = \sigma_{ji}(u) = \lambda \left(\sum_{k=1}^2 \epsilon_{kk}(u)\right) \delta_{ij} + 2\mu\epsilon_{ij}(u).$$

The variational formulation of the problem (3.30),(3.31) is: For a given $g \in H^{1/2}(\partial\Omega)$, find $u \in V$, such that $u|_{\partial\Omega} = g$, and

$$a(u, v) = 0, \quad \forall v \in V_0. \tag{3.32}$$

For simplicity we consider linear triangular elements here. The infinite element formulation is: Find $u \in (S(\Omega))^2$, such that $u|_{\partial\Omega} = g_I$, and

$$a(u, v) = 0, \quad \forall v \in (S_0(\Omega))^2. \tag{3.33}$$

Let the number of nodes on Γ_k be n. According to the anticlockwise direction the displacements on the nodes are $u_{1k}^{(1)}, u_{2k}^{(1)}, u_{1k}^{(2)}, u_{2k}^{(2)}, \cdots, u_{1k}^{(n)}, u_{2k}^{(n)}$ successively, and they form a $(2n)$- dimensional column vector y_k. Now we can evaluate the stiffness matrix and get

$$a(u,v) = \sum_{k=1}^{\infty} \left(y_{k-1}^T \; y_k^T \right) \begin{pmatrix} K_0 & -A^T \\ -A & K_0' \end{pmatrix} \begin{pmatrix} z_{k-1} \\ z_k \end{pmatrix},$$

the form of which is the same as that for the Laplace equation.

In parallel with the propositions in Section 3.1, we establish the following lemmas and theorems. The proof of some propositions is the same as the previous one, thus omitted.

Lemma 54. K_0 and K_0' are positive definite matrices.

Proof. We take $y_0 \neq 0$ and $y_k = 0$, $k = 1, 2, \cdots$, then the corresponding interpolating function is not a rigid body motion, hence we get $a(u,u) > 0$. It follows that K_0 is a positive definite matrix. The proof for K_0' is the same. $\qquad\square$

Lemma 55. *If λ is an eigenvalue of the transfer matrix X and $|\lambda| \geq 1$, then $\lambda = 1$.*

Lemma 56. *The elementary divisor associated with an eigenvalue $\lambda = 1$ of the transfer matrix X is linear.*

Lemma 57. *As $k \to \infty$, $X^k y_0 \to y_\infty$ for any complex vector y_0, where*

$$y_\infty = \alpha g_1 + \beta g_2,$$

where $g_1 = (1, 0, 1, 0, \cdots, 1, 0)^T$, $g_2 = (0, 1, 0, 1, \cdots, 0, 1)^T$, and α, β are constants.

Lemma 58. *If an eigenvalue λ of the transfer matrix X is equal to 1, then the associated eigenvector of it is a linear combination of g_1 and g_2.*

Theorem 55. *The necessary and sufficient conditions to determine the transfer matrix X are*

(a) *X satisfies the equation,*

$$A^T X^2 - KX + A = 0;$$

(b) *There are two eigenvalues $\lambda = 1$ of X, the associated elementary divisors are linear, and the corresponding eigenvectors are g_1, g_2;*

(c) *The absolute values of all other eigenvalues of X are less than 1.*

For three dimensional elastic problems the approach in Section 3.3 can be applied. The algorithm is similar.

3.6   Bi-harmonic equations

According to Section 1.6 the bi-harmonic equations are

$$\triangle^2 u = f, \tag{3.34}$$

and the corresponding bilinear form is

$$
a(u,v) = \int_\Omega \tilde{\nu} \triangle u \cdot \triangle v \, dx
$$
$$
+ \int_\Omega (1 - \tilde{\nu}) \left(\frac{\partial^2 u}{\partial x_1^2} \frac{\partial^2 v}{\partial x_1^2} + 2 \frac{\partial^2 u}{\partial x_1 \partial x_2} \frac{\partial^2 v}{\partial x_1 \partial x_2} + \frac{\partial^2 u}{\partial x_2^2} \frac{\partial^2 v}{\partial x_2^2} \right) dx. \tag{3.35}
$$

We consider homogeneous equation here. For inhomogeneous equations the algorithm in Section 3.4 can be applied here without significant change. Then variational formulation of the Dirichlet problem is: Given $g_1 \in H^{3/2}(\partial\Omega)$, and $g_2 \in H^{1/2}(\partial\Omega)$, find $u \in H^{2,*}(\Omega)$, such that $u|_{\partial\Omega} = g_1, \frac{\partial u}{\partial \nu}|_{\partial\Omega} = g_2$, and

$$a(u,v) = 0, \qquad \forall v \in H_0^{2,*}(\Omega). \tag{3.36}$$

For definiteness we use the Morley triangular elements here [Ciarlet, P.G. (1978)]. The values of u are given at the vertices and the normal derivatives of u are given at the middle points. Quadratic functions are applied as the interpolation functions and six coefficients are determined uniquely by the six nodal values. Here it should be noticed that at each middle point we must fix one normal direction and all layers must be similar, including the direction. Let the six nodes of an element e_i be $X_j^{(i)}$, $j = 1, 2, 3$, and $X_{jl}^{(i)'}$, $1 \le j < l \le 3$. Let the set of all $X_j^{(i)}$'s be Σ_1, and let the set of all $X_{jl}^{(i)'}$'s be Σ_2. Then the infinite element space is defined as

$$S(\Omega) = \{u \in P_2(e_i), \forall i; u \text{ continuous on } \Sigma_1,$$
$$\nabla u \text{ continuous on } \Sigma_2, \sum_i |u|_{2,e_i}^2 < \infty\}.$$

$$S_0(\Omega) = \{u \in S(\Omega); u(x) = 0, \forall x \in \Gamma_0 \cap \Sigma_1, \frac{\partial u(x)}{\partial \nu} = 0, \forall x \in \Gamma_0 \cap \Sigma_2\}.$$

The corresponding bilinear form is

$$a_h(u, v) = \sum_i \int_{e_i} \tilde{\nu} \triangle u \cdot \triangle v \, dx$$

$$+ \int_\Omega (1 - \tilde{\nu}) \left(\frac{\partial^2 u}{\partial x_1^2} \frac{\partial^2 v}{\partial x_1^2} + 2 \frac{\partial^2 u}{\partial x_1 \partial x_2} \frac{\partial^2 v}{\partial x_1 \partial x_2} + \frac{\partial^2 u}{\partial x_2^2} \frac{\partial^2 v}{\partial x_2^2} \right) dx.$$

Lemma 59. *Let* $\|u\|_{S(\Omega)} = \left(\sum_i |u|_{2,e_i}^2 \right)^{\frac{1}{2}}$, *then it is a norm over* $S_0(\Omega)$.

Proof. If $\|u\|_{S(\Omega)} = 0$, then $u \in P_1(e_i)$ for all i. By the continuity of u and ∇u at nodes, $u \in P_1(\Omega)$. Then by the zero boundary condition on Γ_0, $u \equiv 0$. □

Let g_{1I} and g_{2I} be the approximation of g_1 and g_2. The infinite element formulation is: Find $u_h \in S(\Omega)$, such that $u(x) = g_{1I}(x)$, $\forall x \in \Gamma_0 \cap \Sigma_1$, $\frac{\partial u(x)}{\partial \nu} = g_2(x)$, $\forall x \in \Gamma_0 \cap \Sigma_2$, and

$$a_h(u, v) = 0, \qquad \forall v \in S_0(\Omega). \tag{3.37}$$

Theorem 56. *The problem (3.37) admits a unique solution.*

Proof. It is routine to transfer the problem into a problem with homogeneous boundary condition. Applying the Lax-Milgram theorem we need to verify that $a_h(\cdot, \cdot)$ is coercive. Clearly $a_h(u, u)$ is equivalent to $\|u\|_{S(\Omega)}$, thus the conclusion follows. □

We turn now to the implementation. We consider the k-th layer between Γ_{k-1} and Γ_k. If the number of triangular elements is N, then there are N nodes on Γ_{k-1} (or Γ_k) and inside the k-th layer there are other N nodes. Let y_{k-1} (or y_k) be a vector with components of $\frac{u}{|x|}$ or $\frac{\partial u}{\partial \nu}$ at the nodes on Γ_{k-1} (or Γ_k), and $y_{k-\frac{1}{2}}$ be a vector with components of $\frac{\partial u}{\partial \nu}$ at the interior nodes. Notice that we use the value of $\frac{u}{|x|}$ then by (3.35) the total stiffness matrices of all layers are just the same. Upon introducing some matrices, we have the following, which is in the same form as (3.21),

$$a_h(u, v) = \sum_{k=1}^{\infty} \begin{pmatrix} y_{k-1}^T & y_{k-\frac{1}{2}}^T & y_k^T \end{pmatrix} \begin{pmatrix} K_{11} & K_{12} & K_{13} \\ K_{21} & K_{22} & K_{23} \\ K_{31} & K_{32} & K_{33} \end{pmatrix} \begin{pmatrix} z_{k-1} \\ z_{k-\frac{1}{2}} \\ z_k \end{pmatrix} = 0. \tag{3.38}$$

Then following the same lines as those of Section 3.2, $y_{k-\frac{1}{2}}$ can be eliminated. Then there is a transfer matrix X.

Lemma 60. *All eigenvalues of X which are not equal to 1 have a magnitude less than 1. When $\lambda = 1$, λ is a double eigenvalue and its eigenvectors correspond to $u = x$ and $u = y$.*

Proof. Let λ be an eigenvalue of X with an associated eigenvector g. suppose that $|\lambda| \geq 1$, then $y_k = \lambda^k y_0$. $y_{k-\frac{1}{2}}$ are then derived, which depend linearly on y_k and y_{k-1}. If

$$W_1 \equiv \left(y_0^T \; y_{\frac{1}{2}}^T \; y_1^T \right) \begin{pmatrix} K_{11} & K_{12} & K_{13} \\ K_{21} & K_{22} & K_{23} \\ K_{31} & K_{32} & K_{33} \end{pmatrix} \begin{pmatrix} y_0 \\ y_{\frac{1}{2}} \\ y_1 \end{pmatrix} \neq 0,$$

then

$$a_h(u_h, u_h) = \sum_{k=1}^{\infty} \lambda^{k-1} W_1 = \infty,$$

which is impossible. If $W_1 = 0$, then $\|u_h\|_{S(\Omega)} = 0$, which implies that $u_h = x$, or $u_h = y$, or $u_h = 1$, or their linear combination. It can be seen that $u_h = 1$ corresponds to $\lambda = \frac{1}{\xi} < 1$ by noting that the ratio of $|x|$ between Γ_k and Γ_{k-1} is ξ and $\frac{u}{|x|}$ is the variable defined on nodes. If u_h is not the linear combination of x, y, then the magnitude of λ has to be less than 1. This completes the proof. $\qquad\square$

3.7 Stokes equation

We study two dimensional problems first. As before we consider homogeneous equations only. The system of equations is:

$$- \triangle u + \nabla p = 0, \tag{3.39}$$

$$\nabla \cdot u = 0, \tag{3.40}$$

with the boundary condition,

$$u = g, \qquad x \in \partial\Omega. \tag{3.41}$$

The corresponding bilinear forms are:

$$a(u, v) = \int_\Omega \nabla u : \nabla v \, dx, \quad b(u, p) = (p, \nabla u).$$

The corresponding variational problem is: For a given $g \in \left(H^{1/2}(\Omega)\right)^2$, find $u \in \left(H^{1,*}(\Omega)\right)^2$ and $p \in L^2(\Omega)$, such that $u|_{\partial\Omega} = g$, and

$$a(u, v) - b(v, p) = 0, \qquad \forall v \in \left(H_0^{1,*}(\Omega)\right)^2, \tag{3.42}$$

$$b(u, q) = 0, \qquad \forall q \in L^2(\Omega). \tag{3.43}$$

The problem is essentially equivalent to (1.64) and (1.67) with a slightly difference, where the equation is inhomogeneous and the boundary condition is homogeneous. In Section 1.7 it has been explained how a problem with inhomogeneous boundary condition can be reduced to a problem with homogeneous boundary condition. Therefore the problem (3.42),(3.43) admits a unique solution.

We consider the infinite element method for the problem (3.42),(3.43). We use $P_2 - P_0$ elements as an example, where u is approximated by piecewise quadratic polynomials and p is approximated by piecewise constants. The stability and convergence proof of this kind of elements for interior problems can be found in [Girault, V. and Raviart, P.A. (1988)]. The infinite element spaces are

$$S(\Omega) = \left\{ u \in \left(H^{1,*}(\Omega)\right)^2 ; u|_{e_i} \in (P_2(e_i))^2, \forall i \right\},$$

$$M(\Omega) = \left\{ p \in L^2(\Omega); p|_{e_i} \in P_0(e_i), \forall i \right\},$$

$$S_0(\Omega) = \{ u \in S(\Omega); u|_{\partial\Omega} = 0 \},$$

$$S(\Gamma_k) = \left\{ u|_{\Gamma_k} ; u \in S(\Omega) \right\}.$$

Let $g_I \in S(\Gamma_0)$ be the approximation of g. The infinite element formulation is: Find $u_h \in S(\Omega)$ and $p_h \in M(\Omega)$, such that $u_h|_{\partial\Omega} = g_I$, and

$$a(u_h, v) - b(v, p_h) = 0, \qquad \forall v \in S_0(\Omega), \tag{3.44}$$

$$b(u_h, q) = 0, \qquad \forall q \in M(\Omega). \tag{3.45}$$

To prove the well-posedness of (3.44),(3.45) we need to verify the inf-sup condition for this problem.

Lemma 61. *There exists a constant $\beta > 0$ independent of h such that*

$$\inf_{p \in M(\Omega), p \neq 0} \sup_{u \in S_0(\Omega), u \neq 0} \frac{b(u,p)}{\|u\|_{1,*}\|p\|_0} \geq \beta. \tag{3.46}$$

Proof. By Lemma 20, for any $p \in M(\Omega)$, there exists $u \in \left(H_0^{1,*}(\Omega)\right)^2$, such that $\nabla \cdot u = p$, and $\|u\|_{1,*} \leq C\|p\|_0$ with a constant C depending on Ω. We apply the Clement interpolation operator P_h [Clement, P. (1975)] on u. $P_h u$ satisfies the following: Let e_i be an arbitrary element, and E_i be a "macro element" consisting of e_i and all elements neighboring e_i, then

$$|u - P_h u|_{1,e_i} + H^{-1}\|u - P_h u\|_{0,e_i} \leq C|u|_{1,E_i},$$

where H is the longest length of diameters of elements in E_i. Let X_j, $j = 1, 2, 3$ be the vertices of e_i and X_{jl}, $1 \leq j < l \leq 3$, be the middle points of the edges s_{jl}. One interpolation $u \to w$ is defined (see [Girault, V. and Raviart, P.A. (1988)]) such that $w \in S(\Omega)$ and

$$w(X_j) = P_h u(X_j), j = 1, 2, 3, \qquad \int_{s_{jl}} (u - w)\, ds = 0, 1 \leq j < l \leq 3.$$

Clearly

$$\int_{e_i} \nabla \cdot w\, dx = \int_{\partial e_i} w \cdot \nu\, ds = \int_{\partial e_i} u \cdot \nu\, ds = \int_{e_i} \nabla \cdot u\, dx.$$

It is proved in [Girault, V. and Raviart, P.A. (1988)] that

$$|P_h u - w|_{1,e_i} \leq C(h^{-1}\|u - P_h u\|_{0,e_i} + |u - P_h u|_{1,e_i}),$$

where h is the diameter of e_i. Then

$$|u - w|_{1,e_i} \leq C(1 + \frac{H}{h})|u|_{1,E_i}.$$

The constant C depends on the shape of elements. However the number of elements in one layer is finite, and the elements are similar to each other on different layers, so there is a uniform constant C for all elements. Certainly C depends on the mesh. Being the same $\frac{H}{h}$ is bounded by a uniform constant. Hence $|w|_1 \leq C|u|_1$, and

$$\frac{b(w,p)}{\|w\|_{1,*}\|p\|_0} = \frac{b(u,p)}{\|w\|_{1,*}\|p\|_0} \geq \frac{1}{C},$$

which gives (3.46). $\square$

By Theorem 39, we obtain that

Theorem 57. *The problem (3.44),(3.45) admits a unique solution.*

We consider the k-th layer Ω_k. If the number of triangular elements is N, then there are N nodes on Γ_{k-1} (or Γ_k). Inside the k-th layer there are other N nodes. The total number of nodes on one layer is $3N$. There are $2N$ degrees of freedom of the velocity components on Γ_{k-1} (or Γ_k), which are arranged as a $2N$ dimensional vector y_{k-1} (or y_k). There is an analogous $2N$ dimensional vector $y_{k-\frac{1}{2}}$ associated with the interior nodes. The values of pressure in the k-th layer form a N dimensional vector, which is denoted by q.

We impose an additional constraint for the implementation of (3.44),(3.45):

$$\int_{\partial\Omega} g_I \cdot \nu \, ds = 0. \tag{3.47}$$

The reason is interpreted as follows. Let Ω_0 be a sub-domain consisting of a finite number of elements. We take $q = 1$ in Ω_0 and $q = 0$ outside Ω_0, then by (3.45) $\int_{\Omega_0} \nabla \cdot u_h \, dx = 0$, and then by the Green's formula $\int_{\partial\Omega_0} u_h \cdot \nu \, ds = 0$. That is, the total flux on $\partial\Omega_0$ vanishes and the nodal values are constrained by a linear relationship. Therefore the basis functions in V do not have local supports. In order to get a block tri diagonal Toeplitz matrix like (3.4) the property of local supports of the basis functions is essential. That is the difficulty. On the other hand, if (3.47) is satisfied, then one can see that there is a vector γ such that $\gamma^T y_0 = 0$. We take a particular function $q \in M(\Omega)$ at the equation (3.45) as follows: $q = 0$ on $\xi^k\Omega$, and $q = 1$ on $\Omega\backslash\xi^k\Omega$, then we get

$$\gamma^T y_k = 0. \tag{3.48}$$

We normalize γ to a unit vector, then construct an orthogonal matrix $T_r = (\gamma G)$, and set $\tilde{y}_k = G^T y_k$, then there is an one to one correspondence between $\tilde{y}_k$ and y_k. There is no restriction between $\tilde{y}_{k-1}$ and $\tilde{y}_k$, so it is easy to construct stiffness matrices on Ω_k and get a block tri-diagonal Toeplitz matrix with respect to $\tilde{y}_k$.

Given y_0 and y_1, satisfying (3.48), we solve the finite element approximation of the Stokes equation on Ω_1 with boundary data y_0, y_1 and the given mesh. Let the approximate solution be u_h, then there are matrices

K_0, K_0', and A, such that

$$\int_{\Omega_1} \nabla u_h \cdot \nabla v \, dx = (\, \tilde{y}_0^T \; \tilde{y}_1^T \,) \begin{pmatrix} K_0 & -A^T \\ -A & K_0' \end{pmatrix} \begin{pmatrix} \tilde{z}_0 \\ \tilde{z}_1 \end{pmatrix}, \tag{3.49}$$

where $\tilde{z}_k = G^T z_k$, and z_k is associated with $v \in V$.

The above expression is valid for all layers Ω_k, hence the infinite element scheme is deduced to the a system of infinite equations, which takes the same form as (3.4). Then the algorithm is the same as that for the Laplace equation. The transfer matrix X and the combined stiffness matrix K_z can be evaluated.

If $\int_{\partial\Omega} g \cdot \nu \, ds = 0$ then it is easy to get an approximation g_I to satisfy $\int_{\partial\Omega} g_I \cdot \nu \, ds = 0$. If $\int_{\partial\Omega} g \cdot \nu \, ds \neq 0$, we can set

$$u_1 = u + \frac{1}{2\pi} \nabla \ln |x| \int_{\partial\Omega} g \cdot \nu \, ds,$$

then u_1 satisfies the same equation and the condition (3.47).

The link between the combined elements and other conventional elements is as follows: If the conventional elements occupy a domain Ω^*, then we can define

$$a(u, v) = \int_{\Omega^*} \nabla u : \nabla v \, dx + \tilde{y}_0^T K_z \tilde{z}_0,$$

and

$$b(u, p) = \int_{\Omega^*} p \nabla \cdot u \, dx,$$

where $\tilde{y}_0 = G^T y_0$ and $\tilde{z}_0 = G^T z_0$, and Γ_0 is the boundary of the domain Ω. Define spaces

$$S(\Omega^*) = \left\{ u \in \left(H^1(\Omega^*) \right)^2 ; u|_{e_i} \in (P_2(e_i))^2, \forall i \right\},$$

$$M(\Omega^*) = \left\{ p \in L^2(\Omega^*); p|_{e_i} \in P_0(e_i), \forall i, \int_{\Omega^*} p \, dx = 0 \right\}.$$

Let the union of Ω and Ω^* be Ω_∞. The numerical computation for the problem,

$$-\triangle u + \nabla p = 0, \quad x \in \Omega_\infty,$$

$$\nabla \cdot u = 0, \quad x \in \Omega_\infty,$$

$$u = g, \qquad x \in \partial\Omega_\infty,$$

is reduced to a conventional finite element scheme on Ω^*.

The above algorithm can be extended to three dimensional problems and the axial symmetric flow. For the later the system of equations is

$$-\frac{1}{x_1}\nabla(x_1\nabla u_1) + \frac{1}{x_1^2}u_1 + \frac{\partial p}{\partial x_1} = 0,$$

$$-\frac{1}{x_1}\nabla(x_1\nabla u_2) + \frac{\partial p}{\partial x_2} = 0,$$

$$\frac{\partial}{\partial x_1}(x_1 u_1) + \frac{\partial}{\partial x_2}(x_1 u_2) = 0.$$

The corresponding bilinear forms are

$$a(u,v) = \int x_1 \left(\nabla u_1 \cdot \nabla v_1 + \nabla u_2 \cdot \nabla v_2 + \frac{u_1 v_1}{x_1^2} \right) dx,$$

$$b(u,p) = \int p \left(\frac{\partial}{\partial x_1}(x_1 u_1) + \frac{\partial}{\partial x_2}(x_1 u_2) \right) dx.$$

The corresponding Hilbert spaces are equipped with norms

$$\|p\|_0^2 = \int_\Lambda x_1 p^2(x)\,dx,$$

$$\|u_2\|_1^2 = \int_\Omega x_1 \left(|\nabla u_2(x)|^2 + \frac{u_2^2(x)}{|x|^2} \right) dx,$$

$$\|u_1\|_{1,*}^2 = \int_\Omega x_1 \left(|\nabla u_1(x)|^2 + \frac{u_1^2(x)}{x_1^2} \right) dx.$$

For details, see [Ying, L. (1986)].

3.8 Darwin model

Let us recall some definitions in Section 1.12.

$$H(\mathrm{curl}, \mathrm{div}; \Omega_0) = \{u \in (L^2(\Omega_0))^2; \nabla \cdot u, \nabla \times u \in L^2(\Omega_0)\},$$

$$\|u\|_{0,\mathrm{curl},\mathrm{div}} = \left(\|u\|_0^2 + \|\nabla \cdot u\|_0^2 + \|\nabla \times u\|_0^2 \right)^{\frac{1}{2}},$$

$$H_{0c}(\Omega) = \{u \in \mathcal{D}'; \|\zeta u\|_{0,\mathrm{curl,div}} < \infty, (1-\zeta)u \in (H_0^{1,*(\Omega)})^2, u \times \nu|_{\partial\Omega} = 0\},$$

$$\|u\|_* = \{\|\nabla \times u\|_0^2 + \|\nabla \cdot u\|_0^2 + <u \cdot \nu, 1>_{\partial\Omega}^2\}^{\frac{1}{2}}.$$

$$V = \{v \in H_{0c}(\Omega); <v \cdot \nu, 1>_{\partial\Omega} = 0\}$$

Let the closure of the quotient space V/V_0 with respect to the norm $\|\cdot\|_*$ be W. Let $\mathcal{Q} = \{p \in L^2(\Omega); \mathrm{supp} p \subset\subset \Omega\}$, equipped with norm

$$\|p\|_\sharp = \left(\|p\|_0^2 + |\int_\Omega p\, dx|^2\right)^{\frac{1}{2}}.$$

Then take closure to obtain a Hilbert space Q. If $p \in Q$ and $\lim p_n = p, p_n \in \mathcal{Q}$, then define $\int_\Omega p\, dx = \lim \int_\Omega p_n\, dx$. The subspace of Q, $\{p \in Q; \int_\Omega p\, dx = 0\}$ is denoted by Q_0. The bilinear forms are

$$d(u,v) = \int_\Omega (\nabla \times u) \cdot (\nabla \times v)\, dx + \int_\Omega (\nabla \cdot u)(\nabla \cdot v)\, dx$$
$$+ <u \cdot \nu, 1>_{\partial\Omega}<v \cdot \nu, 1>_{\partial\Omega}, \quad u,v \in W,$$

and

$$b(u,p) = \int_\Omega (\nabla \cdot u)p\, dx, \qquad u \in W, p \in Q_0.$$

The problem: Find $u \in W$, $p \in Q_0$, such that

$$d(u,v) + b(v,p) = \int_\Omega B \cdot (\nabla \times v)\, dx, \qquad \forall v \in W, \qquad (3.50)$$

$$b(u,q) = 0, \qquad \forall q \in Q_0, \qquad (3.51)$$

admits a unique solution.

We consider the infinite element method for (3.50),(3.51). Being the same as the Stokes problem, $P_2 - P_0$ elements are applied for u and p. The mesh restricted on a bounded domain $\Omega \setminus \xi^k\Omega$ is also a mesh with finite number of elements. In general let Ω_0 be a union of element, finite or infinite, we define

$$E(\Omega_0) = \{u \in (C(\overline{\Omega_0}))^2, u|_{e_i} \in (P_2(e_i))^2, \forall e_i \subset \Omega_0\}.$$

Let

$$D(\Omega \setminus \xi^k\Omega) = \left\{u \in E(\Omega \setminus \xi^k\Omega); u \times \nu|_{\Gamma_0} = 0, \int_{\Gamma_0} u \cdot \nu\, ds = 0\right\},$$

and let

$$\|u\|_{*,\Omega\setminus\xi^k\Omega} = \left(\int_{\Omega\setminus\xi^k\Omega} |\nabla \cdot u|^2 \, dx + \int_{\Omega\setminus\xi^k\Omega} |\nabla \times u|^2 \, dx\right)^{\frac{1}{2}}.$$

Lemma 62. $\| \cdot \|_{*,\Omega\setminus\xi^k\Omega}$ *is a norm on* $D(\Omega \setminus \xi^k\Omega)$.

Proof. If $\|u\|_{*,\Omega\setminus\xi^k\Omega} = 0$, then $\triangle u = 0$, so u is a harmonic function. $u \in P_2$ on individual elements, hence $u \in P_2$ on $\Omega \setminus \xi^k\Omega$. We extend u to the interior of Γ_0 analytically. Then we define the stream function ψ, such that $u = \left(\frac{\partial\psi}{\partial x_2}, -\frac{\partial\psi}{\partial x_1}\right)$, then $\triangle\psi = 0$. By the boundary condition of u, $\frac{\partial\psi}{\partial\nu}|_{\Gamma_0} = 0$, so ψ is a constant in the interior of Γ_0. Therefore $u = 0$. $\square$

We return now to the exterior domain Ω.

Lemma 63. *The space* $W_h(\Omega) = \{u \in E(\Omega); \nabla \cdot u, \nabla \times u \in L^2(\Omega), u \times \nu|_{\partial\Omega} = 0, <u \cdot \nu, 1 >_{\partial\Omega}= 0\}$ *is a Hilbert space under the norm* $\| \cdot \|_*$.

Proof. because $V_0 \cap E(\Omega) = \{0\}$, $\| \cdot \|_*$ is a norm on $W_h(\Omega)$. Let us prove it is complete. Let $\{u_n\}$ be a Cauchy sequence with limit u. We are going to prove $u \in E(\Omega)$. $\{u_n\}$ is also a Cauchy sequence on $D(\Omega \setminus \xi^k\Omega)$. By Lemma 62, $u \in D(\Omega \setminus \xi^k\Omega)$. Since k is arbitrary, $u \in E(\Omega)$. $\square$

Lemma 64. $W_h(\Omega) \subset H^{1,*}(\Omega)$.

Proof. Let $u \in W_h(\Omega)$. We define a function $\zeta \in C^\infty(\Omega)$, such that $\zeta \equiv 0$ near Γ_0, $\zeta \equiv 1$ near the infinity, and $0 \le \zeta \le 1$. We take an arbitrary $\varphi \in C_0^\infty(\Omega)$, then by the Green's formula,

$$\int_\Omega \nabla\varphi \cdot \nabla(\zeta u) \, dx = \int_\Omega (\nabla \cdot \varphi)(\nabla \cdot (\zeta u)) \, dx + \int_\Omega (\nabla \times \varphi) \cdot (\nabla \times (\zeta u)) \, dx$$

$$\le \|\varphi\|_* \|\zeta u\|_* = |\varphi|_1 \|\zeta u\|_*.$$

Since $u \in H^1$ on any bounded domain, $\|\zeta u\|_*$ is bounded. Consequently

$$|\zeta u|_1 = \sup_{\varphi \in C_0^\infty(\Omega)} \frac{\int_\Omega \nabla\varphi \cdot \nabla(\zeta u) \, dx}{|\varphi|_1} \le \|\zeta u\|_*,$$

therefore $\zeta u \in H_0^{1,*}(\Omega)$, that is, $u \in H^{1,*}(\xi^k\Omega)$ for large k. Then by $u \in H^1$ on any bounded domain we have $u \in H^{1,*}(\Omega)$. $\square$

Lemma 65. *The norm* $\| \cdot \|_*$ *is equivalent to* $\| \cdot \|_{1,*}$ *on* $W_h(\Omega)$.

Proof. It is the direct consequence of the closed graph theorem and Lemma 64. $\square$

We assume an inhomogeneous boundary condition, $u \times \nu|_{\partial\Omega} = g$, and let g_I be an approximation of g, then the formulation of the infinite element method is: Find $u_h \in E(\Omega) \cap H^{1,*}(\Omega)$, $p_h \in M(\Omega)$, such that $u_h \times \nu|_{\Gamma_0} = g_I$, $\int_{\Gamma_0} u_h \cdot \nu\, ds = 0$, and

$$d(u_h, v) + b(v, p_h) = \int_\Omega B \cdot (\nabla \times v)\, dx, \qquad \forall v \in W_h(\Omega), \qquad (3.52)$$

$$b(u_h, q) = 0, \qquad \forall q \in M(\Omega). \qquad (3.53)$$

Theorem 58. *The problem (3.52),(3.53) admits a unique solution.*

Proof. $d(\cdot, \cdot)$ is coercive. To verify the inf-sup condition we apply the inf-sup condition of the infinite element method for the Stokes equation,

$$\sup_{\substack{v \in S_0(\Omega) \\ v \neq 0}} \frac{b(v, p)}{|v|_1} \geq \beta\|p\|_0, \qquad \forall p \in M(\Omega).$$

Now $W_h(\Omega) \supset E(\Omega) \cap H_0^{1,*}(\Omega)$, thus

$$\sup_{\substack{v \in W_h(\Omega) \\ v \neq 0}} \frac{b(v, p)}{|v|_1} \geq \beta\|p\|_0, \qquad \forall p \in M(\Omega).$$

Moreover we have $\|v\|_*^2 \leq 2|v|_1^2$, thus

$$\sup_{\substack{v \in W_h(\Omega) \\ v \neq 0}} \frac{b(v, p)}{\|v\|_*} \geq \frac{\beta}{\sqrt{2}}\|p\|_0, \qquad \forall p \in M(\Omega).$$

Then the conclusion follows, and the proof is the same as the that in Theorem 40. $\qquad \square$

The relationship between the solutions of the infinite element method to the Stokes equation and to the Darwin model is the following:

Theorem 59. *The solution to (3.52),(3.53) is a solution to (3.44),(3.45) with appropriate boundary value and inhomogeneous term.*

Proof. We take v in (3.52) such that $v|_{\Gamma_0} = 0$, then $v \in H_0^{1,*}(\Omega)$. Let a series $\{v_n\} \subset C_0^\infty(\Omega)$ tend to v in $H_0^{1,*}(\Omega)$, then by the Green's formula, $d(u_h, v_n) = a(u_h, v_n)$. Letting $n \to \infty$ we verify that $u_h, \dot{p}_h$ satisfy (3.44),(3.45). $\qquad \square$

Now let us go to the algorithm for the Darwin model. For the formulation (3.52),(3.53), we take a boundary value γ_h for the Stokes equation such that $\gamma_h \times \nu = g_I$, and $\int_{\Gamma_0} \gamma_h \cdot \nu \, ds = 0$. Then solve the following problem: Find $w_h \in S(\Omega)$, $r_h \in M(\Omega)$, such that $w_h|_{\Gamma_0} = \gamma_h$, and

$$a(w_h, v) + b(v, r_h) = \int_\Omega B \cdot (\nabla \times v) \, dx, \qquad \forall v \in S_0(\Omega), \tag{3.54}$$

$$b(w_h, q) = 0, \qquad \forall q \in M(\Omega). \tag{3.55}$$

The solutions to (3.52),(3.53) also satisfy

$$a(u_h, v) + b(v, r_h) = \int_\Omega B \cdot (\nabla \times v) \, dx, \qquad \forall v \in S_0(\Omega), \tag{3.56}$$

$$b(u_h, q) = 0, \qquad \forall q \in M(\Omega). \tag{3.57}$$

Let $u_h - w_h = U_h$ and $p_h - r_h = P_h$, then we have

$$a(U_h, v) + b(v, P_h) = 0, \qquad \forall v \in S_0(\Omega), \tag{3.58}$$

$$b(U_h, q) = 0, \qquad \forall q \in M(\Omega). \tag{3.59}$$

Let the vectors $y_k = B_k U_h$ and $z_k = G^T y_k$, then it holds that $z_{k+1} = X z_k$, which yields

$$y_{k+1} = G X G^T y_k. \tag{3.60}$$

Let the function form of (3.60) be $U_h|_{\Gamma_1} = Y U_h|_{\Gamma_0}$, where Y is an operator.

So far the problem (3.52),(3.53) is deduced to a finite element approximation on a bounded domain Ω_1 for the following system of equations:

$$-\triangle u = \nabla \times B, \qquad x \in \Omega_1,$$

$$\nabla \cdot u = 0, \qquad x \in \Omega_1,$$

$$u \times \nu|_{\Gamma_0} = g_h,$$

$$\int_{\Gamma_0} u \cdot \nu \, ds = 0,$$

$$(u - w_h)|_{\Gamma_1} = Y(u - w_h)|_{\Gamma_0},$$

where w_h is given. We solve the finite element problem on Ω_1 with the same mesh, and get the solution u_h to (3.52),(3.53) on Ω_1. Then we use the formula (3.60) and the solution w_h to (3.54),(3.55) to get the solution u_h on the entire domain Ω.

Finally we need to show that (3.52),(3.53) with homogeneous boundary condition is an approximation to (3.50),(3.51). If the solution to (3.50),(3.51) satisfies $u \in H^{1,*}(\Omega)$, then we can modify the formulation to: Find $u \in H^{1,*}(\Omega) \cap W$, $p \in M(\Omega)$, such that

$$d(u,v) + b(v,p) = \int_\Omega B \cdot (\nabla \times v)\,dx, \qquad \forall v \in H^{1,*}(\Omega) \cap W, \qquad (3.61)$$

$$b(u,q) = 0, \qquad \forall q \in M(\Omega). \qquad (3.62)$$

This is because the solution u satisfies $\nabla \cdot u = 0$, $b(u,p) = 0$ for all $p \in L^2(\Omega)$., Because $H^{1,*}(\Omega) \cap W$ is dense in W, uniqueness holds for the problem (3.61),(3.62). The spaces in (3.52),(3.53) are subspaces of those in (3.61),(3.62), so the solutions to (3.52),(3.53) are indeed approximate solutions.

3.9 Elliptic equations with variable coefficients

3.9.1 *A homogeneous equation*

Let us consider an equation,

$$\nabla \cdot (k(x)\nabla u) = 0 \qquad (3.63)$$

on the domain Ω.

Let $\lambda > 1$ be a parameter. We make a similarity transformation $\lambda x \to x$. Under the transformation the equation (3.63) becomes

$$\nabla \cdot (k(\lambda x)\nabla u) = 0, \qquad (3.64)$$

which is still considered on Ω. The bilinear form associated with (3.64) is thus defined by

$$a_\lambda(u,v) = \int_\Omega k(\lambda x)\nabla u \nabla v\,dx. \qquad (3.65)$$

We assume that (3.63) is elliptic, that is, there is a constant $k_0 > 0$ such that $k(x) \geq k_0$ for all $x \in \Omega$. In addition we assume that $k(x)$ is continuous and approaches to a constant as $x \to \infty$. It has no harm in assuming $\lim_{x \to \infty} k(x) = 1$.

We still use linear triangular elements for the infinite element mesh. The total stiffness matrix on Ω_1 is defined by

$$\begin{pmatrix} K_0 & -A^T \\ -A & K_0' \end{pmatrix},$$

that is,

$$\int_{\Omega_1} k(\lambda x) \nabla u \cdot \nabla v \, dx = \begin{pmatrix} y_0^T & y_1^T \end{pmatrix} \begin{pmatrix} K_0 & -A^T \\ -A & K_0' \end{pmatrix} \begin{pmatrix} z_0 \\ z_1 \end{pmatrix},$$

where $u, v \in S(\Omega)$, and the space $S(\Omega)$ is defined in Section 3.1. The infinite formulation for the equation (3.64) is: for a given $g_I \in S(\Gamma_0)$, find $u_h \in S(\Omega)$, such that

$$a_\lambda(u_h, v) = 0, \qquad \forall v \in S_0(\Omega). \tag{3.66}$$

The problem (3.66) admits a unique solution. Then a combined stiffness matrix $K_z(\lambda)$ is defined by $y_0^T K_z(\lambda) z_0 = a_\lambda(u_h, v)$. $K_z(\lambda)$ is real and symmetric. The transfer matrix $X(\lambda)$ is defined by $y_1 = X(\lambda) y_0$. Let the domain Ω be regarded as a union of Ω_1 and $\xi \Omega$. The bilinear form a_λ can also be written as

$$a_\lambda(u_h, v) = \begin{pmatrix} y_0^T & y_0^T X(\lambda)^T \end{pmatrix} \begin{pmatrix} K_0 & -A^T \\ -A & K_0' \end{pmatrix} \begin{pmatrix} z_0 \\ z_1 \end{pmatrix} + y_0^T X(\lambda)^T K_z(\lambda\xi) z_1$$

Since z_1 is arbitrary, we have

$$y_0^T(-A + X(\lambda)^T K_0' + X(\lambda) K_z(\lambda\xi)) = 0.$$

y_0 is arbitrary too, so

$$X(\lambda) = (K_0' + K_z(\lambda\xi))^{-1} A.$$

On the other hand, being analogous to Theorem 52 it holds that

$$K_z(\lambda) = K_0 - A^T X(\lambda).$$

Therefore

$$K_z(\lambda) = K_0 - A^T (K_0' + K_z(\lambda\xi))^{-1} A. \tag{3.67}$$

To solve an equation with variable coefficients like (3.64), it is convenient to consider the critical case $\xi = 1^+$. We set $y_1 = y_0 + (\xi - 1)\eta_1$, then

$$(y_0^T \; y_1^T) \begin{pmatrix} K_0 & -A^T \\ -A & K_0' \end{pmatrix} \begin{pmatrix} y_0 \\ y_1 \end{pmatrix}$$

$$= (y_0^T \; \eta_1^T) \begin{pmatrix} K_0 - A - A^T + K_0' & -(\xi - 1)(A^T - K_0') \\ -(\xi - 1)(A - K_0') & (\xi - 1)^2 K_0' \end{pmatrix} \begin{pmatrix} y_0 \\ \eta_1 \end{pmatrix}$$

$$\equiv (y_0^T \; \eta_1^T) \begin{pmatrix} \tilde{L}_0 & -\tilde{B}^T \\ -\tilde{B} & \tilde{L}_0' \end{pmatrix} \begin{pmatrix} y_0 \\ \eta_1 \end{pmatrix}.$$

By (3.67) we get

$$K_z(\lambda) = K_z(\lambda\xi) + \tilde{L}_0$$

$$- \left(\frac{\tilde{B}^T}{\xi - 1} - K_z(\lambda\xi) \right) \left(\frac{\tilde{L}_0'}{(\xi - 1)^2} + K_z(\lambda\xi) \right)^{-1} \left(\frac{\tilde{B}}{\xi - 1} - K_z(\lambda\xi) \right).$$

Dividing it by $\xi - 1$, letting $\xi \to 1$ and defining

$$L_0 = \lim_{\xi \to 1} \frac{\tilde{L}_0}{\xi - 1}, \qquad B = \lim_{\xi \to 1} \frac{\tilde{B}}{\xi - 1}, \qquad L_0' = \lim_{\xi \to 1} \frac{\tilde{L}_0'}{\xi - 1},$$

to obtain

$$\lambda \frac{dK_z(\lambda)}{d\lambda} = -L_0 + (B^T - K_z(\lambda))(L_0')^{-1}(B - K_z(\lambda)), \qquad (3.68)$$

which is the ordinary differential equation for $K_z(\lambda)$.

Letting $\lambda \to \infty$, the equation (3.64) approaches to the Laplace equation, $\triangle u = 0$ asymptotically. We denote by K_{z0} the combined stiffness matrix associated with the Laplace equation, which is the initial data of (3.68) at $\lambda = \infty$. If we know more properties of the function $k(x)$ near $|x| = \infty$ we can get more accurate asymptotic expression for $K_z(\lambda)$. Let us assume that

$$k(x) = 1 + \frac{c_1(\theta)}{|x|} + o\left(\frac{1}{|x|} \right), \qquad \theta = \tan^{-1} \frac{x_2}{x_1},$$

then

$$k(\lambda x) = 1 + \frac{c_1(\theta)}{|x|} \lambda^{-1} + o\left(\frac{\lambda^{-1}}{|x|} \right),$$

accordingly we have

$$L_0 = L_{00} + L_{01}\lambda^{-1} + o(\lambda^{-1}),$$

$$L_0' = L_{00}' + L_{01}'\lambda^{-1} + o(\lambda^{-1}),$$

$$B = B_0 + B_1\lambda^{-1} + o(\lambda^{-1}).$$

Formally we set $K_z(\lambda) = K_{z0} + K_{z1}\lambda^{-1}$ and substitute it into (3.68), then we obtain a linear equation for K_{z1}:

$$\begin{aligned}
&K_{z1} - K_{z1}(L_{00}')^{-1}(B_0 - K_{z0}) - (B_0^T - K_{z0})(L_{00}')^{-1}K_{z1}\\
=&L_{01} - (B_0^T - K_{z0})(L_{00}')^{-1}L_{01}'(L_{00}')^{-1}(B_0 - K_{z0}) \qquad (3.69)\\
&- B_1^T(L_{00}')^{-1}(B_0 - K_{z0}) - (B_0^T - K_{z0})(L_{00}')^{-1}B_1.
\end{aligned}$$

We can take a constant ξ slightly larger than 1, then get an approximation of K_{z0}. Using (3.69) we get K_{z1}. $K_{z0} + K_{z1}\lambda_0^{-1}$ is taken as an approximation to $K_z(\lambda_0)$ for a large λ_0. Then we solve (3.68) backward to get $K_z(\lambda)$ for all $\lambda \in [1, \lambda_0]$. $K_z(1)$ is our desired combined stiffness matrix for the equation (3.63).

3.9.2 *An inhomogeneous equation*

Next let us consider an equation,

$$-\nabla \cdot (k(x)\nabla u) = f(x). \qquad (3.70)$$

To guarantee well-posedness it is assumed that $f(x)|x|^2 \log^2 |x|$ is bounded for large $|x|$. Under the transformation $\lambda x \to x$ it becomes

$$-\nabla \cdot (k(\lambda x)\nabla u) = \lambda^2 f(\lambda x). \qquad (3.71)$$

We define

$$F_\lambda(v) = \lambda^2 \int_\Omega f(\lambda x)v \, dx,$$

and

$$J_\lambda(u, v) = a_\lambda(u, v) - F_\lambda(v).$$

The infinite element formulation of the Dirichlet boundary value problem to (3.71) is: Find $u_h \in S(\Omega)$ for given $g_I \in S(\Gamma_0)$ such that $u_h|_{\Gamma_0} = g_I$ and

$$J_\lambda(u_h, v) = 0, \qquad \forall v \in S_0(\Omega). \qquad (3.72)$$

If $v_1, v_2 \in S(\Omega)$ and $v_1|_{\Gamma_0} = v_2|_{\Gamma_0} = z_0$, then $v = v_1 - v_2 \in S_0(\Omega)$. (3.72) implies that $J_\lambda(u, v_1) = J_\lambda(u, v_2)$. Therefore as a functional on the space $S(\Omega)$, $J_\lambda(u, \cdot)$ depends on z_0 only. We write J_λ as the sum of a bilinear part and a linear part:

$$J_\lambda(u, v) = y_0^T K_z(\lambda) z_0 - H(\lambda) z_0, \tag{3.73}$$

where $H(\lambda)$ is a row vector. Because (3.73) is valid for all homogeneous equations, $K_z(\lambda)$ is just the combined stiffness matrix associated with (3.64).

Let us derive the equation for $H(\lambda)$. We divide the domain Ω into two parts: Ω_1 and $\xi\Omega$ for given $\xi > 1$, then we have

$$y_0^T K_z(\lambda) z_0 - H(\lambda) z_0 = y_1^T K_z(\lambda\xi) z_1$$
$$+ (\, y_0^T \ y_1^T \,) \begin{pmatrix} K_0 & -A^T \\ -A & K_0' \end{pmatrix} \begin{pmatrix} z_0 \\ z_1 \end{pmatrix} - H(\lambda\xi) z_1 - \lambda^2 \int_{\Omega_1} f(\lambda x) v \, dx.$$

The last term depends on $\begin{pmatrix} z_0 \\ z_1 \end{pmatrix}$ linearly. We define

$$(\, \Phi_0(\lambda) \ \Phi_1(\lambda) \,) \begin{pmatrix} z_0 \\ z_1 \end{pmatrix} = \lambda^2 \int_{\Omega_1} f(\lambda x) v \, dx.$$

Noting (3.67) we get

$$y_0^T (K_0 - A^T (K_0' + K_z(\lambda\xi))^{-1} A) z_0 - H(\lambda) z_0 = y_1^T K_z(\lambda\xi) z_1$$
$$+ (\, y_0^T \ y_1^T \,) \begin{pmatrix} K_0 & -A^T \\ -A & K_0' \end{pmatrix} \begin{pmatrix} z_0 \\ z_1 \end{pmatrix} - H(\lambda\xi) z_1 - (\, \Phi_0(\lambda) \ \Phi_1(\lambda) \,) \begin{pmatrix} z_0 \\ z_1 \end{pmatrix}.$$

$$\tag{3.74}$$

Since u_h is the solution, (3.74) holds for all z_1, we have

$$y_1^T K_z(\lambda\xi) - y_0^T A + y_1^T K_0' - H(\lambda\xi) - \Phi_1(\lambda) = 0,$$

which gives

$$y_1 = (K_0' + K_z(\lambda\xi))^{-1} (A y_0 + H^T(\lambda\xi) + \Phi_1^T(\lambda)). \tag{3.75}$$

We substitute (3.75) into (3.74) to obtain

$$-H(\lambda)z_0 = -H(\lambda\xi)(K_0' + K_z(\lambda\xi))^{-1}Az_0$$
$$- \Phi_1(\lambda)(K_0' + K_z(\lambda\xi))^{-1}Az_0 - \Phi_0(\lambda)z_0.$$

Since z_0 is arbitrary, it holds that

$$-H(\lambda) = -(H(\lambda\xi) + \Phi_1(\lambda))(K_0' + K_z(\lambda\xi))^{-1}A - \Phi_0(\lambda).$$

We set $z_1 = z_0 + (\xi - 1)\zeta_1$ then

$$(\Phi_0(\lambda)\ \Phi_1(\lambda)) \begin{pmatrix} z_0 \\ z_1 \end{pmatrix} = (\Phi_0(\lambda) + \Phi_1(\lambda)\ (\xi - 1)\Phi_1(\lambda)) \begin{pmatrix} z_0 \\ \zeta_1 \end{pmatrix}.$$

We define

$$\tilde{\Psi}_0(\lambda) = \Phi_0(\lambda) + \Phi_1(\lambda), \quad \tilde{\Psi}_1(\lambda) = (\xi - 1)\Phi_1(\lambda),$$

then it holds that

$$H(\lambda) = \left(H(\lambda\xi) + \frac{\tilde{\Psi}_1(\lambda)}{\xi - 1} \right)$$
$$\cdot \left\{ I + \left(\frac{\tilde{L}_0'}{(\xi - 1)^2} + K_z(\lambda\xi) \right)^{-1} \left(\frac{\tilde{B}}{\xi - 1} - K_z(\lambda\xi) \right) \right\} + \tilde{\Psi}_0 - \frac{\tilde{\Psi}_1}{\xi - 1}.$$

$$(3.76)$$

Let

$$\psi_0(\lambda) = \lim_{\xi \to 1} \frac{\tilde{\Psi}_0}{\xi - 1}, \quad \psi_1(\lambda) = \lim_{\xi \to 1} \frac{\tilde{\Psi}_1}{(\xi - 1)^2}.$$

We divide (3.76) by $\xi - 1$ and let $\xi \to 1$ to obtain

$$\lambda\frac{dH(\lambda)}{d\lambda} = -H(\lambda)(L_0')^{-1}(B - K_z(\lambda)) - \psi_0(\lambda), \qquad (3.77)$$

which is the ordinary differential equation for $H(\lambda)$.

If $f(x) = f_0(\theta)|x|^{-3} + o(|x|^{-3})$, then $\psi_0(\lambda) = \psi_{00}\lambda^{-1} + o(\lambda^{-1})$. Formally we set $H(\lambda) = H_0 + H_1\lambda^{-1}$, then by (3.77) we have

$$-H_0(L_{00}')^{-1}(B_0 - K_{z0}) = 0,$$

$$-H_1 = -H_0(L_{00}')^{-1}L_{01}'(L_{00}')^{-1}(B_0 - K_{z0})$$
$$- H_1(L_{00}')^{-1}(B_0 - K_{z0}) - H_0(L_{00}')^{-1}(B_1 - K_{z1}) - \psi_{00},$$

which gives

$$H_0 = 0,$$

$$H_1 = \psi_{00}\{I - (L'_{00})^{-1}(B_0 - K_{z0})\}^{-1}.$$

3.9.3 General multiply connected domains

Let $\bar{D}_j \subset \Omega$, $j = 1, \cdots, N$, where D_j are simply-connected domains with boundary γ_j. We assume that $\bar{D}_j$ are disjoint to each other and set $\Omega_0 = \Omega \setminus \bigcup_{j=1}^{N} \bar{D}_j$, then Ω_0 is a multiply connected domain. Let us consider the equation (3.70) on Ω_0.

First of all we consider the Neŭmann boundary condition:

$$\left. \frac{\partial u}{\partial \nu} \right|_{\gamma_j} = 0, \qquad j = 1, \cdot, N. \tag{3.78}$$

The boundary condition on Γ_0 is flexible as usual.

If the boundaries γ_j are suitable regular, say, satisfying the Lipschitz condition, then functions in $H^1_{\mathrm{loc}}(\Omega_0)$ can be extended into D_j, still in H^1. Therefore if we assume

$$\tilde{k}(x) = \begin{cases} k(x), \; x \in \Omega_0, \\ 0, \quad x \in \Omega \setminus \Omega_0, \end{cases}$$

$$\tilde{f}(x) = \begin{cases} f(x), \; x \in \Omega_0, \\ 0, \quad x \in \Omega \setminus \Omega_0, \end{cases}$$

and consider the equation

$$-\nabla \cdot (\tilde{k}(x)\nabla u) = \tilde{f}(x)$$

on Ω, then it is equivalent to (3.70) on Ω_0. We define a seminorm

$$\|u\|_{1,*} = \left(\int_\Omega \tilde{k}(x) \left(|\nabla u(x)|^2 + \frac{u^2(x)}{|x|^2 \log^2 |x|} \right) dx \right)^{1/2} \tag{3.79}$$

on $H^{1,*}(\Omega)$ and let

$$T = \{u \in H^{1,*}(\Omega); \|u\|_{1,*} = 0\},$$

then for a given function u_0 on Γ_0 we consider the variational problem: find $u \in H^{1,*}(\Omega)/T$, such that $u|_{\Gamma_0} = g_I$, and

$$J_\lambda(u, v) = 0, \qquad \forall v \in H_0^{1,*}(\Omega).$$

The problem is thus defined on Ω, and the approaches in the previous section can be applied here. We provide more details about the mathematical formulations in the next subsection.

Secondly we consider the Dirichlet boundary conditions

$$u|_{\gamma_j} = 0, \qquad j = 1, \cdots, N. \tag{3.80}$$

We use penalty method and consider the problem: find $u \in H^{1,*}(\Omega)/T$ such that $u|_{\Gamma_0} = g_I$ and

$$\int_\Omega \tilde{k}(x)\nabla u \cdot \nabla v \, dx - \int_\Omega \tilde{f}(x)v \, dx + \frac{1}{\varepsilon} \sum_{j=1}^N \oint_{\gamma_j} uv \, ds = 0, \quad \forall v \in H_0^{1,*}(\Omega),$$

where $\varepsilon > 0$ is a small constant. We denote by $\lambda^{-1}\gamma_j$ the image of γ_j under the transformation $\lambda x \to x$, then define

$$J_\lambda(u, v) = \int_\Omega \tilde{k}(\lambda x)\nabla u \cdot \nabla v \, dx - \lambda^2 \int_\Omega \tilde{f}(\lambda x)v \, dx + \frac{\lambda}{\varepsilon} \sum_{j=1}^N \int_{(\lambda^{-1}\gamma_j)\cap\Omega} uv \, ds,$$

$$\tag{3.81}$$

where we notice that $(\lambda^{-1}\gamma_j) \cap \Omega$ may not be a closed curve. Then we use the approach in the previous section to solve this problem. Concerning the last term of (3.81) the scheme to evaluate $K_z(\lambda)$ should be modified. Let us consider one domain D_j as an example. Let Γ_λ be the similar figures of Γ_0 with the center O and the constant of proportionality $\lambda > 1$. We assume that γ_j consists of some parts, some of which lie in Γ_{λ_1} and Γ_{λ_2}, and some are transversal to Γ_λ (Fig. 5).

The matrix $K_z(\lambda)$ is discontinuous at $\lambda = \lambda_1$ and λ_2:

$$y_0^T K_z(\lambda_2 - 0)z_0 = y_0^T K_z(\lambda_2 + 0)z_0 + \frac{\lambda_2}{\varepsilon} \int_{(\lambda_2^{-1}\gamma_j)\cap\Gamma_0} uv \, ds,$$

$$y_0^T K_z(\lambda_1 - 0)z_0 = y_0^T K_z(\lambda_1 + 0)z_0 + \frac{\lambda_1}{\varepsilon} \int_{(\lambda_1^{-1}\gamma_j)\cap\Gamma_0} uv \, ds.$$

The other parts of γ_j have some contribution to the matrix L_0. Let us consider "element stiffness matrices".

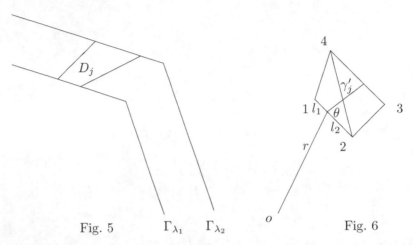

Fig. 5 Γ_{λ_1} Γ_{λ_2} Fig. 6

Suppose $1 - 2 - 3 - 4$ is a combination of two elements. the intersection of $\lambda^{-1}\gamma_j$ with it is approximated by a line segment γ'_j with an oblique angle θ.(Fig 6) γ'_j divides the line segment $1 - 2$ into two parts with lengths l_1, l_2 respectively. Let $\varphi_i, i = 1, \cdots, 4$ be the basis functions associated with y_0 and η_1, and let f_k be the value of φ_i at the node k, then

$$\varphi_1 : f_1 = f_4 = 1, f_2 = f_3 = 0,$$

$$\varphi_2 : f_2 = f_3 = 1, f_1 = f_4 = 0,$$

$$\varphi_3 : f_4 = \xi - 1, f_1 = f_2 = f_3 = 0,$$

$$\varphi_3 : f_3 = \xi - 1, f_1 = f_2 = f_4 = 0.$$

φ_3 and φ_4 have no contribution to the stiffness matrices, and the length of γ'_j is $r(\xi - 1)/\sin\theta$, therefore

$$\lim_{\xi \to 1} \frac{1}{\xi - 1} \int_{\gamma'_j} \varphi_1^2 \, dx = \lim_{\xi \to 1} \frac{1}{\xi - 1} \left(\frac{l_2}{l_1 + l_2} \right)^2 \frac{r(\xi - 1)}{\sin\theta} = \frac{r}{\sin\theta} \left(\frac{l_2}{l_1 + l_2} \right)^2.$$

The other entries can be evaluated in the same way. We get

$$K = \frac{\lambda r}{\varepsilon \sin\theta} \begin{pmatrix} \left(\frac{l_2}{l_1+l_2} \right)^2 & \frac{l_1 l_2}{(l_1+l_2)^2} \\ \frac{l_1 l_2}{(l_1+l_2)^2} & \left(\frac{l_1}{l_1+l_2} \right)^2 \end{pmatrix}.$$

We assemble all matrices K of one layer then add the sum to L_0 in the equation (3.68).

3.9.4 *Transfer matrices*

To solve a boundary value problem after getting $K_z(\lambda)$ we define transfer matrices $X(\lambda_0, \lambda)$ and $X_1(\lambda_0, \lambda)$ by

$$y_\lambda = X(\lambda_0, \lambda)y_{\lambda_0} + X_1(\lambda_0, \lambda), \quad \lambda > \lambda_0, \tag{3.82}$$

where $y_\lambda = u|_{\Gamma_\lambda}$. If y_{λ_0} is given, then we regard it as a boundary data. By the well-posedness of the problem y_λ is determined uniquely, hence the matrix X and X_1 are well defined. By some calculation we get from (3.75) that

$$y_1 = \left\{ I + \left(\frac{\tilde{L}_0'}{(\xi - 1)^2} + K_z(\lambda\xi) \right)^{-1} \left(\frac{\tilde{B}}{\xi - 1} - K_z(\lambda\xi) \right) \right\} y_0$$

$$+ \left(\frac{\tilde{L}_0'}{(\xi - 1)^2} + K_z(\lambda\xi) \right)^{-1} \left(H^T(\lambda\xi) + \frac{\tilde{\Psi}_1(\lambda)}{\xi - 1} \right).$$

Therefore

$$\lim_{\xi \to 1} \frac{y_1 - y_0}{\xi - 1} = (L_0')^{-1}(B - K_z(\lambda))y_0 + (L_0')^{-1}H^T(\lambda).$$

We define an infinitesimal transfer matrix $Y(\lambda) = (L_0')^{-1}(B - K_z(\lambda))$ and an infinitesimal transfer vector $Y_1(\lambda) = (L_0')^{-1}H^T(\lambda)$. We then take $\lambda_0 \geq 1$, $\Delta\lambda_0 > 0$, and make a similarity transformation $\lambda_0 x \to x$, then $(\lambda_0 + \Delta\lambda_0)x \to (1 + \frac{\Delta\lambda_0}{\lambda_0})x$. Let $\xi = 1 + \frac{\Delta\lambda_0}{\lambda_0}$, then approximately

$$y_{\lambda_0 + \Delta\lambda_0} = y_{\lambda_0} + (\xi - 1)(Y(\lambda_0)y_{\lambda_0} + Y_1(\lambda_0)) = y_{\lambda_0} + \frac{\Delta\lambda_0}{\lambda_0}(Y(\lambda_0)y_{\lambda_0} + Y_1(\lambda_0)),$$

and

$$y_\lambda = X(\lambda_0, \lambda)y_{\lambda_0} + X_1(\lambda_0, \lambda)$$

$$= X(\lambda_0 + \Delta\lambda_0, \lambda)y_{\lambda_0 + \Delta\lambda_0} + X_1(\lambda_0 + \Delta\lambda_0, \lambda)$$

$$= X(\lambda_0 + \Delta\lambda_0, \lambda)\left(y_{\lambda_0} + \frac{\Delta\lambda_0}{\lambda_0}(Y(\lambda_0)y_{\lambda_0} + Y_1(\lambda_0)) \right) + X_1(\lambda_0 + \Delta\lambda_0, \lambda).$$

Since y_{λ_0} is arbitrary, we get approximately

$$X(\lambda_0, \lambda) = X(\lambda_0 + \Delta\lambda_0, \lambda)\left(I + \frac{\Delta\lambda_0}{\lambda_0}Y(\lambda_0) \right),$$

$$X_1(\lambda_0, \lambda) = \frac{\Delta\lambda_0}{\lambda_0}X(\lambda_0 + \Delta\lambda_0, \lambda)Y_1(\lambda_0) + X_1(\lambda_0 + \Delta\lambda_0, \lambda).$$

Dividing them by $\Delta\lambda_0$ and letting $\Delta\lambda_0 \to 0$ we get

$$\frac{\partial}{\partial\lambda_0}X(\lambda_0, \lambda) = -\lambda_0^{-1}X(\lambda_0, \lambda)Y(\lambda_0), \tag{3.83}$$

$$\frac{\partial}{\partial \lambda_0} X_1(\lambda_0, \lambda) = -\lambda_0^{-1} X(\lambda_0, \lambda) Y_1(\lambda_0). \qquad (3.84)$$

We have initial value

$$X(\lambda, \lambda) = I, \qquad X_1(\lambda, \lambda) = 0,$$

so we can solve (3.83),(3.84) to get X and X_1, then get the solution from a given boundary condition.

3.10 Convergence

All schemes in this chapter can be proved convergent. We investigate one of them for example. Let us consider the Poisson equation,

$$-\triangle u = f, \qquad x \in \Omega, \qquad (3.85)$$

with homogeneous boundary condition,

$$u|_{\partial \Omega} = 0, \qquad (3.86)$$

where dim $(\Omega) = 2$.

We assume that $f \in H^{m-1}(\Omega)$ for a positive integer m, and f is compactly supported. Moreover we assume that $u \in H_{loc}^{m+1}(\Omega)$. The partition of the domain Ω is described in Section 3.1. Let h be a reference length of the mesh, for example we can set h to be the maximum length of the diameters of all elements in Ω_1. Some assumptions are made on the mesh:

(A) The partition is regular, that is the interior angles of all elements possess a common lower bound $\theta_0 > 0$.

(B) $O \notin \bar{e}_i, \quad \forall i$. And let $d(O, e_i)$ be the distance from e_i to the point O, then there is a constant χ, such that

$$\text{meas } e_i \leq \chi h^2 (d(O, e_i))^2, \qquad \forall i.$$

For instance the similar partition given in Section 3.1 satisfies the condition (B).

Let m-th order elements are employed in the scheme. Then the infinite element subspaces are

$$S(\Omega) = \{u \in H^{1,*}(\Omega); |u|_{e_i} \in P_m(e_i), \forall i\},$$

and

$$S_0(\Omega) = \{u \in S(\Omega); u|_{\partial \Omega} = 0\}.$$

The infinite element formulation of the problem is: Find $u_h \in S_0(\Omega)$, such that

$$a(u_h, v) = (f, v), \qquad \forall v \in S_0(\Omega). \tag{3.87}$$

By the expressions of the solutions of the Laplace equation in Section 1.1, we find the weighted semi-norm

$$|u|_{m+1(m)} = \left\{ \int_\Omega |x|^{2m} |D^{m+1} u(x)|^2 \, dx \right\}^{\frac{1}{2}} \tag{3.88}$$

is bounded, where $D^{m+1}u$ denotes the $(m+1)$-th order derivatives of u. Let Π be the interpolating operator with respect to $S(\Omega)$, then we have

Theorem 60. *If the conditions* (A),(B) *hold,* $u \in H_{loc}^{m+1}(\Omega)$, *and* $|u|_{m+1(m)} < \infty$, *then we have the interpolating estimate,*

$$|u - \Pi u|_1 \leq C h^m |u|_{m+1(m)}, \tag{3.89}$$

$$\left\{ \int_\Omega \frac{1}{|x|^2} (u(x) - \Pi u(x))^2 \, dx \right\}^{\frac{1}{2}} \leq C h^{m+1} |u|_{m+1(m)}. \tag{3.90}$$

Proof. By the embedding theorem of Sobolev spaces, u is continuous, so Πu makes sense. According to the interpolating estimate on triangular elements [Ciarlet, P.G. (1978)],

$$|u - \Pi u|_{s,e_i} \leq C h_i^{m+1-s} |u|_{m+1,e_i}, \quad s = 0, 1, \tag{3.91}$$

where h_i is the diameter of e_i.

By the condition (A) we have

$$\text{meas } e_i \geq h_i^2 \sin^3 \theta_0 \cos \theta_0,$$

then by the condition (A),(B) we have

$$h_i^2 \leq \chi \frac{h^2 (d(O, e_i))^2}{\sin^3 \theta_0 \cos \theta_0}.$$

Substituting it into (3.91) we get

$$|u - \Pi u|_{s,e_i} \leq C h^{m+1-s} (d(O, e_i))^{m+1-s} |u|_{m+1,e_i}.$$

We have $d(O, e_i) \leq |x|$ on the triangle e_i, therefore

$$(d(O, e_i))^{m+1-s} |u|_{m+1,e_i} \leq (d(O, e_i))^{1-s} |u|_{m+1(m),e_i}.$$

We obtain

$$(d(O, e_i))^{s-1} |u - \Pi u|_{s,e_i} \leq C h^{m+1-s} |u|_{m+1(m),e_i}.$$

Upon summing them up with respect to i we get (3.89) and (3.90). $\qquad \square$

We turn to estimate the error $u - u_h$.

Theorem 61. *If the conditions* (A),(B) *hold, and the solution u to the problem* (3.85),(3.86) *satisfies* $|u|_{m+1(m)} < \infty$, *then*

$$|u - u_h|_1 \leq Ch^m |u|_{m+1(m)}. \tag{3.92}$$

Proof. The proof is routine. We take an arbitrary $v \in S(\Omega)$, then

$$a(u - u_h, v) = 0.$$

Thus

$$a(u - u_h, u - u_h) = a(u - u_h, u - \Pi u),$$

which gives

$$|u - u_h|_1^2 \leq |u - u_h|_1 |u - \Pi u|_1.$$

(3.92) follows by applying Theorem 60. $\qquad\square$

Thus the error estimates are optimal. Using Nitsche's trick, we can prove L^2-norm estimate. However, because the exterior domain Ω can not be convex, and the boundary $\partial\Omega$ is not smooth, the result is not optimal. There are some approaches to deal with reentrant corners, but it is another story.

Let $R > 0$ be large enough, such that $B(O, R) \supset \Omega^c$. We have the theorem:

Theorem 62. *Under the assumptions of Theorem 61 it holds that*

$$\|u - u_h\|_{0,\Omega \cap B(O,R)} \leq Ch^s |u - u_h|_1, \tag{3.93}$$

where $0 < s < 1$, depending on Ω, and the constant C depends on Ω and R.

Proof. We set

$$f = \begin{cases} u - u_h, & |x| < R, \\ 0, & |x| > R, \end{cases}$$

and consider an auxiliary problem: find $\varphi \in H_0^{1,*}(\Omega)$, such that

$$a(v, \varphi) = (f, v), \qquad v \in H_0^{1,*}(\Omega). \tag{3.94}$$

Let us estimate φ first. By Theorem 14

$$\|\varphi\|_{1,*} \leq C\|f\|_0.$$

If x_0 is a corner point on $\partial\Omega$, we consider two small domains $\Omega_\delta = \Omega \cap B(x_0, \delta)$, $\Omega_{2\delta} = \Omega \cap B(x_0, 2\delta)$ with $\delta > 0$. Then by the separation of variables, the solution φ in $\Omega_{2\delta}$ can be expressed explicitly, and it can be estimated that

$$\|\varphi\|_{1+s,\Omega_\delta} \le C(\|f\|_{0,\Omega_\delta} + \|\varphi\|_{0,\Omega_{2\delta}}),$$

where $0 < s < 1$, depending on the angle. Let $\Omega_1 = \Omega \cap B(O, R+1)$, and $\Omega_2 = \{x; |X| > R+1\}$, then using the interior estimate of elliptic equations [Gilbarg, D. and Trudinger, N.S. (1977)], we obtain

$$\|\varphi\|_{1+s,\Omega_1} \le C(\|f\|_0 + \|\varphi\|_{1,*}).$$

Using the results in Section 1.1 the solution φ in Ω_2 can be expressed in terms of potentials, hence

$$\left(\int_{\Omega_2} |x|^2 |D^2\varphi(x)|^2 \, dx\right)^{\frac{1}{2}} \le C \max_{|x|=R} |\varphi(x)| \le C\|\varphi\|_{1+s,\Omega_1}.$$

The second inequality is due to the embedding theorem of Sobolev spaces. Thus we obtain

$$\|\varphi\|_{1+s,\Omega_1} + \left(\int_{\Omega_2} |x|^2 |D^2\varphi(x)|^2 \, dx\right)^{\frac{1}{2}} \le C\|f\|_0.$$

Letting $v = u - u_h$ in (3.94) and noting that

$$a(u - u_h, \Pi\varphi) = 0,$$

we get

$$|u - u_h|_1 |\varphi - \Pi\varphi|_1 \ge a(u - u_h, \varphi - \Pi\varphi) = (f, u - u_h).$$

Using the normal estimate for interpolating operators on Ω_1 and Theorem 60 on Ω_2, we have

$$|\varphi - \Pi\varphi|_1 \le Ch^s \|f\|_0.$$

Noting that $(f, u - u_h) = \|f\|_0^2$, we obtain finally

$$\|f\|_0 \le Ch^s |u - u_h|_1,$$

and the proof is complete. □

Other kinds of error estimates can be found in [Ying, L. (1995)].

Chapter 4

Artificial Boundary Conditions

Some artificial boundary conditions, exact or approximate, have been given in the previous chapters, for example, the application of "Dirichlet to Neŭmann" operator or the combined stiffness matrix. In this chapter we will investigate more artificial boundary conditions. A pseudo-differential operator is introduced for the wave equation to reduce the boundary condition at the infinity to an interface. Since this condition is global in space, some works have been done to localize the boundary condition on the analytic level. The problem of stability for some approximations is still open, because there are some higher order derivatives in the conditions. In this chapter we will also show some works on the artificial boundary conditions of the Navier-Stokes equations.

4.1 Absorbing boundary conditions

Absorbing boundary conditions are applied for wave problems. Let Ω be an exterior domain, and let Ω_0 be a bounded domain such that $\Omega_0 \supset \overline{\Omega^c}$. The truncated domain is $\Omega_1 = \Omega \cap \Omega_0$ and $\partial\Omega_0$ is the artificial boundary. Suppose there is no incoming waves from the exterior of Ω_0. The artificial boundary condition on $\partial\Omega_0$ is defined to absorb all outgoing waves from Ω_1, so that the waves do not reflect from $\partial\Omega_0$. It is named as an "absorbing boundary condition".

We start from one dimensional case and consider the wave equation,

$$\frac{\partial^2 u}{\partial t^2} = \frac{\partial^2 u}{\partial x^2}. \tag{4.1}$$

Suppose the domain is $(0, \infty)$, and the initial and boundary conditions are

$$u|_{t=0} = u_0, \qquad \frac{\partial u}{\partial t}\bigg|_{t=0} = u_1, \qquad x > 0, \qquad (4.2)$$

$$u|_{x=0} = 0, \qquad t > 0, \qquad (4.3)$$

where $u_0(0) = 0$, $u_1(0) = 0$.

The equation (4.1) can be factorized to

$$\left(\frac{\partial}{\partial t} - \frac{\partial}{\partial x}\right)\left(\frac{\partial}{\partial t} + \frac{\partial}{\partial x}\right) u = 0. \qquad (4.4)$$

It yields two equations

$$\frac{\partial u}{\partial t} - \frac{\partial u}{\partial x} = 0, \qquad (4.5)$$

and

$$\frac{\partial u}{\partial t} + \frac{\partial u}{\partial x} = 0. \qquad (4.6)$$

Obviously the solutions to (4.5),(4.6) are solutions to (4.1). Conversely, if we make a transform of variables $\xi = x - t$ and $\eta = x + t$, then the equation (4.1) is transformed to

$$\frac{\partial^2 u}{\partial \xi \partial \eta} = 0,$$

and the general solution is $u = f(\xi) + g(\eta) = f(x - t) + g(x + t)$, where f, g are arbitrary functions. Therefore the general solution to (4.1) is the linear combination of the solutions to (4.5) and (4.6). Using the expression of general solutions one can get the solution to the initial-boundary value problem (4.1),(4.2),(4.3):

$$u(x,t) = \begin{cases} \frac{1}{2}(u_0(x-t) + u_0(x+t)) + \frac{1}{2}\int_{x-t}^{x+t} u_1(\xi)\,d\xi, & x > t, \\ \frac{1}{2}(-u_0(t-x) + u_0(x+t)) + \frac{1}{2}\int_{t-x}^{x+t} u_1(\xi)\,d\xi, & x < t. \end{cases}$$

If u_0 and u_1 vanish for $x > a$, then on the domain $\{(x,t); x > a\}$ the solution is

$$u(x,t) = \begin{cases} \frac{1}{2}u_0(x-t) + \frac{1}{2}\int_{x-t}^{a} u_1(\xi)\,d\xi, & x > t, \\ -\frac{1}{2}u_0(t-x) + \frac{1}{2}\int_{t-x}^{a} u_1(\xi)\,d\xi, & x < t. \end{cases}$$

Thus only outgoing waves appear in this domain. The solution u satisfies (4.5) on the line $x = a$. (4.5) is the exact absorbing boundary condition on $x = a$. With this boundary condition the exterior problem is reduced to an initial boundary value problem on the interval $(0, a)$, and the DtN (Dirichlet to Neŭmann) operator is thus $K = \frac{\partial}{\partial t}$.

We turn now to investigate two dimensional problems and consider the equation

$$\frac{\partial^2 u}{\partial t^2} = \triangle u. \tag{4.7}$$

Let the plane $x_1 = a$ be the artificial boundary, and there is no incoming waves from $x_1 > a$. Like (4.4), formally we can make the factorization:

$$\left(\frac{\partial}{\partial x_1} - \frac{\partial}{\partial t} \sqrt{1 - \frac{\frac{\partial^2}{\partial x_2^2}}{\frac{\partial^2}{\partial t^2}}} \right) \left(\frac{\partial}{\partial x_1} + \frac{\partial}{\partial t} \sqrt{1 - \frac{\frac{\partial^2}{\partial x_2^2}}{\frac{\partial^2}{\partial t^2}}} \right) u = 0. \tag{4.8}$$

Then the absorbing boundary condition at $x = a$ is

$$\left(\frac{\partial}{\partial x_1} - \frac{\partial}{\partial t} \sqrt{1 - \frac{\frac{\partial^2}{\partial x_2^2}}{\frac{\partial^2}{\partial t^2}}} \right) u = 0, \tag{4.9}$$

and the DtN operator is

$$K = \frac{\partial}{\partial t} \sqrt{1 - \frac{\frac{\partial^2}{\partial x_2^2}}{\frac{\partial^2}{\partial t^2}}}.$$

In order to interpret (4.9) precisely, we introduce some basic definitions of pseudo-differential operators briefly.[Taylor, M. (1974)]

The Fourier transform is defined on the spaces:

$$\mathcal{S}(\mathbb{R}^n) = \{\varphi \in C^\infty(\mathbb{R}^n); \lim_{|x| \to \infty} x^\alpha D^\beta \varphi(x) - 0, \forall \alpha, \beta\},$$

where α, β are multiple indices, and $C_0^\infty(\mathbb{R}^n) \subset \mathcal{S}(\mathbb{R}^n) \subset C^\infty(\mathbb{R}^n)$.

The dual space of $C_0^\infty(\mathbb{R}^n)$, $\mathcal{S}(\mathbb{R}^n)$, and $C^\infty(\mathbb{R}^n)$ are denote by $\mathcal{D}'(\mathbb{R}^n)$, $\mathcal{S}'(\mathbb{R}^n)$, and $\mathcal{E}'(\mathbb{R}^n)$ respectively. They are spaces of distributions, and the relation of them is

$$\mathcal{D}'(\mathbb{R}^n) \supset \mathcal{S}'(\mathbb{R}^n) \supset \mathcal{E}'(\mathbb{R}^n).$$

Letting $f \in \mathcal{S}(\mathbb{R}^n)$, the Fourier transform $\tilde{f}$ of it is defined by

$$\tilde{f}(\xi) = \int_{\mathbb{R}^n} f(x) e^{-ix \cdot \xi} \, dx,$$

where $\xi \in \mathbb{R}^n$. The inverse Fourier transform is

$$f(x) = \frac{1}{(2\pi)^n} \int_{\mathbb{R}^n} \tilde{f}(\xi) e^{ix \cdot \xi} \, d\xi.$$

Using duality the Fourier transform $F : \mathcal{S}(\mathbb{R}^n) \to \mathcal{S}(\mathbb{R}^n)$ can be extended to: $F : \mathcal{S}'(\mathbb{R}^n) \to \mathcal{S}'(\mathbb{R}^n)$.

Differentiating by parts one can find that if $\tilde{f} = F(f)$, then $i\xi_j \tilde{f} = F\left(\frac{\partial f}{\partial x_j}\right)$, that is

$$\frac{\partial f(x)}{\partial x_j} = \frac{1}{(2\pi)^n} \int_{\mathbb{R}^n} i\xi_j \tilde{f}(\xi) e^{ix \cdot \xi} \, d\xi.$$

In general, if $P(\xi)$ is a polynomial of the independent variables ξ with constant coefficients, then

$$P(D)f(x) = \frac{1}{(2\pi)^n} \int_{\mathbb{R}^n} P(i\xi) \tilde{f}(\xi) e^{ix \cdot \xi} \, d\xi,$$

where $D = \frac{1}{i}\left(\frac{\partial}{\partial x_1}, \frac{\partial}{\partial x_2}, \cdots, \frac{\partial}{\partial x_n}\right)$.

If $p(x, \xi)$ is a function defined on $\Omega \times \mathbb{R}^n$, satisfying

$$|D_x^\beta D_\xi^\alpha p(x, \xi)| \le C(1 + |\xi|)^{m - |\alpha|},$$

where m is a real number. The pseudo-differential operator is defined by

$$P(x, D)f(x) = \frac{1}{(2\pi)^n} \int_{\mathbb{R}^n} p(x, \xi) \tilde{f}(\xi) e^{ix \cdot \xi} \, d\xi,$$

which is a linear continuous mapping from $\mathcal{E}'(\mathbb{R}^n)$ to $\mathcal{D}'(\mathbb{R}^n)$. $p(x, \xi)$ is called the "symbol" of the pseudo-differential operator $P(x, D)$.

Let us now return to the equation (4.7). Let $\tilde{u}$ be the Fourier transform of u, then

$$u(x_1, x_2, t) = \frac{1}{(2\pi)^n} \int_{\mathbb{R}^3} \tilde{u}(\xi_1, \xi_2, \omega) e^{i(\xi_1 x_1 + \xi_2 x_2 + \omega t)} \, d\xi_1 d\xi_2 d\omega. \tag{4.10}$$

The formula (4.10) can be explained as follows. For a fixed constant vector (ξ_1, ξ_2, ω), the function

$$w(x, t) = \tilde{u}(\xi_1, \xi_2, \omega) e^{i(\xi_1 x_1 + \xi_2 x_2 + \omega t)}$$

is a particular solution to the equation (4.7). w is a constant on the plane $\xi_1 x_1 + \xi_2 x_2 + \omega t = c$, where c is a constant, so it represents a plane wave traveling along a direction (ξ_1, ξ_2) on the x plane with a speed of $s = \omega / |\xi|$. The formula (4.10) shows that u is the linear combination of all plane waves.

Applying the Fourier transform operator to the equation (4.7), we get

$$(\omega^2 - \xi_1^2 - \xi_2^2)\tilde{u} = 0. \tag{4.11}$$

Therefore u vanishes if $\omega^2 - \xi_1^2 - \xi_2^2 \neq 0$. The support of u lies in a cone: $\xi_1 \pm \omega\sqrt{1 - \left(\frac{\xi_2}{\omega}\right)^2} = 0$. Since $\left|\frac{\xi_2}{\omega}\right| = 1$ on this cone, the wave speed is 1. Restricted on this cone the integral (4.10) can be expressed by

$$u(x_1, x_2, t) = \frac{1}{(2\pi)^n} \int_{|\xi_2| \leq |\omega|} \tilde{f}_1(\xi_2, \omega) e^{i(\omega\sqrt{1-\left(\frac{\xi_2}{\omega}\right)^2}x_1 + \xi_2 x_2 + \omega t)} \, d\xi_2 d\omega$$
$$+ \frac{1}{(2\pi)^n} \int_{|\xi_2| \leq |\omega|} \tilde{f}_2(\xi_2, \omega) e^{i(-\omega\sqrt{1-\left(\frac{\xi_2}{\omega}\right)^2}x_1 + \xi_2 x_2 + \omega t)} \, d\xi_2 d\omega.$$

In the first term $\xi_1/\omega \geq 0$, so it represents the combination of those waves traveling towards to the positive x_1 direction, the outgoing waves. On the other hand the second term represents the combination of incoming waves. The equation (4.11) can be factorized to

$$\left(\xi_1 - \omega\sqrt{1 - \left(\frac{\xi_2}{\omega}\right)^2}\right)\left(\xi_1 + \omega\sqrt{1 - \left(\frac{\xi_2}{\omega}\right)^2}\right)\tilde{u} = 0.$$

It yields two equations

$$\left(\xi_1 - \omega\sqrt{1 - \left(\frac{\xi_2}{\omega}\right)^2}\right)\tilde{u} = 0, \tag{4.12}$$

and

$$\left(\xi_1 + \omega\sqrt{1 - \left(\frac{\xi_2}{\omega}\right)^2}\right)\tilde{u} = 0. \tag{4.13}$$

They are the equations for outgoing waves and incoming waves respectively. Returning to the physical variables, they are

$$\frac{\partial u}{\partial x_1} - Ku = 0, \qquad \frac{\partial u}{\partial x_1} + Ku = 0, \tag{4.14}$$

respectively, where

$$Ku = \frac{1}{(2\pi)^n} \int_{\mathbb{R}^3} \tilde{u}(\xi_1, \xi_2, i\omega)\omega\sqrt{1 - \left(\frac{\xi_2}{\omega}\right)^2} e^{i(\xi_1 x_1 + \xi_2 x_2 + \omega t)} \, d\xi_1 d\xi_2 d\omega. \tag{4.15}$$

The first equation in (4.14) is the absorbing boundary condition for the artificial boundary on the right, while the second one is that for the artificial boundary on the left.

4.2   Some approximations

The conventional differential operator is local in the sense that the derivatives of a function at one point depends on the function value in a neighborhood, as small as desired. On the other hand the DtN operator K defined above is a pseudo-differential operator, and it is global. It is not so convenient to carry out any numerical computation for a pseudo-differential operator. Many approximations have been developed to localize the operator (4.15). The idea is the following: The symbol $\omega\sqrt{1 - \left(\frac{\xi_2}{\omega}\right)^2}$ is approximated by a rational function $p(s)/q(s)$, where $s = \frac{\xi_2}{\omega}$, then the equation (4.12) is reduced to

$$\left(\xi_1 - \omega\left(p(\frac{\xi_2}{\omega})/q(\frac{\xi_2}{\omega})\right)\right)\tilde{u} = 0,$$

that is

$$\left(\xi_1 q(\frac{\xi_2}{\omega}) - \omega p(\frac{\xi_2}{\omega})\right)\tilde{u} = 0.$$

Multiply the both sides by ω^m, where m is a suitable power, then it becomes a conventional differential equation applied on the artificial boundary. If the domain Ω_0 is rectangular, then similar conditions are applied corresponding to x_2 on the other two line segments .

1. Clayton-Engquist-Majda's boundary conditions [Engquist, B. and Majda, A. (1977)][Engquist, B. and Majda, A. (1979)]

Using Padé's approximation, the function $\sqrt{1 - s^2}$ is replaced by rational functions,

$$r^{(1)}(s) = 1, \qquad r^{(k+1)}(s) = 1 - \frac{s^2}{1 + r^{(k)}(s)}, k \geq 1.$$

The absorbing boundary conditions of different orders are:

$$B_1 u = \frac{\partial u}{\partial t} - \frac{\partial u}{\partial x_1} = 0,$$

$$B_2 u = \frac{\partial^2 u}{\partial t^2} - \frac{\partial^2 u}{\partial t \partial x_1} - \frac{1}{2}\frac{\partial^2 u}{\partial x_2^2} = 0,$$

$$B_{N+1} u = \frac{\partial}{\partial t} B_N u - \frac{1}{4}\frac{\partial^2}{\partial x_2^2} B_{N-1} u = 0.$$

Padé's approximation is developed near the origin, so it gives satisfactory results if s is small, that is, if the incidence angles of the outgoing waves are small.

2. Halpern-Trefethen's boundary conditions [Halpern, L. and Trefethen, L.N. (1988)]

Approximations are made for different $s's$ corresponding to different incidence angles, which can be large. Take $2K$ symmetric points on the interval $[-1, 1]$:

$$\pm s_1 = \pm \sin\theta_1, \cdots, \pm s_K = \pm\sin\theta_K,$$

then define a rational function $r(s)$, such that $r(s) = \sqrt{1-s^2}$ at the $2K$ points. The formulas are the following:

$$p(\eta) = (\eta - \sqrt{1-s_1^2})\cdots(\eta - \sqrt{1-s_K^2}),$$

$$r(s) = \frac{(p(\eta) + p(-\eta))\eta}{-p(\eta) + p(-\eta)},$$

where $\eta = \sqrt{1-s^2}$. Because the numerator and denominator are even functions of η, $r(s)$ is a rational function. Besides, because $|p(-\eta)| > |p(\eta)|$ for $\eta > 0$ and $p'(0) \neq 0$, $r(s)$ has neither poles nor zero points on $[-1, 1]$. Therefore $r(s)$ is indeed an interpolation function of $\sqrt{1-s^2}$.

There are several ways to define θ_k:

a. Padé; b. Chebyshev; c. Least squares; d. Chebyshev-Padé; e. Interpolation in Newman points.

We omit the details here.

3. Lindman's boundary conditions [Lindman, E.L. (1975)]

A rational function $r(s)$ is introduced to approximate $\frac{1}{\sqrt{1-s^2}}$:

$$r(s) = 1 + \sum_{k=1}^{N} \frac{\alpha_k s^2}{1 - \beta_k s^2}.$$

Applying this approximation in (4.12) yields

$$\left(\xi_1 \left(1 + \sum_{k=1}^{N} \frac{\alpha_k \xi_2^2}{\omega^2 - \beta_k \xi_2^2} \right) - \omega \right) \tilde{u} = 0.$$

Then applying the inverse Fourier transform to get the boundary conditions:

$$\frac{\partial u}{\partial t} - \frac{\partial u}{\partial x_1} = \sum_{k=1}^{N} h_k,$$

$$\frac{\partial^2 h_k}{\partial t^2} - \beta_k \frac{\partial^2 h_k}{\partial x_2^2} = \alpha_k \frac{\partial^2}{\partial x_2^2} \left(\frac{\partial u}{\partial x_1} \right).$$

4. Higdon's boundary conditions [Higdon, R.L. (1986)][Higdon, R.L. (1987)]

A general boundary condition is

$$\prod_{k=1}^{N} \left(\cos \alpha_k \frac{\partial}{\partial t} - \frac{\partial u}{\partial x_1} \right) u = 0,$$

and all the above boundary conditions are particular cases. Plane waves with incidence angles α_k satisfy this condition.

5. Safjan's boundary conditions [Safjan, A.J. (1998)]

In order to implement some high order absorbing boundary conditions, some approximations are introduced. For example, the second order Clayton-Engquist-Majda's boundary condition,

$$\frac{\partial^2 u}{\partial t^2} - \frac{\partial^2 u}{\partial t \partial x_1} - \frac{1}{2} \frac{\partial^2 u}{\partial x_2^2} = 0$$

is approximated by a two steps scheme:

$$\frac{\partial u^{(1)}}{\partial t} - \frac{\partial u^{(1)}}{\partial x_1} = 0,$$

$$\frac{\partial^2 u^{(2)}}{\partial t^2} - \frac{\partial^2 u^{(2)}}{\partial t \partial x_1} = \frac{1}{2} \frac{\partial^2 u^{(1)}}{\partial x_2^2}.$$

That is

$$\frac{\partial u^{(1)}}{\partial t} - \frac{\partial u^{(1)}}{\partial x_1} = 0,$$

$$\frac{\partial u^{(2)}}{\partial t} - \frac{\partial u^{(2)}}{\partial x_1} = \frac{1}{2} \frac{\partial^2 \tilde{u}^{(1)}}{\partial x_2^2},$$

where

$$\tilde{u}(x_1, x_2, t) = \int_0^t u(x_1, x_2, \tau) \, d\tau.$$

The operator on the left hand side is always first order.

Likely the third order boundary condition,

$$\frac{\partial^3 u}{\partial x_1 \partial t^2} - \frac{\partial^3 u}{\partial t^3} + \frac{3}{4} \frac{\partial^3 u}{\partial x_2^2 \partial t} - \frac{1}{4} \frac{\partial^3 u}{\partial x_1 \partial x_2^2} = 0,$$

can be approximated by

$$\frac{\partial u^{(1)}}{\partial t} - \frac{\partial u^{(1)}}{\partial x_1} = 0,$$

$$\frac{\partial u^{(2)}}{\partial t} - \frac{\partial u^{(2)}}{\partial x_1} = \frac{1}{2} \frac{\partial^2 \tilde{u}^{(1)}}{\partial x_2^2},$$

$$\frac{\partial u^{(3)}}{\partial t} - \frac{\partial u^{(3)}}{\partial x_1} = \frac{3}{4} \frac{\partial^2 \tilde{u}^{(2)}}{\partial x_2^2} - \frac{1}{4} \frac{\partial^3 \hat{u}^{(2)}}{\partial x_1 \partial x_2^2},$$

where

$$\hat{u}(x_1, x_2, t) = \int_0^t \tilde{u}(x_1, x_2, \tau) \, d\tau.$$

4.3 Bayliss-Turkel radiation boundary conditions

Consider the equation (4.7) for dim $(\Omega) = 2, 3$. The radiation boundary conditions in [Bayliss, A. and Turkel, E. (1980)] is based on the expansions of solutions near the infinity. For three dimensional cases the solutions u can be expressed in terms of independent variables $t - r, r, \theta, \phi$, where (r, θ, ϕ) is the spherical coordinate system. For large r, the solutions propagating in directions that are outward from the origin can be expanded in a convergent series of the form

$$u = \sum_{j=1}^{\infty} \frac{u_j(t - r, \theta, \phi)}{r^j}. \tag{4.16}$$

For two dimensional cases, the expansion expression is

$$u = \sum_{j=0}^{\infty} \frac{u_j(t - r, \theta)}{r^{j+\frac{1}{2}}}. \tag{4.17}$$

The leading term for the solutions are $O(r^{-1})$ and $O(r^{\frac{1}{2}})$ respectively. This fact can be seen from the fundamental solutions of the wave equations, which are

$$u = \frac{1}{4\pi t}\delta(|x| - t), \quad u = \frac{U(t - |x|)}{2\pi\sqrt{t^2 - |x|^2}}, \quad U(t - |x|) = \begin{cases} 0, & |x| > t, \\ 1, & |x| < t, \end{cases}$$

for dimension three or two, where δ is the Dirac function. The fundamental solutions satisfy (4.7) and initial data,

$$u|_{t=0} = 0, \qquad \frac{\partial u}{\partial t}\Big|_{t=0} = \delta(x).$$

For the proof of (4.16) and (4.17), see [Friedlander, F.G. (1962)][Wilcox, C.H. (1956)] [Karp, S.N. (1961)].

Define an operator

$$L = \frac{\partial}{\partial t} + \frac{\partial}{\partial x},$$

then substitute (4.16) in Lu to get $Lu = O(r^{-2})$. To improve the error let $B_1 = L + \frac{1}{r}$, then $B_1 u = O(r^{-3})$. $B_1 u = 0$ is then applied as the absorbing boundary condition on $r = R$ with R large.

To improve the operator B_1, let $B_2 = (L + \frac{3}{r})B_1$, then $B_2 u = O(r^{-5})$. In general

$$B_j = \left(L + \frac{2j - 1}{r}\right)B_{j-1}, \quad B_j u = O(r^{-2j-1}).$$

For two dimensional problems substitute (4.17) in Lu to get $Lu = O(r^{-\frac{3}{2}})$. Then define $B_1 = L + \frac{1}{2r}$, and

$$B_j = \left(L + \frac{4j - 3}{2r}\right)B_{j-1}, \quad B_j u = O(r^{-2j-\frac{1}{2}}).$$

4.4   A lower order absorbing boundary condition

In the previous sections some absorbing boundary conditions are of higher order. That is, the order of the boundary condition is higher than that of the equation. Some of them may cause unstable problems. One second order in derivatives, but higher order in precision absorbing boundary condition is given in the following. [Zhang, G. (1985)][Zhang, G. (1993)][Zhang, G. and Wei, S.(1998)]

Noting that

$$\sqrt{1 - \xi^2} = 1 - \frac{1}{\pi} \int_{-1}^{1} \sqrt{1 - r^2} \frac{\xi^2}{1 - r^2 \xi^2} \, dr,$$

the expression (4.15) can be written as

$$Ku = \frac{1}{(2\pi)^n} \int_{\mathbb{R}^3} \tilde{u}(\xi_1, \xi_2, \omega) i\omega \left(1 - \frac{1}{\pi} \int_{-1}^{1} \sqrt{1 - r^2} \frac{\xi_2^2}{\omega^2 - r^2 \xi_2^2} \, dr \right)$$
$$\cdot e^{i(\xi_1 x_1 + \xi_2 x_2 + \omega t)} \, d\xi_1 d\xi_2 d\omega.$$

Let

$$\tilde{v}(\xi_1, \xi_2, \omega, r) = \frac{\xi_2^2}{\omega^2 - r^2 \xi_2^2} \tilde{u},$$

then the first equation of (4.14) is equivalent to

$$\frac{\partial u}{\partial x_1} - \frac{\partial u}{\partial t} = \frac{\partial}{\partial t} \left\{ \frac{1}{\pi} \int_{-1}^{1} \sqrt{1 - r^2} v(x, t, r) \, dr \right\},$$

$$\frac{\partial^2 v}{\partial t^2} - r^2 \frac{\partial^2 v}{\partial x_2^2} = \frac{\partial^2 u}{\partial x_2^2}, \qquad r \in (-1, 1).$$

One can get an approximate boundary condition with n-th order accuracy:

$$\frac{\partial u}{\partial x_1} - \frac{\partial u}{\partial t} = \frac{\partial}{\partial t} \sum_{k=1}^{n} a_k v_k(x, t, r_k),$$

$$\frac{\partial^2 v_k}{\partial t^2} - r_k^2 \frac{\partial^2 v_k}{\partial x_2^2} = \frac{\partial^2 u}{\partial x_2^2},$$

where

$$r_k = \cos \frac{k\pi}{n+1}, \qquad a_k = \frac{1}{n+1} \sin^2 \frac{k\pi}{n+1}.$$

4.5   Liao extrapolation in space and time

Suppose that the solution for $t \leq t_0$ and $x_1 < a$ is known and $x_1 = a$ is an artificial boundary. Liao introduced an extrapolation scheme to evaluate $u(a, x_2, x_3, t_0 + \Delta t)$. [Liao, Z., Wong, H., Yang, B. and Yuan, Y. (1984)][Liao, Z. and Wong, H. (1984)]

Let $u_0 = u(a, x_2, x_3, t_0 + \Delta t)$, $u_1 = u(a - h, x_2, x_3, t_0)$, $u_2 = u(a - 2h, x_2, x_3, t_0 - \Delta t)$, $\cdots$, $u_L = u(a - Lh, x_2, x_3, t_0 - (L-1)\Delta t)$, where $h = \alpha \Delta t$ and $0 \leq \alpha \leq 2$. The m-th backward difference of u_1 can be written in terms of the underlying fields as follows:

$$\Delta^m u_1 = \sum_{k=1}^{m+1} (-1)^{k+1} C_{k-1}^m u(a - kh, x_2, x_3 t_0 - (k-1)\Delta t),$$

where C_k^m is the binomial coefficient defined by

$$C_k^m = \frac{m!}{(m-k)!k!}.$$

u_0 is unknown and evaluated by means of extrapolation.

$$u_0 = u_1 + \Delta^1 u_1 + \Delta^2 u_1 + \cdots + \Delta^{L-1} u_1.$$

Shao and Lan introduced a modified scheme of this kind.[Shao, X. and Lan, Z.][Shao, X.]

4.6   Maxwell equations

For the initial-boundary value problems of Maxwell equations (1.106) we notice that

$$A^2 = \varepsilon^{-1} \mu^{-1} \begin{pmatrix} \nabla \times \nabla \times & 0 \\ 0 & \nabla \times \nabla \times \end{pmatrix}.$$

We assume that $\rho = 0$ and $j = 0$ in (1.101),(1.104). Then $\nabla \cdot E = 0$ and $\nabla \cdot H = 0$, and we get

$$A^2 = -\varepsilon^{-1} \mu^{-1} \begin{pmatrix} \triangle & 0 \\ 0 & \triangle \end{pmatrix}.$$

For the homogeneous equation (1.106) we apply the operator $\frac{\partial}{\partial t}$ to obtain

$$\frac{\partial^2 u}{\partial t^2} = \varepsilon^{-1} \mu^{-1} \triangle u,$$

which is the wave equation. Therefore all the above artificial boundary conditions for wave equations can be applied to Maxwell equations.

For two dimensional cases we assume that E and $\hat{H}$ are independent of x_3, then the system of equation (1.106) are separated to two independent system of equations:

1. The transverse magnetic mode (TM):

$$\frac{\partial H_1}{\partial t} = -\mu^{-1}\frac{\partial E_3}{\partial x_2},$$
$$\frac{\partial H_2}{\partial t} = \mu^{-1}\frac{\partial E_3}{\partial x_1}, \tag{4.18}$$
$$\frac{\partial E_3}{\partial t} = \varepsilon^{-1}\left(\frac{\partial H_2}{\partial x_1} - \frac{\partial H_1}{\partial x_2}\right).$$

2. The transverse electric mode (TE):

$$\frac{\partial E_1}{\partial t} = \varepsilon^{-1}\frac{\partial H_3}{\partial x_2},$$
$$\frac{\partial E_2}{\partial t} = -\varepsilon^{-1}\frac{\partial H_3}{\partial x_1}, \tag{4.19}$$
$$\frac{\partial H_3}{\partial t} = \mu^{-1}\left(\frac{\partial E_1}{\partial x_2} - \frac{\partial E_2}{\partial x_1}\right).$$

We set $u = (H_1\ H_2\ E_3)^T$ and $u = (E_1\ E_2\ H_3)^T$ for the TM mode and the TE mode respectively, then both of the systems of equations can be written in the form of

$$\frac{du}{dt} + iAu = 0, \tag{4.20}$$

where

$$A = -i\begin{pmatrix} \mu^{-1} & 0 & 0 \\ 0 & \mu^{-1} & 0 \\ 0 & 0 & \varepsilon^{-1} \end{pmatrix}\begin{pmatrix} 0 & 0 & \frac{\partial}{\partial x_2} \\ 0 & 0 & -\frac{\partial}{\partial x_1} \\ \frac{\partial}{\partial x_2} & -\frac{\partial}{\partial x_1} & 0 \end{pmatrix},$$

and

$$A = i\begin{pmatrix} \varepsilon^{-1} & 0 & 0 \\ 0 & \varepsilon^{-1} & 0 \\ 0 & 0 & \mu^{-1} \end{pmatrix}\begin{pmatrix} 0 & 0 & \frac{\partial}{\partial x_2} \\ 0 & 0 & -\frac{\partial}{\partial x_1} \\ \frac{\partial}{\partial x_2} & -\frac{\partial}{\partial x_1} & 0 \end{pmatrix}.$$

We eliminate H_1, H_2 in (4.18) by apply the operator

$$\left(\frac{\partial}{\partial x_2}, \frac{\partial}{\partial x_1}, \frac{\partial}{\partial t}\right)$$

on the right, then we get the wave equation for E_3,

$$\frac{\partial^2 E_3}{\partial t^2} = \varepsilon^{-1}\mu^{-1}\triangle E_3.$$

Analogously the system (4.19) yields

$$\frac{\partial^2 H_3}{\partial t^2} = \varepsilon^{-1}\mu^{-1}\triangle H_3.$$

Therefore the above artificial boundary conditions can be applied here as well.

We turn now to investigate the initial-boundary value problems of (4.20). Let the initial conditions be

$$u|_{t=0} = u_0, \tag{4.21}$$

satisfying $\nabla \cdot E = 0$ and $\nabla \cdot \hat{H} = 0$. Being the same as that in Section 1.11 we consider the total reflection boundary condition, $E \times \nu|_{\partial\Omega} = 0$, $\frac{\partial}{\partial t} B \cdot \nu|_{\partial\Omega} = 0$. Then for the TM mode (4.18)

$$\left(-\frac{\partial E_3}{\partial x_2}, \frac{\partial E_3}{\partial x_1}\right) \cdot \nu\bigg|_{\partial\Omega} = 0,$$

and

$$E_3\nu_1 = E_3\nu_2 = 0.$$

The combination of these two conditions leads to the Dirichlet boundary value,

$$E_3|_{\partial\Omega} = 0. \tag{4.22}$$

Being analogous the Neŭmann boundary condition,

$$\frac{\partial H_3}{\partial \nu}\bigg|_{\partial\Omega} = \left(\frac{\partial H_3}{\partial x_2}, -\frac{\partial H_3}{\partial x_1}\right) \times \nu\bigg|_{\partial\Omega} = 0, \tag{4.23}$$

is generated for the TE mode (4.19) .

To study well-posedness, we introduce some spaces,

$$S(\text{curl}; \Omega) = \left\{E = (E_1, E_2) \in (L^2(\Omega))^2; \frac{\partial E_1}{\partial x_2} - \frac{\partial E_2}{\partial x_1} \in L^2(\Omega)\right\},$$

$$S_0(\text{curl}; \Omega) = \{E \in S(\text{curl}; \Omega); E \times \nu|_{\partial\Omega} = 0\}.$$

For the TM mode, we work in the space

$$V_{TM} = S(\text{curl}; \Omega) \times H_0^1(\Omega),$$

and for the TE mode the space is

$$V_{TE} = S_0(\text{curl}; \Omega) \times H^1(\Omega).$$

Then the initial-boundary value problems are (4.20),(4.21),(4.22), where the domain of A is V_{TM}, and (4.20),(4.21),(4.23), where the domain of A is V_{TE}, respectively.

Following the argument in Section 1.11 let us introduce weighted L^2-spaces H with the weights

$$M = \begin{pmatrix} \mu\ 0\ 0 \\ 0\ \mu\ 0 \\ 0\ 0\ \varepsilon \end{pmatrix} \quad \text{or} \quad M = \begin{pmatrix} \varepsilon\ 0\ 0 \\ 0\ \varepsilon\ 0 \\ 0\ 0\ \mu \end{pmatrix}$$

and prove that

Lemma 66. $A : V_{TM} \to (L^2(\Omega))^3$ and $A : V_{TE} \to (L^2(\Omega))^3$ are self-adjoint operators in H.

Proof. We consider the TM mode first. Clearly A is closed and symmetric. By definition

$$D(A^*) = \{v \in (L^2(\Omega))^3; \exists g \in (L^2(\Omega))^3, (Au, v)_H - (u, g)_H, \forall u \in V_{TM}\}.$$

We denote $v = (v_1, v_2)^T$ and $g = (g_1, g_2)^T$. Letting $H_1 = H_2 = 0$ in u, we have

$$-i\mu^{-1}\left(\left(\frac{\partial E_3}{\partial x_2}, -\frac{\partial E_3}{\partial x_1}\right), v\right) = (E_3, g_3), \quad \forall E_3 \in H_0^1(\Omega).$$

Therefore $g_3 = i\mu^{-1}(\frac{\partial v_1}{\partial x_2} - \frac{\partial v_2}{\partial x_1})$ and the domain for v is $S(\text{curl}; \Omega)$. Similarly letting $E_3 = 0$, we have

$$-i\varepsilon^{-1}\left(\frac{\partial H_2}{\partial x_1} - \frac{\partial H_1}{\partial x_2}, v_3\right) = ((H_1, H_2), g), \quad \forall (H_1, H_2) \in S(\text{curl}; \Omega).$$

Therefore $g = i\varepsilon^{-1}(\frac{\partial v_3}{\partial x_1}, -\frac{\partial v_3}{\partial x_2})$ and the domain for v_3 is $H_0^1(\Omega)$. Hence

$$D(A^*) = S(\text{curl}; \Omega) \times H_0^1(\Omega) = V_{TM}.$$

The proof for the TE mode is analogous. $\qquad\square$

Following the same lines as that in Section 1.11, we can prove that the problems (4.20),(4.21),(4.22) and (4.20),(4.21),(4.23) admit unique solutions. The details are thus omitted.

4.7 Finite difference schemes

After truncating the domain Ω and imposing some artificial boundary conditions, numerical schemes for bounded domains can be applied. For wave problems one finite difference scheme is the second order Yee's scheme.[Yee, K.S. (1966)] For example the TM mode can be approximated by the following scheme: Let $(E_3)^j_{k,l} = E_3(k\Delta x_1, l\Delta x_2, j\Delta t)$, and the same notations for H_1 and H_2, then the approximate solutions are given by

$$\frac{(H_1)^{j+1}_{k+\frac{1}{2},l} - (H_1)^j_{k+\frac{1}{2},l}}{\Delta t} = -\mu^{-1}\frac{(E_3)^{j+\frac{1}{2}}_{k+\frac{1}{2},l+\frac{1}{2}} - (E_3)^{j+\frac{1}{2}}_{k+\frac{1}{2},l-\frac{1}{2}}}{\Delta x_2},$$

$$\frac{(H_2)^{j+1}_{k,l+\frac{1}{2}} - (H_2)^j_{k,l+\frac{1}{2}}}{\Delta t} = \mu^{-1}\frac{(E_3)^{j+\frac{1}{2}}_{k+\frac{1}{2},l+\frac{1}{2}} - (E_3)^{j+\frac{1}{2}}_{k-\frac{1}{2},l+\frac{1}{2}}}{\Delta x_1},$$

$$\frac{(E_3)^{j+\frac{1}{2}}_{k-\frac{1}{2},l+\frac{1}{2}} - (E_3)^{j-\frac{1}{2}}_{k-\frac{1}{2},l+\frac{1}{2}}}{\Delta t}$$
$$= \varepsilon^{-1}\left(\frac{(H_2)^j_{k,l+\frac{1}{2}} - (H_2)^j_{k-1,l+\frac{1}{2}}}{\Delta x_1} - \frac{(H_1)^j_{k-\frac{1}{2},l+1} - (H_1)^j_{k-\frac{1}{2},l}}{\Delta x_2}\right).$$

The finite difference schemes for three dimensional problems and for the TE mode are the same.

Mur introduced a finite difference scheme on the artificial boundary which is consistent with the Yee's scheme .[Mur, G. (1981)] For example for the Clayton-Engquist-Majda's boundary condition,

$$B_2 u = \frac{\partial^2 u}{\partial t^2} - \frac{\partial^2 u}{\partial t \partial x_1} - \frac{1}{2}\frac{\partial^2 u}{\partial x_2^2} = 0,$$

the derivatives are approximated by

$$\frac{\partial^2 u}{\partial t \partial x_1} \doteq \frac{u^{j+1}_{k+1,l+\frac{1}{2}} - u^{j+1}_{k,l+\frac{1}{2}} - u^{j-1}_{k+1,l+\frac{1}{2}} + u^{j-1}_{k,l+\frac{1}{2}}}{2\Delta t \Delta x_1},$$

$$\frac{\partial^2 u}{\partial t^2} \doteq \frac{1}{2} \left(\frac{u_{k+1,l+\frac{1}{2}}^{j+1} - 2u_{k+1,l+\frac{1}{2}}^{j} + u_{k+1,l+\frac{1}{2}}^{j-1}}{\Delta t^2} + \frac{u_{k,l+\frac{1}{2}}^{j+1} - 2u_{k,l+\frac{1}{2}}^{j} + u_{k,l+\frac{1}{2}}^{j-1}}{\Delta t^2} \right),$$

$$\frac{\partial^2 u}{\partial x_2^2} \doteq \frac{1}{2} \left(\frac{u_{k+1,l+\frac{3}{2}}^{j} - 2u_{k+1,l+\frac{1}{2}}^{j} + u_{k+1,l-\frac{1}{2}}^{j}}{\Delta x_2^2} + \frac{u_{k,l+\frac{3}{2}}^{j} - 2u_{k,l+\frac{1}{2}}^{j} + u_{k,l-\frac{1}{2}}^{j}}{\Delta x_2^2} \right).$$

4.8 Stationary Navier-Stokes equations

4.8.1 *Homogeneous boundary condition at the infinity*

The governing system of equations with boundary conditions for stationary incompressible viscous flow is the following:

$$u \cdot \nabla u + \nabla p = \tilde{\nu} \triangle u + f, \tag{4.24}$$

$$\nabla \cdot u = 0, \tag{4.25}$$

$$u = g, \qquad x \in \partial\Omega, \tag{4.26}$$

$$\lim_{|x| \to \infty} u = u_\infty. \tag{4.27}$$

An artificial boundary condition for three dimensional problems with homogeneous boundary condition at the infinity, $u_\infty = 0$, is investigated by [Nazarov, S.A. and Specovius-Neugebauer, M. (2003)], where the incompressible condition (4.25) is generalized to

$$-\nabla \cdot u = f_1. \tag{4.28}$$

For $u_\infty = 0$, the system behaves like the Stokes equation near the infinity. A "power-law solution" for the associated homogeneous system is defined by $(u, p) = (|x|^{-1}U(s), |x|^{-2}P(s))$, where $x = |x|s$, $s \in S^3$, and S^3 is the unit sphere in $\mathbb{R}^3$. It is proved that there exists a "power-law solution", which is the leading term of the solution for the problem under some smallness conditions. The "power-law solution" is applied as the artificial boundary condition.

Let the artificial boundary be $\partial B(O, R)$ with R large enough, so that $B(O, R) \supset \overline{\Omega^c}$. Introducing spherical coordinates,

$$
\begin{cases}
x_1 = r \cos \phi \sin \theta, \\
x_2 = r \sin \phi \sin \theta, \\
x_3 = r \cos \theta,
\end{cases}
$$

and letting $u = (u_r, u_\phi, u_\theta)$, the homogeneous system of equations in the coordinates is

$$
u_r \frac{\partial u_r}{\partial r} + \frac{u_\theta}{r} \frac{\partial u_r}{\partial \theta} + \frac{u_\phi}{r \sin \theta} \frac{\partial u_r}{\partial \phi} - \frac{u_\theta^2 + u_\phi^2}{r}
$$
$$
= -\frac{\partial p}{\partial r} + \tilde{\nu} \left(\triangle u_r - \frac{2}{r^2} u_r - \frac{2}{r^2 \sin \theta} \frac{\partial(u_\theta \sin \theta)}{\partial \theta} - \frac{2}{r^2 \sin \theta} \frac{\partial u_\phi}{\partial \phi} \right),
$$

$$
\frac{1}{r^2} \frac{\partial(r^2 u_r)}{\partial r} + \frac{1}{r \sin \theta} \left(\frac{\partial(u_\theta \sin \theta)}{\partial \theta} + \frac{\partial u_\phi}{\partial \phi} \right) = 0,
$$

where we write down the radial part only, which is needed later on. We plug the power-law solution $(u, p) = (r^{-1} U(s), r^{-2} P(s))$, $U = (U_r, U_\theta, U_\phi)$, into them and multiply them by r^3 and r^2 respectively to obtain

$$
-U_r^2 - U_\theta^2 - U_\phi^2 + U_\theta \frac{\partial U_r}{\partial \theta} + \frac{U_\phi}{\sin \theta} \frac{\partial U_r}{\partial \phi}
$$
$$
= 2P + \tilde{\nu} \left(\tilde{\triangle} U_r - 2 \left(U_r + \frac{1}{\sin \theta} \frac{\partial(U_\theta \sin \theta)}{\partial \theta} + \frac{1}{\sin \theta} \frac{\partial U_\phi}{\partial \phi} \right) \right),
\tag{4.29}
$$

$$
U_r + \frac{1}{\sin \theta} \left(\frac{\partial(U_\theta \sin \theta)}{\partial \theta} + \frac{\partial U_\phi}{\partial \phi} \right) = 0,
\tag{4.30}
$$

where $\tilde{\triangle}$ is the angular part of the Laplacian— the Laplace-Beltrami operator on the unit sphere,

$$
\tilde{\triangle} = \frac{\partial^2}{\partial \theta^2} + \cot \theta \frac{\partial}{\partial \theta} + \frac{1}{\sin^2 \theta} \frac{\partial^2}{\partial \phi^2}.
$$

Applying (4.30) the equation (4.29) turns into

$$
2P = -\tilde{\nu} \tilde{\triangle} U_r - |U|^2 + U_\theta \frac{\partial U_r}{\partial \theta} + \frac{U_\phi}{\sin \theta} \frac{\partial U_r}{\partial \phi}
$$
$$
= -\tilde{\nu} \tilde{\triangle} U_r - |U|^2 + (U^T \cdot \tilde{\nabla}) U_r,
\tag{4.31}
$$

where

$$\tilde{\nabla} = \left(\frac{\partial}{\partial \theta}, \frac{1}{\sin \theta} \frac{\partial}{\partial \phi} \right).$$

The Stokes equation gives a hint of the Neŭmann operator,

$$T(u, p) = \begin{pmatrix} -\tilde{\nu} \frac{\partial}{\partial r} u_r + p \\ -\tilde{\nu} \frac{\partial}{\partial r} u_\theta \\ -\tilde{\nu} \frac{\partial}{\partial r} u_\phi \end{pmatrix},$$

and for the power-law solution,

$$-\frac{\partial}{\partial r} u = \frac{1}{r} u.$$

Then (4.31) is applied to replace p in the Neŭmann operator to define an artificial Neŭmann boundary condition at $r = R$, namely,

$$-\tilde{\nu} \frac{\partial u_r}{\partial r} + p - \frac{\tilde{\nu}}{R} u_r + \frac{\tilde{\nu}}{2} \tilde{\triangle} u_r + \frac{1}{2} \left(|u|^2 - (u^T \cdot \tilde{\nabla}) u_r \right) = 0, \qquad (4.32)$$

$$-\tilde{\nu} \frac{\partial u_\theta}{\partial r} = \frac{\tilde{\nu}}{R} u_\theta, \qquad -\tilde{\nu} \frac{\partial u_\phi}{\partial r} = \frac{\tilde{\nu}}{R} u_\phi. \qquad (4.33)$$

This defines an approximate DtN operator.

Introduce weighted Hölder spaces $\Lambda_\beta^{l;\alpha}(\Omega)$, $\alpha \in (0,1)$, equipped with the norm,

$$\|f\|_{\Lambda_\beta^{l,\alpha}} = \sum_{k=0}^{l} \sup_{x \in \Omega} |x|^{\beta - l - \alpha + k} |D^k f(x)|$$

$$+ \sup_{\substack{x, y \in \Omega \\ 2|x - y| \le |x|}} |x - y|^{-\alpha} ||x|^\beta D^k f(x) - |y|^\beta D^k f(y)|.$$

We denote by (u, p) the solution to (4.24)-(4.27), and by (u_h, p_h) the solution to (4.24)-(4.26),(4.32),(4.33). Uniqueness and error estimates are summarized in the following theorem.

Theorem 63. *Let l be a positive integer, $\alpha \in (0, 1)$, $\beta \in (l+1+\alpha, l+2+\alpha)$, and $\gamma \in (l+2+\alpha, l+3+\alpha)$. Then there exist positive constants ρ, C_1 and R_1, such that if*

$$\|f\|_{\Lambda_\gamma^{l,\alpha}} + \|g\|_{\Lambda_\gamma^{l,\alpha}} < \rho,$$

and $R > R_1$, then the problem (4.24)-(4.26),(4.32),(4.33) admits a unique solution (u_h, p_h) in the ball

$$\|u - u_h\|_{\Lambda_\beta^{l,\alpha}} \leq C_1 R^{\beta - 2 - l - \alpha},$$

and u_h satisfies the error estimate,

$$\|u - u_h\|_{\Lambda_\beta^{l,\alpha}} \leq C_2 R^{\beta - \gamma}(\|f\|_{\Lambda_\gamma^{l,\alpha}} + \|g\|_{\Lambda_\gamma^{l,\alpha}}),$$

Where C_2 is independent of $R > R_1$ and (f, g).

4.8.2 *Inhomogeneous boundary conditions at the infinity*

An artificial boundary condition for three dimensional problems (4.24)-(4.27) with inhomogeneous boundary condition at the infinity, $u_\infty \neq 0$, is investigated by [Deuring, P. and Kracmar, S (2004)]. Without losing generality we choose coordinates so that u_∞ is parallel to the x_1-axis. Replacing u by $u - u_\infty$ and introducing the Reynolds number $\tilde{R}$, the Navier-Stokes equations can be transformed into a dimensionless form

$$\tilde{R}\frac{\partial u}{\partial x_1} + \tilde{R}u \cdot \nabla u + \nabla p = \triangle u + f. \tag{4.34}$$

For $u_\infty \neq 0$, the flow behaves like the Oseen flow near the infinity. A more general equation,

$$\tilde{R}\frac{\partial u}{\partial x_1} + \tau u \cdot \nabla u + \nabla p = \triangle u + f, \tag{4.35}$$

is investigated, where $\tau \in [0, \tilde{R}]$, and if $\tau = 0$, it is the Oseen flow.

The following artificial boundary condition is imposed at $|x| = R$ with R large enough:

$$\frac{\partial u}{\partial r} - p\frac{x}{R} - \frac{\tau}{2R}(x \cdot u)u + \left(\frac{1}{R} + \frac{\tilde{R}}{2}\frac{R - x_1}{R}\right)u = 0. \tag{4.36}$$

This defines an approximate DtN operator.

Letting $\Omega_1 = \Omega \cap B(O, R)$, two bilinear forms are defined for the problem:

$$a(u, v) = \int_{\Omega_1}\left(\nabla u : \nabla v + \tilde{R}\frac{\partial u}{\partial x_1} \cdot v + \tilde{R}(u \cdot \nabla)u \cdot v\right)dx$$

$$+ \int_{|x|=R}\left(\left(\frac{1}{R} + \frac{\tilde{R}}{2}\frac{R - x_1}{R}\right)u \cdot v - \frac{\tau}{2R}(u \cdot x)(u \cdot v)\right)ds,$$

$$b(u, p) = -\int_{\Omega_1}\nabla u \cdot p\,dx.$$

The following results are proved:

Theorem 64. *There is a constant $\delta > 0$, such that for $g \in \left(H^{\frac{1}{2}}(\partial\Omega)\right)^3$, $f \in \left(L^{\frac{6}{5}}_{loc}(\Omega)\right)^3$ with $|\int_{\partial\Omega} g \cdot \nu\, ds| < \delta$, there exists a pair of functions $(u, p) \in \left(H^1(\Omega_1)\right)^3 \times L^2(\Omega_1)$ with $\nabla \cdot u = 0$, $u|_{\partial\Omega} = g$, and satisfying*

$$a(u, v) + b(v, p) = (f, v)_{\Omega_1}, \quad \forall v \in \left(H^1(\Omega_1)\right)^3, v|_{\partial\Omega} = 0. \tag{4.37}$$

We denote by (u, p) the solution to (4.24)-(4.27), and by (u_h, p_h) the solution to (4.24)-(4.26),(4.36). Uniqueness and error estimates require stronger conditions.

Theorem 65. *We assume the hypotheses of Theorem 64 to be satisfied. Moreover, assume that there are positive constants γ and $\sigma > 4$, such that $f \in \left(L^{\frac{6}{5}}(\Omega)\right)^3$, and $|f(x)| \leq \gamma|x|^{-\sigma}$. Let $\Gamma = \|f\|_{0,6/5} + \gamma + \|b\|_{1/2}$ and $\tau_m = \max(1, \tilde{R})$. Then there exist positive constants $\alpha_1, \alpha_2, C_1, C_2$, such that if $\tau \max(\Gamma, \Gamma^2) \leq C_1 \tau_m^{-\alpha_1}$, then the solution to (4.37) is unique, and the following estimates are satisfied:*

$$\|\nabla(u - u_h)\|_0 + (R^{-1} + \tilde{R})^{1/2}\|u - u_h\|_{0,\partial B(O,R)}$$
$$\leq C_2 \tau_m^{\alpha_2} \Gamma \cdot R^{-1/2} \cdot \min(1, (\tilde{R}R)^{-1/2}).$$

4.8.3 A linear boundary condition

For two dimensional symmetric flow with inhomogeneous boundary conditions at the infinity a linear artificial boundary condition is proposed by [Bönisch,S., Heuveline, V. and Wittwer, P. (2005)], which is based on a theorem about the asymptotic behavior of solutions at the infinity [Wittwer, P. (2002)]. The flow is governed by the problem (4.24)-(4.27). Consider the homogeneous equation, that is $f = 0$, and assume that $u_\infty \neq 0$. The coordinates are chosen so that u_∞ is parallel to the x_1-axis as before. By a scaling argument the problem can be transformed into a dimensionless form,

$$u \cdot \nabla u + \nabla p = \triangle u, \tag{4.38}$$

$$\nabla \cdot u = 0, \tag{4.39}$$

$$u = 0, \qquad x \in \partial\Omega, \tag{4.40}$$

$$\lim_{|x|\to\infty} u = (1,0). \tag{4.41}$$

Regard the x_1 variable as "time", then the equations behave like parabolic equations at long times. Consider an "initial condition"

$$u|_{x_1=1} = (1,0) + g, \tag{4.42}$$

and the corresponding "initial value problem" on $x_1 > 1$. Let $v = (v_1, v_2) = u - (1,0)$, then the following holds:

Theorem 66. *Under some smallness conditions there exists a (locally unique) solution to (4.38),(4.39),(4.42) and the boundary condition (4.41) on $x_1 = \infty$. Furthermore there exist functions*

$$v_{01}(x) = \frac{c}{2\sqrt{\pi}} \frac{1}{\sqrt{x_1}} e^{-\frac{x_2^2}{4x_1}} + \frac{d}{\pi} \frac{x_1}{|x|^2},$$

$$v_{02}(x) = \frac{c}{4\sqrt{\pi}} \frac{x_2}{x_1^{3/2}} e^{-\frac{x_2^2}{4x_1}} + \frac{d}{\pi} \frac{x_2}{|x|^2},$$

such that

$$\lim_{x_1\to\infty} x_1^{1/2} \left(\sup_{x_2\in\mathbb{R}} |(v_1 - v_{01})(x_1, x_2)| \right) = 0,$$

$$\lim_{x_1\to\infty} x_1 \left(\sup_{x_2\in\mathbb{R}} |(v_2 - v_{02})(x_1, x_2)| \right) = 0,$$

and

$$\int_{-\infty}^{\infty} v_1(x)\, dx_2 = \int_{-\infty}^{\infty} v_{01}(x)\, dx_2 = c + d.$$

The first terms in v_{01} and v_{02} represent a potential flow with a source or sink at the origin. The second terms represent a wake, within which the vorticity of the fluid is concentrated. The asymptotic behavior $v_0 = (v_{01}, v_{02})$ is universal in the sense that this asymptotic behavior is independent of the geometry of the domain Ω and Ω^c, except for the amplitudes c and d. In the original coordinates the parameter c should be multiplied by a factor $\sqrt{\tilde{\nu}|u_\infty|}$, and d by $\tilde{\nu}$.

We now consider the exterior domain Ω. Suppose the body Ω^c is symmetric with respect to the x_1-axis, and the flow is also symmetric with respect to the x_1-axis, that is, $u_1(x_1, x_2) = u_1(x_1, -x_2)$, $u_2(x_1, x_2) = -u_2(x_1, -x_2)$, and $p(x_1, x_2) = p(x_1, -x_2)$. Let a square, $\Omega_0 = \{x; |x_1| < $

$a, |x_2| < a\}$, be large enough, so that $\Omega_0 \supset \overline{\Omega^c}$. The system is solved on $\Omega_1 = \Omega \cap \Omega_0$, with the following artificial boundary condition:

$$u = v_0 + (1,0), \qquad x_1 > 0,$$

$$u = \frac{d}{\pi} \frac{x}{|x|^2} + (1,0), \qquad x_1 < 0.$$

There are two parameters c and d, so two more equations are needed, which are obtained by integrating the equations and using Green's formula.

$$\int_{\partial \Omega} \left(\frac{\partial u}{\partial \nu} - p\nu \right) ds = \lim_{a \to \infty} \int_{\partial \Omega_0} (u_0 u_0^T + p)\nu \, ds, \qquad (4.43)$$

$$\int_{\partial \Omega} u \cdot \nu \, ds = \lim_{a \to \infty} \int_{\partial \Omega_0} u_0 \cdot \nu \, ds, \qquad (4.44)$$

where $u_0 = v_0 + (1,0)$, and we notice that $u = 0$ on $\partial \Omega$ and the viscous term tends to zero as $a \to \infty$. Noting that the flow is symmetric, the x_2 component of the equation (4.43) makes no contribution, so there are exact two equations.

We remark that for high Reynolds number symmetric flow is unstable, so the above algorithm can be applied to low Reynolds number flow only.

Chapter 5

Perfectly Matched Layer Method

The artificial boundary conditions in the previous chapter can be regarded as an artificial material with zero thickness. Another approach to absorb the outgoing waves is to put a layer with a positive thickness. A special medium, not necessary existing in the real world, is designed to be filled in this layer, which can absorb most of the waves. This layer should bear two important features: 1. The interface must be penetrable so that no wave is reflected to the real underlying domain. 2. The special medium must have a damping behavior. We investigate some kinds of layers in this chapter. Functions in this chapter are complex.

5.1 Wave equations

Like Section 4.1, we start from one dimensional case and consider the wave equation,

$$\frac{\partial^2 u}{\partial t^2} = \frac{\partial^2 u}{\partial x^2}. \tag{5.1}$$

Suppose the domain is $(0, \infty)$, and the initial and boundary conditions are

$$u|_{t=0} = u_0, \qquad \left.\frac{\partial u}{\partial t}\right|_{t=0} = u_1, \qquad x > 0, \tag{5.2}$$

$$u|_{x=0} = 0, \qquad t > 0, \tag{5.3}$$

where $u_0(0) = 0$ and $u_1(0) = 0$. Let $\frac{\partial u}{\partial x} = p$, and $\frac{\partial u}{\partial t} = q$, then the equation (5.1) is equivalent to

$$\frac{\partial p}{\partial t} = \frac{\partial q}{\partial x}, \quad \frac{\partial q}{\partial t} = \frac{\partial p}{\partial x}. \tag{5.4}$$

We see that

$$\frac{\partial(p+q)}{\partial t} - \frac{\partial(p+q)}{\partial x} = 0, \quad \frac{\partial(p-q)}{\partial t} + \frac{\partial(p-q)}{\partial x} = 0.$$

Therefore $p + q$ are constants along characteristics $x + t = $ constant, which are left going waves. On the other hand, $p - q$ are constants along characteristics $x - t = $ constant, which are right going waves. The solutions to (5.4) are the superposition of the right going waves $(p, q) = (f(x-t), -f(x-t))$ and the left going waves $(p, q) = (g(x+t), g(x+t))$, where f and g are arbitrary functions. By the initial condition (5.2), $p|_{t=0} = p_0 = u_0'$ and $q|_{t=0} = q_0 = u_1$. We have $f = \frac{1}{2}(p_0 - q_0)$ and $g = \frac{1}{2}(p_0 + q_0)$. The left going waves touches the boundary $x = 0$, then the boundary reflects the wave to create a right going wave $(p, q) = (g(t-x), -g(t-x))$. The complete solution is:

$$(p, q) = \begin{cases} (f(x-t) + g(x+t), -f(x-t) + g(x+t)), & x > t, \\ (g(x+t) + g(t-x), g(x+t) - g(t-x)), & x < t. \end{cases} \quad (5.5)$$

We assume that u_0 and u_1 vanish for $x > a$ and put some damping medium on the interval $(a, a+d)$. Artificial boundary condition is imposed at $x = a + d$. Then the approximate problem is:

$$\frac{\partial p}{\partial t} = \frac{\partial q}{\partial x}, \quad \frac{\partial q}{\partial t} = \frac{\partial p}{\partial x}, \quad x \in (0, a),$$

$$\frac{\partial p}{\partial t} + \sigma p = \frac{\partial q}{\partial x}, \quad \frac{\partial q}{\partial t} + \sigma q = \frac{\partial p}{\partial x}, \quad x \in (a, a+d),$$

$$p|_{t=0} = p_0, \quad q|_{t=0} = q_0, \quad x \in (0, a),$$

$$p|_{t=0} = q|_{t=0} = 0, \quad x \in (a, a+d),$$

$$q|_{x=0} = 0, \quad p|_{x=a-0} = p|_{x=a+0}, \quad q|_{x=a-0} = q|_{x=a+0},$$

$$p|_{x=a+d} = 0,$$

where $\sigma > 0$ is the damping constant. Noting that the interface $x = a$ is penetrable, because p and q are continuous across the interface, so are the right going waves $p - q$ and left going waves $p + q$. Therefore a wave can go through the interface without any reflection.

We set $\tilde{p} = pe^{\sigma t}$ and $\tilde{q} = qe^{\sigma t}$, then the equations on $(a, a + d)$ become

$$\frac{\partial \tilde{p}}{\partial t} = \frac{\partial \tilde{q}}{\partial x}, \quad \frac{\partial \tilde{q}}{\partial t} = \frac{\partial \tilde{p}}{\partial x},$$

which are the same as (5.4). The right going waves satisfy

$$(\tilde{p} - \tilde{q})(a + d, t) = (\tilde{p} - \tilde{q})(a, t - d), \quad t \geq d,$$

that is

$$(p - q)(a + d, t) = e^{-\sigma d}(p - q)(a, t - d), \quad t \geq d.$$

Consequently a right going wave passing through the medium decays by a factor $e^{-\sigma d}$. The same is true for a left going wave. Let $(p, q) = (f(x - t), -f(x - t))$, $t \geq 0$, be a right going wave on $(0, a)$. The interface $x = a$ is penetrable, so $(\tilde{p}, \tilde{q}) = e^{\sigma t}(f(a - t), -f(a - t))$, $t \geq 0$, at $x = a + 0$. As the wave is reflected at $x = a + d$, and the left going wave reaches $x = a$, it is $(\tilde{p}, \tilde{q}) = e^{\sigma(t-2d)}(-f(a + 2d - t), -f(a + 2d - t))$, $t \geq 2d$, then $(p, q) = e^{-2\sigma d}(-f(a + 2d - t), -f(a + 2d - t))$, $t \geq 2d$, at $x = a - 0$. We see that the amplitude of the wave decays by a factor $e^{-2\sigma d}$.

By (5.5) the exact solution at $x = 0$ is $p(0, t) = 2g(t) = p_0(t) + q_0(t)$. Since $p_0(t) = q_0(t) = 0$ for $t > a$, the exact solution $p(0, t)$ vanishes for $t > a$. The value of $p(0, t)$ is not disturbed by the reflection wave from the artificial boundary $x = a + d$ when $t < a + 2d$, so still we have $p(0, t) = 2g(t)$, but it is disturbed when $t > a + 2d$. By induction we can see

$$p(0, t) = (-1)^m e^{-2m\sigma d} p(0, t - 2m(a+d)), \quad t \in (2m(a+d), 2(m+1)(a+d)),$$

for $m = 1, 2, \cdots$. It seems the bigger σd the better, but if d large, the truncated domain is large, and the computing cost is high, while if σ is too large, the difference of equations between two sides of $x = a$ is too large, which would cause a large computational error and instability.

There is another point of view. Let the Fourier transform of p, q with respect to t be $\tilde{p}(x, -\omega) = F(p(x, t))$, and $\tilde{q}(x, -\omega) = F(q(x, t))$, then the equations are

$$-i\omega \tilde{p} = \frac{\partial \tilde{q}}{\partial x}, \quad -i\omega \tilde{q} = \frac{\partial \tilde{p}}{\partial x}, \quad x \in (0, a), \tag{5.6}$$

and

$$(-i\omega + \sigma)\tilde{p} = \frac{\partial \tilde{q}}{\partial x}, \quad (-i\omega + \sigma)\tilde{q} = \frac{\partial \tilde{p}}{\partial x}, \quad x \in (a, a + d), \tag{5.7}$$

respectively. They are known as the time harmonic forms. By eliminating $\tilde{q}$ (or $\tilde{p}$) they become second order equations

$$\frac{\partial^2 \tilde{p}}{\partial x^2} + \omega^2 \tilde{p} = 0, \qquad x \in (0, a),$$

and

$$\frac{\partial^2 \tilde{p}}{\partial x^2} - (-i\omega + \sigma)^2 \tilde{p} = 0, \qquad x \in (a, a + d).$$

The general solutions for them are

$$\tilde{p} = c_1(\omega)e^{i\omega x} + c_2(\omega)e^{-i\omega x}, \qquad x \in (0, a), \tag{5.8}$$

and

$$\tilde{p} = c_3(\omega)e^{-(-i\omega+\sigma)x} + c_4(\omega)e^{(-i\omega+\sigma)x}, \qquad x \in (a, a + d). \tag{5.9}$$

By multiplying a factor $e^{-i\omega t}$ we find that the first term is a right going wave and the second term is a left going wave. From (5.9) we see that the first term decays to zero exponentially as $x \to +\infty$, and so does the second term as $x \to -\infty$. By (5.8) and (5.6) we get

$$\tilde{q} = -c_1(\omega)e^{i\omega x} + c_2(\omega)e^{-i\omega x},$$

and analogously by (5.9) and (5.7) we get

$$\tilde{q} = -c_3(\omega)e^{-(-i\omega+\sigma)x} + c_4(\omega)e^{(-i\omega+\sigma)x}.$$

When a right going wave reaches $x = a$, it is equal to

$$(\tilde{p}, \tilde{q}) = (c_1(\omega)e^{i\omega a}, -c_1(\omega)e^{i\omega a}).$$

Simply we set $c_3(\omega) = c_1(\omega)e^{\sigma a}$, then the right going wave on $x > a$ is

$$(\tilde{p}, \tilde{q}) = (c_1(\omega)e^{\sigma a}e^{-(-i\omega+\sigma)x}, -c_1(\omega)e^{\sigma a}e^{(-i\omega+\sigma)x}).$$

$\tilde{p}$ and $\tilde{q}$ are continuous across the interface $x = a$ without any reflection.

Another observation is useful to design the damping layer. If we define a transform of independent variables:

$$\tilde{x} = \begin{cases} x, & x < a, \\ a + \left(1 + i\frac{\sigma}{\omega}\right)(x - a), & x > a, \end{cases} \tag{5.10}$$

then the equation (5.6) and (5.7) can be written in a uniform form

$$-i\omega\tilde{p} = \frac{\partial \tilde{q}}{\partial \tilde{x}}, \qquad -i\omega\tilde{q} = \frac{\partial \tilde{p}}{\partial \tilde{x}}. \tag{5.11}$$

The solutions to (5.8),(5.9) can be expressed by

$$\tilde{p} = c_1(\omega)e^{i\omega\tilde{x}} + c_2(\omega)e^{-i\omega\tilde{x}},$$

which satisfies the equation (5.11). $\tilde{p}$ is certainly continuous at $x = a$, therefore a right going wave (5.8) can pass across the interface $x = a$ without any reflection, which again leads to the conclusion that the interface is penetrable.

We turn now to investigate the two dimensional wave equation

$$\frac{\partial^2 u}{\partial t^2} = \frac{\partial^2 u}{\partial x_1^2} + \frac{\partial^2 u}{\partial x_2^2}. \tag{5.12}$$

Let $p = (p_1, p_2) = \nabla u$ and $q = \frac{\partial u}{\partial t}$, then the equation is equivalent to

$$\frac{\partial q}{\partial t} = \nabla \cdot p, \qquad \frac{\partial p}{\partial t} = \nabla q. \tag{5.13}$$

We define the Fourier transform with respect to t and x_2, $\tilde{p}(x_1, \xi_2, -\omega) = F(p(x_1, x_2, t))$ and $\tilde{q}(x_1, \xi_2, -\omega) = F(q(x_1, x_2, t))$. The equations for $\tilde{p}$, $\tilde{q}$ are

$$-i\omega\tilde{q} = \frac{\partial\tilde{p}_1}{\partial x_1} + i\xi_2\tilde{p}_2,$$

$$-i\omega\tilde{p} = \left(\frac{\partial\tilde{q}}{\partial x_1}, i\xi_2\tilde{q}\right).$$

By eliminating $\tilde{p}$ we get a second order equation for $\tilde{q}$:

$$\frac{\partial^2\tilde{q}}{\partial x_1^2} + (\omega^2 - \xi_2^2)\tilde{q} = 0. \tag{5.14}$$

The general solution is

$$\tilde{q} = c_1(\omega, \xi_2)e^{i\sqrt{\omega^2-\xi_2^2}\,x_1} + c_2(\omega, \xi_2)e^{-i\sqrt{\omega^2-\xi_2^2}\,x_1}, \tag{5.15}$$

and

$$\tilde{p}_1 = \sqrt{1 - \frac{\xi_2^2}{\omega^2}}\left(-c_1(\omega, \xi_2)e^{i\sqrt{\omega^2-\xi_2^2}\,x_1} + c_2(\omega, \xi_2)e^{-i\sqrt{\omega^2-\xi_2^2}\,x_1}\right), \tag{5.16}$$

$$\tilde{p}_2 = -\frac{\xi_2}{\omega}\left(c_1(\omega, \xi_2)e^{i\sqrt{\omega^2-\xi_2^2}\,x_1} + c_2(\omega, \xi_2)e^{-i\sqrt{\omega^2-\xi_2^2}\,x_1}\right). \tag{5.17}$$

The first term is a right going wave and the second term is left going.

In the damping layer $x_1 \in (a, a+d)$ the equations are

$$\frac{\partial q}{\partial t} + \sigma q = \nabla \cdot p, \qquad \frac{\partial p}{\partial t} + \sigma p = \nabla q,$$

$$(-i\omega + \sigma)\tilde{q} = \frac{\partial \tilde{p}_1}{\partial x_1} + i\xi_2 \tilde{p}_2,$$

$$(-i\omega + \sigma)\tilde{p} = \left(\frac{\partial \tilde{q}}{\partial x_1}, i\xi_2 \tilde{q} \right),$$

$$\frac{\partial^2 \tilde{q}}{\partial x_1^2} + ((-i\omega + \sigma)^2 - \xi_2^2)\tilde{q} = 0. \qquad (5.18)$$

The general solution is

$$\tilde{q} = c_3(\omega, \xi_2) e^{i\sqrt{(\omega+i\sigma)^2 - \xi_2^2} x_1} + c_4(\omega, \xi_2) e^{-i\sqrt{(\omega+i\sigma)^2 - \xi_2^2} x_1}, \qquad (5.19)$$

$$\tilde{p}_1 = \sqrt{1 - \frac{\xi_2^2}{(\omega+i\sigma)^2}} \left(-c_3(\omega, \xi_2) e^{i\sqrt{(\omega+i\sigma)^2 - \xi_2^2} x_1} \right.$$
$$\left. + c_4(\omega, \xi_2) e^{-i\sqrt{(\omega+i\sigma)^2 - \xi_2^2} x_1} \right), \qquad (5.20)$$

$$\tilde{p}_2 = -\frac{\xi_2}{\omega + i\sigma} \left(c_3(\omega, \xi_2) e^{i\sqrt{(\omega+i\sigma)^2 - \xi_2^2} x_1} + c_4(\omega, \xi_2) e^{-i\sqrt{(\omega+i\sigma)^2 - \xi_2^2} x_1} \right). \qquad (5.21)$$

We see the system for $\tilde{p}$, $\tilde{q}$ consists of one algebraic equation and two first order differential equations for $\tilde{q}$ and $\tilde{p}_1$, thus the interface condition is

$$\tilde{q}|_{x_1=a-0} = \tilde{q}|_{x_1=a+0}, \qquad \tilde{p}_1|_{x_1=a-0} = \tilde{p}_1|_{x_1=a+0}.$$

For a right going wave on $x_1 < a$, the necessary condition for a right going wave on $x_1 > a$ is

$$c_3(\omega, \xi_2) e^{i\sqrt{(\omega+i\sigma)^2 - \xi_2^2} a} = c_1(\omega, \xi_2) e^{i\sqrt{\omega^2 - \xi_2^2} a},$$

$$-\sqrt{1 - \frac{\xi_2^2}{(\omega+i\sigma)^2}} c_3(\omega, \xi_2) e^{i\sqrt{(\omega+i\sigma)^2 - \xi_2^2} a} = -\sqrt{1 - \frac{\xi_2^2}{\omega^2}} c_1(\omega, \xi_2) e^{i\sqrt{\omega^2 - \xi_2^2} a}.$$

This is an over-determined system for the coefficient c_3, which causes difficulty for two dimensional problems. The interface $x_1 = a$ is not fully penetrable in this setting, and there must be a reflection wave going to the left.

On the other hand if we introduce a new variable $\tilde{x}_1$ as the one dimensional case,

$$\tilde{x}_1 = \begin{cases} x_1, & x_1 < a, \\ a + \left(1 + i\frac{\sigma}{\omega}\right)(x_1 - a), & x_1 > a, \end{cases} \tag{5.22}$$

and replace x_1 in (5.15) by $\tilde{x}_1$,

$$\tilde{q} = c_1(\omega, \xi_2)e^{i\sqrt{\omega^2 - \xi_2^2}\,\tilde{x}_1},$$

which satisfies

$$\frac{\partial^2 \tilde{q}}{\partial \tilde{x}_1^2} + (\omega^2 - \xi_2^2)\tilde{q} = 0. \tag{5.23}$$

Then the interface is fully penetrable. Taking inverse Fourier transform, we can get the equation on the physical domain. However the equation (5.23) leads to a convolution product. To prevent this inconvenience some approaches have been given. We will discuss them at Section 5.2 and Section 5.4.

If $\xi_2 = 0$, these two equations(5.18) and (5.23) are the same. By the discussion in Section 4.1 we know that ξ_2 represents the incidence angle. So if the incidence angle is small, the reflection in (5.18) is weak, otherwise the reflection is strong.

5.2 Bérenger's perfectly matched layers

We investigate the two dimensional wave equation again. Following Bérenger's idea [Berenger, J.P. (1994)], q is divided into two parts: $q = q_1 + q_2$, and the system of equations (5.13) is equivalent to

$$\frac{\partial q_1}{\partial t} = \frac{\partial p_1}{\partial x_1}, \qquad \frac{\partial q_2}{\partial t} = \frac{\partial p_2}{\partial x_2}, \tag{5.24}$$

$$\frac{\partial p_1}{\partial t} = \frac{\partial(q_1 + q_2)}{\partial x_1}, \qquad \frac{\partial p_2}{\partial t} = \frac{\partial(q_1 + q_2)}{\partial x_2}. \tag{5.25}$$

In the perfectly matched layer $x_1 \in (a, a + d)$ the system of equations is

$$\frac{\partial q_1}{\partial t} + \sigma q_1 = \frac{\partial p_1}{\partial x_1}, \qquad \frac{\partial q_2}{\partial t} = \frac{\partial p_2}{\partial x_2}, \tag{5.26}$$

$$\frac{\partial p_1}{\partial t} + \sigma p_1 = \frac{\partial(q_1 + q_2)}{\partial x_1}, \qquad \frac{\partial p_2}{\partial t} = \frac{\partial(q_1 + q_2)}{\partial x_2}. \tag{5.27}$$

Applying the Fourier transform to the system, we get

$$(-i\omega + \sigma)\tilde{q}_1 = \frac{\partial \tilde{p}_1}{\partial x_1}, \qquad -i\omega\tilde{q}_2 = i\xi_2\tilde{p}_2, \tag{5.28}$$

$$(-i\omega + \sigma)\tilde{p}_1 = \frac{\partial(\tilde{q}_1 + \tilde{q}_2)}{\partial x_1}, \qquad -i\omega\tilde{p}_2 = i\xi_2(\tilde{q}_1 + \tilde{q}_2). \tag{5.29}$$

If we apply the transform of independent variables (5.22), we find (5.28),(5.29) is just the Fourier transform of (5.24),(5.25) with x_1 replaced by $\tilde{x}_1$. Following the argument in the previous section, we conclude that the interface $x_1 = a$ is penetrable.

Now let Ω be an exterior domain as usual. Suppose a square $\Omega_0 = \{x \in \mathbb{R}^2, |x_1| < a, |x_2| < a\}$ is large enough, so that $\Omega_0 \supset \overline{\Omega^c}$. We extend it to a larger square, $\Omega_d = \{x \in \mathbb{R}^2, |x_1| < a + d, |x_2| < a + d\}$. Then $\Omega_d \setminus \overline{\Omega_0}$ is the PML domain. The system of equations in the PML domain is designed by

$$\frac{\partial q_1}{\partial t} + \sigma_1 q_1 = \frac{\partial p_1}{\partial x_1}, \qquad \frac{\partial q_2}{\partial t} + \sigma_2 q_2 = \frac{\partial p_2}{\partial x_2}, \tag{5.30}$$

$$\frac{\partial p_1}{\partial t} + \sigma_1 p_1 = \frac{\partial(q_1 + q_2)}{\partial x_1}, \qquad \frac{\partial p_2}{\partial t} + \sigma_2 p_2 = \frac{\partial(q_1 + q_2)}{\partial x_2}. \tag{5.31}$$

Two damping parameters σ_1 and σ_2 are designed by

$$\left.\begin{array}{l} \sigma_j = 0, \qquad \text{for } |x_j| < a \\ \sigma_j = \sigma > 0, \text{ for } |x_j| > a \end{array}\right\} \text{ where } j = 1, 2.$$

If Ω_0 is a disk, $\{x; |x| < a\}$, polar coordinates can be used. The equation is

$$\frac{\partial^2 u}{\partial t^2} = \frac{1}{r}\frac{\partial}{\partial r}\left(r\frac{\partial u}{\partial r}\right) + \frac{1}{r^2}\frac{\partial^2 u}{\partial \theta^2}. \tag{5.32}$$

Introducing

$$q = \frac{\partial u}{\partial t}, \quad p_1 = \frac{\partial u}{\partial r}, \quad p_2 = \frac{1}{r}\frac{\partial u}{\partial \theta},$$

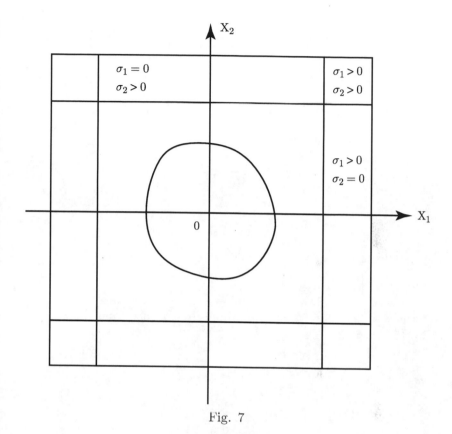

Fig. 7

we get the equivalent system of equations,

$$\frac{\partial q}{\partial t} = \frac{1}{r}\left(\frac{\partial}{\partial r}(rp_1) + \frac{\partial p_2}{\partial \theta}\right),$$

$$\frac{\partial p_1}{\partial t} = \frac{\partial q}{\partial r}, \quad \frac{\partial p_2}{\partial t} = \frac{1}{r}\frac{\partial q}{\partial \theta}.$$

Define the Fourier transform of (q, p_1, p_2) by $(\tilde{q}, \tilde{p}_1, \tilde{p}_2)(-\omega, r, \theta)$, then the system is

$$-i\omega\tilde{q} = \frac{1}{r}\left(\frac{\partial}{\partial r}(r\tilde{p}_1) + \frac{\partial \tilde{p}_2}{\partial \theta}\right),$$

$$-i\omega\tilde{p}_1 = \frac{\partial\tilde{q}}{\partial r}, \quad -i\omega\tilde{p}_2 = \frac{1}{r}\frac{\partial\tilde{q}}{\partial\theta}.$$

Introducing a new independent variable,

$$\tilde{r} = \begin{cases} r, & r < a, \\ a + \left(1 + i\frac{\sigma}{\omega}\right)(r - a), & r > a, \end{cases}$$

we replace the independent variable r by $\tilde{r}$ to obtain

$$-i\omega\tilde{q} = \frac{1}{\tilde{r}}\left(\frac{\partial}{\partial\tilde{r}}(\tilde{r}\tilde{p}_1) + \frac{\partial\tilde{p}_2}{\partial\theta}\right), \tag{5.33}$$

$$-i\omega\tilde{p}_1 = \frac{\partial\tilde{q}}{\partial\tilde{r}}, \quad -i\omega\tilde{p}_2 = \frac{1}{\tilde{r}}\frac{\partial\tilde{q}}{\partial\theta}. \tag{5.34}$$

The variable $\tilde{q}$ is divided into two parts: $\tilde{q} = \tilde{q}_1 + \tilde{q}_2$, and $\tilde{q}_1$ satisfies

$$-i\omega\tilde{q}_1 = \frac{\partial\tilde{p}_1}{\partial\tilde{r}}. \tag{5.35}$$

Returning to the independent variable r, we deduce from the equations (5.33)-(5.35) to get

$$-i\omega\left(1 + i\frac{r-a}{r}\frac{\sigma}{\omega}\right)\tilde{q} = \frac{1}{r}\left(\frac{\partial}{\partial r}(r\tilde{p}_1) + \frac{\partial\tilde{p}_2}{\partial\theta}\right) + \frac{\tilde{r} - (1 + i\frac{\sigma}{\omega})r}{r}\frac{\partial\tilde{p}_1}{\partial\tilde{r}}$$
$$= \frac{1}{r}\left(\frac{\partial}{\partial r}(r\tilde{p}_1) + \frac{\partial\tilde{p}_2}{\partial\theta}\right) - i\omega\frac{\tilde{r} - (1 + i\frac{\sigma}{\omega})r}{r}\tilde{q}_1,$$

$$-i\omega\left(1 + i\frac{\sigma}{\omega}\right)\tilde{p}_1 = \frac{\partial\tilde{q}}{\partial r}, \quad -i\omega\left(1 + i\frac{r-a}{r}\frac{\sigma}{\omega}\right)\tilde{p}_2 = \frac{1}{r}\frac{\partial\tilde{q}}{\partial\theta},$$

$$-i\omega\left(1 + i\frac{\sigma}{\omega}\right)\tilde{q}_1 = \frac{\partial\tilde{p}_1}{\partial r},$$

for $r > a$. We notice that

$$i\omega\frac{\tilde{r} - (1 + i\frac{\sigma}{\omega})r}{r} = \frac{\sigma a}{r}.$$

Returning to the original independent variable (t, r, θ), we get the equations in the PML domain,

$$\frac{\partial q}{\partial t} + \frac{r-a}{r}\sigma q + \frac{\sigma a}{r}q_1 = \frac{1}{r}\left(\frac{\partial}{\partial r}(rp_1) + \frac{\partial p_2}{\partial\theta}\right),$$

$$\frac{\partial p_1}{\partial t} + \sigma p_1 = \frac{\partial q}{\partial r},$$

$$\frac{\partial p_2}{\partial t} + \frac{r-a}{r}\sigma p_2 = \frac{1}{r}\frac{\partial q}{\partial \theta},$$

$$\frac{\partial q_1}{\partial t} + \sigma q_1 = \frac{\partial p_1}{\partial r}.$$

5.3 Stability analysis

The system (5.26),(5.27) is not symmetric. It is not trivial to see if it is stable. Using the energy estimate, we study the problem in this section. We begin with a simpler case: the system (5.24),(5.25) with initial value imposed on $\mathbb{R}^2$. This system is equivalent to the system (5.13) plus an additional equation:

$$\frac{\partial q}{\partial t} = \frac{\partial p_1}{\partial x_1} + \frac{\partial p_2}{\partial x_2}, \tag{5.36}$$

$$\frac{\partial p_1}{\partial t} = \frac{\partial q}{\partial x_1}, \qquad \frac{\partial p_2}{\partial t} = \frac{\partial q}{\partial x_2}. \tag{5.37}$$

$$\frac{\partial q_1}{\partial t} = \frac{\partial p_1}{\partial x_1}. \tag{5.38}$$

Let us define the energy norm. Let $U = (q, p_1, p_2)^T$, and

$$E(t) = \int_{\mathbb{R}^2} \left(\left|\frac{\partial U}{\partial t}\right|^2 + \left|\frac{\partial U}{\partial x_1}\right|^2 + \left|\frac{\partial U}{\partial x_2}\right|^2 \right) dx,$$

and

$$\|q_1(t)\|_0^2 = \int_{\mathbb{R}^2} q_1^2 \, dx.$$

Lemma 67. *If the solutions to the system (5.36)-(5.38) are* $U \in (C([0,\infty); H^1(\mathbb{R}^2)))^3 \cap (C^1([0,\infty); L^2(\mathbb{R}^2)))^3$, $q_1 \in C([0,\infty); L^2(\mathbb{R}^2))$, *then the following are valid:*

$$E(t) = E(0), \tag{5.39}$$

$$\|q_1(t)\|_0 \leq \|q_1(0)\|_0 + \int_0^t E(\tau)^{\frac{1}{2}}\, d\tau. \tag{5.40}$$

Proof. We may assume that the solutions are sufficiently smooth, then use a series of smooth functions tending to the solutions to achieve the conclusion. Taking first order derivatives in the equations (5.36),(5.37) gives

$$\frac{\partial^2 q}{\partial t^2} = \frac{\partial^2 p_1}{\partial x_1 \partial t} + \frac{\partial^2 p_2}{\partial x_2 \partial t} = \frac{\partial^2 q}{\partial x_1^2} + \frac{\partial^2 q}{\partial x_2^2},$$

$$\frac{\partial^2 p_1}{\partial t^2} = \frac{\partial^2 q}{\partial x_1 \partial t} = \frac{\partial^2 p_1}{\partial x_1^2} + \frac{\partial^2 p_2}{\partial x_1 \partial x_2},$$

$$\frac{\partial^2 p_2}{\partial t^2} = \frac{\partial^2 q}{\partial x_2 \partial t} = \frac{\partial^2 p_1}{\partial x_1 \partial x_2} + \frac{\partial^2 p_2}{\partial x_2^2}.$$

Multiplying the equations by $\frac{\partial q}{\partial t}$, $\frac{\partial p_1}{\partial t}$, and $\frac{\partial p_2}{\partial t}$ respectively, then integrating on $\mathbb{R}^2$, we get

$$\frac{1}{2}\frac{d}{dt}\int_{\mathbb{R}^2}\left(\frac{\partial q}{\partial t}\right)^2 dx = -\frac{1}{2}\frac{d}{dt}\int_{\mathbb{R}^2}\left(\frac{\partial q}{\partial x_1}\right)^2 dx - \frac{1}{2}\frac{d}{dt}\int_{\mathbb{R}^2}\left(\frac{\partial q}{\partial x_2}\right)^2 dx,$$

$$\frac{1}{2}\frac{d}{dt}\int_{\mathbb{R}^2}\left(\frac{\partial p_1}{\partial t}\right)^2 dx = -\frac{1}{2}\frac{d}{dt}\int_{\mathbb{R}^2}\left(\frac{\partial p_1}{\partial x_1}\right)^2 dx + \int_{\mathbb{R}^2}\frac{\partial p_1}{\partial t}\frac{\partial^2 p_2}{\partial x_1 \partial x_2}\, dx,$$

$$\frac{1}{2}\frac{d}{dt}\int_{\mathbb{R}^2}\left(\frac{\partial p_2}{\partial t}\right)^2 dx = \int_{\mathbb{R}^2}\frac{\partial p_2}{\partial t}\frac{\partial^2 p_1}{\partial x_1 \partial x_2}\, dx - \frac{1}{2}\frac{d}{dt}\int_{\mathbb{R}^2}\left(\frac{\partial p_2}{\partial x_2}\right)^2 dx.$$

Applying the equations (5.37) and integrating by parts gives

$$\int_{\mathbb{R}^2}\frac{\partial p_1}{\partial t}\frac{\partial^2 p_2}{\partial x_1 \partial x_2}\, dx = -\int_{\mathbb{R}^2}\frac{\partial^2 p_1}{\partial x_2 \partial t}\frac{\partial p_2}{\partial x_1}\, dx = -\int_{\mathbb{R}^2}\frac{\partial^2 q}{\partial x_1 \partial x_2}\frac{\partial p_2}{\partial x_1}\, dx$$

$$= -\int_{\mathbb{R}^2}\frac{\partial^2 p_2}{\partial x_1 \partial t}\frac{\partial p_2}{\partial x_1}\, dx = -\frac{1}{2}\frac{d}{dt}\int_{\mathbb{R}^2}\left(\frac{\partial p_2}{\partial x_1}\right)^2 dx.$$

Analogously it holds that

$$\int_{\mathbb{R}^2}\frac{\partial p_2}{\partial t}\frac{\partial^2 p_1}{\partial x_1 \partial x_2}\, dx = -\frac{1}{2}\frac{d}{dt}\int_{\mathbb{R}^2}\left(\frac{\partial p_1}{\partial x_2}\right)^2 dx.$$

Adding these three equations together yields

$$\frac{1}{2}\frac{dE(t)}{dt} = 0,$$

which implies (5.39). Then (5.40) is a direct consequence of the equation (5.38). □

The estimate (5.40) is classified as a "weak stability", since the L^2-norms of q_1, q_2 are not only bounded by the L^2-norms of the initial data, but also by the L^2-norms of the derivatives of the initial data. This property may cause instabilities in numerical computation.

The energy estimate for the system (5.26),(5.27) is more technical. The Gronwall inequality is needed here:

Theorem 67. *If a function $E(t) \geq 0$ for $t \geq 0$, and satisfies*

$$E(t) \leq a \int_0^t E(\tau)\, d\tau + b, \tag{5.41}$$

with two constants a, b, then $E(t) \leq be^{at}$ for $t \geq 0$.

Proof. We multiply the inequality (5.41) by e^{-at} and get

$$e^{-at}\left(E(t) - a\int_0^t E(\tau)\, d\tau\right) \leq be^{-at}.$$

It is

$$\partial_t\left(e^{-at}\int_0^t E(\tau)\, d\tau\right) \leq be^{-at}.$$

Then we have

$$e^{-at}\int_0^t E(\tau)\, d\tau \leq b\int_0^t e^{-a\tau}\, d\tau = \frac{b}{a}(1 - e^{-at}).$$

We substitute it into (5.41) and the conclusion follows. □

The system (5.26),(5.27) is equivalent to

$$\frac{\partial q}{\partial t} + \sigma q_1 = \frac{\partial p_1}{\partial x_1} + \frac{\partial p_2}{\partial x_2}, \tag{5.42}$$

$$\frac{\partial p_1}{\partial t} + \sigma p_1 = \frac{\partial q}{\partial x_1}, \qquad \frac{\partial p_2}{\partial t} = \frac{\partial q}{\partial x_2}. \tag{5.43}$$

$$\frac{\partial q_1}{\partial t} + \sigma q_1 = \frac{\partial p_1}{\partial x_1}. \tag{5.44}$$

Theorem 68. *If the solutions to the system (5.42)-(5.44) are* $U \in \left(C([0,\infty); H^1(\mathbb{R}^2))\right)^3 \cap \left(C^1([0,\infty); L^2(\mathbb{R}^2))\right)^3$, $q_1 \in C([0,\infty); L^2(\mathbb{R}^2))$, *then the following are valid:*

$$E(t) \le C(\sigma) e^{C\sigma t} (E(0) + \|q_1(0)\|_0^2), \tag{5.45}$$

$$\|q_1(t)\|_0 \le e^{-\sigma t} \|q_1(0)\|_0 + \int_0^t E(\tau)^{\frac{1}{2}} \, d\tau, \tag{5.46}$$

where $C(\sigma)$ *depends on* σ.

Proof. Being analogous to the previous lemma we obtain equations,

$$\frac{\partial^2 q}{\partial t^2} = \frac{\partial^2 q}{\partial x_1^2} + \frac{\partial^2 q}{\partial x_2^2} - 2\sigma \frac{\partial p_1}{\partial x_1} + \sigma^2 q_1,$$

$$\frac{\partial^2 p_1}{\partial t^2} = \frac{\partial^2 p_1}{\partial x_1^2} + \frac{\partial^2 p_2}{\partial x_1 \partial x_2} - \sigma \frac{\partial q_1}{\partial x_1} - \sigma \frac{\partial p_1}{\partial t},$$

$$\frac{\partial^2 p_2}{\partial t^2} = \frac{\partial^2 p_1}{\partial x_1 \partial x_2} + \frac{\partial^2 p_2}{\partial x_2^2} - \sigma \frac{\partial q_1}{\partial x_2}.$$

Multiplying the equations by $\frac{\partial q}{\partial t}$, $\frac{\partial p_1}{\partial t}$, and $\frac{\partial p_2}{\partial t}$ respectively, then integrating on $\mathbb{R}^2$, we get

$$\frac{1}{2} \frac{dE(t)}{dt} = \int_{\mathbb{R}^2} \left\{ -2\sigma \frac{\partial p_1}{\partial x_1} \frac{\partial q}{\partial t} + \sigma^2 q_1 \frac{\partial q}{\partial t} - \sigma \frac{\partial q_1}{\partial x_1} \frac{\partial p_1}{\partial t} \right.$$
$$\left. - \sigma \left(\frac{\partial p_1}{\partial t}\right)^2 - \sigma \frac{\partial q_1}{\partial x_2} \frac{\partial p_2}{\partial t} + \sigma \frac{\partial p_2}{\partial x_1 \partial x_2} \frac{\partial p_1}{\partial x_1 \partial x_2} - \sigma \left(\frac{\partial p_1}{\partial x_2}\right)^2 \right\} dx \cong \sum_{k=1}^{7} I_k.$$

Then we have

$$\frac{1}{2}(E(t) - E(0)) = \sum_{k=1}^{7} \int_0^t I_k. \tag{5.47}$$

Let us estimate the terms on the right hand side of (5.47). We have the following:

$$-2\sigma \int_0^t \int_{\mathbb{R}^2} \frac{\partial p_1}{\partial x_1} \frac{\partial q}{\partial t} \, dx \, dt \leq \sigma \int_0^t \int_{\mathbb{R}^2} \left(\frac{\partial p_1}{\partial x_1}\right)^2 dx \, dt + \sigma \int_0^t \int_{\mathbb{R}^2} \left(\frac{\partial q}{\partial t}\right)^2 dx \, dt,$$

$$\sigma^2 \int_0^t \int_{\mathbb{R}^2} q_1 \frac{\partial q}{\partial t} \, dx \, dt \leq \frac{\sigma^3}{2} \int_0^t \int_{\mathbb{R}^2} q_1^2 \, dx \, dt + \frac{\sigma}{2} \int_0^t \int_{\mathbb{R}^2} \left(\frac{\partial q}{\partial t}\right)^2 dx \, dt,$$

$$\sigma \int_0^t \int_{\mathbb{R}^2} \frac{\partial p_2}{\partial x_1} \frac{\partial p_1}{\partial x_2} \, dx \, dt \leq \frac{\sigma}{2} \int_0^t \int_{\mathbb{R}^2} \left(\frac{\partial p_2}{\partial x_1}\right)^2 dx \, dt + \frac{\sigma}{2} \int_0^t \int_{\mathbb{R}^2} \left(\frac{\partial p_1}{\partial x_2}\right)^2 dx \, dt.$$

Noting the equation (5.44) and taking integration by parts, we obtain

$$- \sigma \int_0^t \int_{\mathbb{R}^2} \frac{\partial q_1}{\partial x_1} \frac{\partial p_1}{\partial t} \, dx \, dt$$

$$= - \sigma \int_{\mathbb{R}^2} \frac{\partial q_1}{\partial x_1} p_1 \, dx \Big|_0^t + \sigma \int_0^t \int_{\mathbb{R}^2} \frac{\partial^2 q_1}{\partial x_1 \partial t} p_1 \, dx \, dt$$

$$= \sigma \int_{\mathbb{R}^2} q_1 \frac{\partial p_1}{\partial x_1} \, dx \Big|_0^t + \sigma \int_0^t \int_{\mathbb{R}^2} \left(-\sigma \frac{\partial q_1}{\partial x_1} + \frac{\partial^2 p_1}{\partial x_1^2}\right) p_1 \, dx \, dt$$

$$= \sigma \int_{\mathbb{R}^2} q_1 \frac{\partial p_1}{\partial x_1} \, dx \Big|_0^t + \sigma \int_0^t \int_{\mathbb{R}^2} \left(\sigma q_1 \frac{\partial p_1}{\partial x_1} - \left(\frac{\partial p_1}{\partial x_1}\right)^2\right) dx \, dt$$

$$\leq \sigma \left\{\frac{1}{2\varepsilon} \int_{\mathbb{R}^2} q_1^2 \, dx + \frac{\varepsilon}{2} \int_{\mathbb{R}^2} \left(\frac{\partial p_1}{\partial x_1}\right)^2 dx\right\} \Big|_t$$

$$+ \left\{\frac{\sigma}{2} \int_{\mathbb{R}^2} q_1^2 \, dx + \frac{\sigma}{2} \int_{\mathbb{R}^2} \left(\frac{\partial p_1}{\partial x_1}\right)^2 dx\right\} \Big|_0 + \frac{\sigma^3}{4} \int_0^t \int_{\mathbb{R}^2} q_1^2 \, dx \, dt,$$

where $\varepsilon > 0$ is a small constant. Analogously we have

$$- \sigma \int_0^t \int_{\mathbb{R}^2} \frac{\partial q_1}{\partial x_2} \frac{\partial p_2}{\partial t} \, dx \, dt$$

$$= - \sigma \int_{\mathbb{R}^2} \frac{\partial q_1}{\partial x_2} p_2 \, dx \Big|_0^t + \sigma \int_0^t \int_{\mathbb{R}^2} \frac{\partial^2 q_1}{\partial x_2 \partial t} p_2 \, dx \, dt$$

$$= \sigma \int_{\mathbb{R}^2} q_1 \frac{\partial p_2}{\partial x_2} \, dx \Big|_0^t + \sigma \int_0^t \int_{\mathbb{R}^2} \left(-\sigma \frac{\partial q_1}{\partial x_2} + \frac{\partial^2 p_1}{\partial x_1 \partial x_2}\right) p_2 \, dx \, dt$$

$$= \sigma \int_{\mathbb{R}^2} q_1 \frac{\partial p_2}{\partial x_2} \, dx \Big|_0^t + \sigma \int_0^t \int_{\mathbb{R}^2} \left(\sigma q_1 \frac{\partial p_2}{\partial x_2} - \frac{\partial p_1}{\partial x_1} \frac{\partial p_2}{\partial x_2}\right) dx \, dt$$

$$\leq \sigma \left\{ \frac{1}{2\varepsilon} \int_{\mathbb{R}^2} q_1^2 \, dx + \frac{\varepsilon}{2} \int_{\mathbb{R}^2} \left(\frac{\partial p_2}{\partial x_2} \right)^2 dx \right\} \Big|_t$$

$$+ \sigma \left\{ \frac{1}{2} \int_{\mathbb{R}^2} q_1^2 \, dx + \frac{1}{2} \int_{\mathbb{R}^2} \left(\frac{\partial p_2}{\partial x_2} \right)^2 dx \right\} \Big|_0 + \frac{\sigma^3}{4} \int_0^t \int_{\mathbb{R}^2} q_1^2 \, dx \, dt$$

$$+ \sigma \int_0^t \int_{\mathbb{R}^2} \left(\frac{\partial p_1}{\partial x_1} \right)^2 dx \, dt + \frac{5\sigma}{4} \int_0^t \int_{\mathbb{R}^2} \left(\frac{\partial p_2}{\partial x_2} \right)^2 dx \, dt.$$

We plug these estimates in (5.47) and obtain

$$(1 - \varepsilon\sigma) E(t) \leq (1 + \sigma) E(0) + 2\sigma \|q_1(0)\|_0^2 + 4\sigma \int_0^t E(\tau) \, d\tau \tag{5.48}$$

$$+ 2\sigma^3 \int_0^t \int_{\mathbb{R}^2} q_1^2 \, dt + \frac{2\sigma}{\varepsilon} \int_{\mathbb{R}^2} q_1^2 \, dx.$$

We apply the equation (5.44) to estimate q_1.

$$q_1(x,t) = q_1(x,0)e^{-\sigma t} + \int_0^t e^{-\sigma(t-\tau)} \frac{\partial p_1(x,\tau)}{\partial x_1} \, d\tau.$$

Therefore we obtain

$$\|q_1(t)\|_0 \leq \|q_1(0)\|_0 e^{-\sigma t} + \int_0^t e^{-\sigma(t-\tau)} \left\| \frac{\partial p_1}{\partial x_1}(\tau) \right\|_0 d\tau, \tag{5.49}$$

and by the Schwarz inequality we get

$$\|q_1(t)\|_0^2 \leq 2\|q_1(0)\|_0^2 e^{-2\sigma t} + 2 \left(\int_0^t e^{-\sigma(t-\tau)} \left\| \frac{\partial p_1}{\partial x_1}(\tau) \right\|_0 d\tau \right)^2$$

$$\leq 2\|q_1(0)\|_0^2 e^{-2\sigma t} + 2 \int_0^t e^{-\sigma(t-\tau)} \, d\tau \int_0^t e^{-\sigma(t-\tau)} \left\| \frac{\partial p_1}{\partial x_1}(\tau) \right\|_0^2 d\tau$$

$$\leq 2\|q_1(0)\|_0^2 e^{-2\sigma t} + \frac{2}{\sigma} \int_0^t e^{-\sigma(t-\tau)} \left\| \frac{\partial p_1}{\partial x_1}(\tau) \right\|_0^2 d\tau.$$

Then we have

$$\int_0^t \|q_1(s)\|_0^2 \, ds \leq \frac{1}{\sigma} \|q_1(0)\|_0^2 + \frac{2}{\sigma} \int_0^t ds \int_0^s e^{-\sigma(s-\tau)} \left\| \frac{\partial p_1}{\partial x_1}(\tau) \right\|_0^2 d\tau$$

$$= \frac{1}{\sigma} \|q_1(0)\|_0^2 + \frac{2}{\sigma} \int_0^t \left\| \frac{\partial p_1}{\partial x_1}(\tau) \right\|_0^2 d\tau \int_\tau^t e^{-\sigma(s-\tau)} \, ds$$

$$\leq \frac{1}{\sigma} \|q_1(0)\|_0^2 + \frac{2}{\sigma^2} \int_0^t \left\| \frac{\partial p_1}{\partial x_1}(\tau) \right\|_0^2 d\tau.$$

We plug these inequalities into (5.48) and set $\varepsilon = 1/2\sigma$ to obtain

$$\frac{1}{2}E(t) \leq (1+\sigma)E(0) + 16\sigma \int_0^t E(\tau)\,d\tau + (2\sigma + 10\sigma^2)\|q_1(0)\|_0^2.$$

Then applying the Gronwall inequality we obtain (5.45), and (5.46) follows from (5.49). □

The fact of weak stability may cause unstable finite difference schemes. The second order Yee's scheme in Section 4.7 is stable for Maxwell equations. However following an analysis of [Abarbanel, S. and Gottlieb, D. (1997)] we will show that it is unstable as it is applied to the system (5.24),(5.25). The system of finite difference equations is

$$p_1^{n+1} = p_1^n + \Delta t \delta_1(q_1^{n+\frac{1}{2}} + q_2^{n+\frac{1}{2}}), \qquad p_2^{n+1} = p_2^n + \Delta t \delta_2(q_1^{n+\frac{1}{2}} + q_2^{n+\frac{1}{2}}),$$

$$q_1^{n+\frac{1}{2}} = q_1^{n-\frac{1}{2}} + \Delta t \delta_1 p_1^n, \qquad q_2^{n+\frac{1}{2}} = q_2^{n-\frac{1}{2}} + \Delta t \delta_2 p_2^n,$$

where $p_\alpha^n = p_\alpha|_{t=n\Delta t}$, $q_\alpha^{n+\frac{1}{2}} = q_\alpha|_{t=(n+\frac{1}{2})\Delta t}$, $\alpha = 1, 2$, and δ_α is the finite difference operator $\Delta/\Delta x_\alpha$. The matrix form of the system is

$$\begin{pmatrix} 1 & 0 & -\Delta t\delta_1 & -\Delta t\delta_1 \\ 0 & 1 & -\Delta t\delta_2 & -\Delta t\delta_2 \\ 0 & 0 & 1 & 0 \\ 0 & 0 & 0 & 1 \end{pmatrix} \begin{pmatrix} p_1^{n+1} \\ p_2^{n+1} \\ q_1^{n+\frac{1}{2}} \\ q_2^{n+\frac{1}{2}} \end{pmatrix} = \begin{pmatrix} 1 & 0 & 0 & 0 \\ 0 & 1 & 0 & 0 \\ \Delta t\delta_1 & 0 & 1 & 0 \\ 0 & \Delta t\delta_2 & 0 & 1 \end{pmatrix} \begin{pmatrix} p_1^n \\ p_2^n \\ q_1^{n-\frac{1}{2}} \\ q_2^{n-\frac{1}{2}} \end{pmatrix}.$$

The Fourier symbols of δ_1 and δ_2 are, respectively

$$\delta_1 \to \frac{2i\sin\theta}{\Delta x_1} = i\frac{k_1}{\Delta t}, \qquad \delta_2 \to \frac{2i\sin\phi}{\Delta x_2} = i\frac{k_2}{\Delta t},$$

where $-\pi/2 \leq \theta, \phi \leq \pi/2$. The amplification matrix is

$$G = \begin{pmatrix} 1 & 0 & -ik_1 & -ik_1 \\ 0 & 1 & -ik_2 & -ik_2 \\ 0 & 0 & 1 & 0 \\ 0 & 0 & 0 & 1 \end{pmatrix}^{-1} \begin{pmatrix} 1 & 0 & 0 & 0 \\ 0 & 1 & 0 & 0 \\ ik_1 & 0 & 1 & 0 \\ 0 & ik_2 & 0 & 1 \end{pmatrix} = \begin{pmatrix} 1 - k_1^2 & -k_1 k_2 & ik_1 & ik_1 \\ -k_1 k_2 & 1 - k_2^2 & ik_2 & ik_2 \\ ik_1 & 0 & 1 & 0 \\ 0 & ik_2 & 0 & 1 \end{pmatrix}$$

A necessary, but not sufficient, condition for the stability of the difference scheme is that all the eigenvalues of G are less than or equal to 1 in magnitude (the von Neŭmann condition). The eigenvalues λ's of G are

$$\lambda = 1, 1, 1 - \frac{\gamma}{2} \pm \sqrt{\frac{\gamma^2}{4} - \gamma} = 1, 1, \lambda_1, \lambda_2,$$

where $\gamma = k_1^2 + k_2^2$. Under the CFL condition, $\gamma \leq 4$, then $\lambda's$ satisfy $|\lambda| = 1$, so the scheme meets the necessary condition of stability. However it is unstable. The Jordan canonical form of G is:

$$J = \begin{pmatrix} \lambda_1 & 0 & 0 & 0 \\ 0 & \lambda_2 & 0 & 0 \\ 0 & 0 & 1 & 1 \\ 0 & 0 & 0 & 1 \end{pmatrix},$$

That is, $G = TJT^{-1}$, where T is the "Jordanizing" matrix. After n steps, we have $G^n = TJ^nT^{-1}$, and

$$J^n = \begin{pmatrix} \lambda_1^n & 0 & 0 & 0 \\ 0 & \lambda_2^n & 0 & 0 \\ 0 & 0 & 1 & n \\ 0 & 0 & 0 & 1 \end{pmatrix}.$$

The L^2-norm of J^n is

$$\|J^n\| = \max_{|\xi|=1} \|J^n\xi\| = (n^2 + 1)^{\frac{1}{2}}.$$

Thus $\|G^n\|$ grows with n without a finite bound, and the scheme is unstable.

5.4 Uniaxial perfectly matched layers

Since Bérenger's perfectly matched layer method yields a system which is weakly stable only, it is desirable to replace it to a form without splitting the field. Let

$$s = \begin{cases} 1 + i\frac{\sigma}{\omega}, & x_1 > a, \\ 1, & x_1 < a, \end{cases}$$

then the equations (5.28),(5.29) are equivalent to

$$-i\omega\tilde{q} = \frac{1}{s}\frac{\partial\tilde{p}_1}{\partial x_1} + i\xi_2\tilde{p}_2, \tag{5.50}$$

$$-i\omega\tilde{p}_1 = \frac{1}{s}\frac{\partial\tilde{q}}{\partial x_1}, \qquad -i\omega\tilde{p}_2 = i\xi_2\tilde{q}, \tag{5.51}$$

where $\tilde{q} = \tilde{q}_1 + \tilde{q}_2$. The important matter is to derive the corresponding equations in the physical domain. To this end, introduce two variables

$$\tilde{P} = \frac{\tilde{p}_1}{s}, \qquad \tilde{Q} = \frac{\tilde{q}}{s},$$

and replace the equations (5.50), (5.51) by

$$-i\omega\tilde{q} = \frac{\partial \tilde{P}}{\partial x_1} + i\xi_2\tilde{p}_2,$$ (5.52)

$$-i\omega\tilde{p}_1 = \frac{\partial \tilde{Q}}{\partial x_1}, \qquad -i\omega\tilde{p}_2 = i\xi_2\tilde{q}.$$ (5.53)

Taking inverse Fourier transform, we get the equations on PML:

$$\frac{\partial q}{\partial t} = \frac{\partial P}{\partial x_1} + \frac{\partial p_2}{\partial x_2},$$ (5.54)

$$\frac{\partial p_1}{\partial t} = \frac{\partial Q}{\partial x_1}, \qquad \frac{\partial p_2}{\partial t} = \frac{\partial q}{\partial x_2},$$ (5.55)

$$\frac{\partial q}{\partial t} = \frac{\partial Q}{\partial t} + \sigma Q, \qquad \frac{\partial p_1}{\partial t} = \frac{\partial P}{\partial t} + \sigma P.$$ (5.56)

This is the "uniaxial perfectly matched layer" (UPML).

We remark that the system (5.52),(5.53) is different from (5.50),(5.51). Although they are the same for $x_1 < a$ and $x_1 > a$, s is a piecewise constant function of x_1, so the interface boundary conditions at $x_1 = a$ for them are different. Let us derive the interface boundary conditions to make a comparison. We multiply (5.50) by $-i\omega$ and obtain

$$-\omega^2\tilde{q} = -i\omega\frac{1}{s}\frac{\partial \tilde{p}_1}{\partial x_1} + \omega\xi_2\tilde{p}_2,$$

then by (5.51) we get the equation for $\tilde{q}$,

$$-\omega^2\tilde{q} = \frac{1}{s}\frac{\partial}{\partial x_1}\left(\frac{1}{s}\frac{\partial \tilde{q}}{\partial x_1}\right) - \xi_2^2\tilde{q}.$$ (5.57)

On the other hand by (5.52),(5.53) we can get

$$-\omega^2\tilde{q} = \frac{\partial}{\partial x_1}\left(\frac{1}{s}\frac{\partial}{\partial x_1}\left(\frac{\tilde{q}}{s}\right)\right) - \xi_2^2\tilde{q}.$$ (5.58)

There are two first order differential equations for $\tilde{p}_1$ and $\tilde{q}$, and one algebraic equation, in (5.50),(5.51), therefore the interface condition is:

$$\tilde{p}_1|_{x_1=a+0} = \tilde{p}_1|_{x_1=a-0}, \qquad \tilde{q}|_{x_1=a+0} = \tilde{q}|_{x_1=a-0}.$$

On the other hand the interface condition for (5.52),(5.53) is:

$$\tilde{P}|_{x_1=a+0} = \tilde{P}|_{x_1=a-0}, \qquad \tilde{Q}|_{x_1=a+0} = \tilde{Q}|_{x_1=a-0}.$$

Therefore (5.52),(5.53) represents a model different from (5.50),(5.51). This fact has to be taken into account when a finite difference scheme, or a finite element scheme, is designed on the entire truncated domain, including the PML domain.

5.5 Maxwell equations

For two dimensional TM mode (4.18), if we replace $(\varepsilon\mu)^{\frac{1}{2}}x$ by x and set $q = \left(\frac{\varepsilon}{\mu}\right)^{\frac{1}{2}} E_3$, $p_1 = H_2$, and $p_2 = -H_1$, then it is identity to (5.13). Analogously, for the TE mode (4.19) we replace $(\varepsilon\mu)^{\frac{1}{2}}x$ by x and set $q = -\left(\frac{\mu}{\varepsilon}\right)^{\frac{1}{2}} H_3$, $p_1 = -E_2$, and $p_2 = E_1$, then it is identity to (5.13), too. Therefore the above approaches can be applied.

We turn now to consider three dimensional Maxwell equations. Let Ω_0 be a cube $\{x; |x_k| < a, k = 1, 2, 3\}$ with the constant a large enough so that $\Omega_0 \supset \overline{\Omega^c}$. Then take $d > 0$ and define a cube, $\Omega_d = \{x; |x_k| < a + d, k = 1, 2, 3\}$. $\Omega_d \setminus \overline{\Omega_0}$ is the PML domain.

For homogeneous Maxwell equations, E and H are split into: $E_1 = E_{12} + E_{13}$, $E_2 = E_{23} + E_{21}$, $E_3 = E_{31} + E_{32}$, $H_1 = H_{12} + H_{13}$, $H_2 = H_{23} + H_{21}$, and $H_3 = H_{31} + H_{32}$. Introducing damping factors $\sigma_k, \sigma_k^* > 0$, $k = 1, 2, 3$, for $|x_k| \in (a, a + d)$, with the relationship of

$$\frac{\sigma_k}{\varepsilon} = \frac{\sigma_k^*}{\mu}.$$

Let us assume $\sigma_k = \sigma_k^* = 0$ for $|x_k| < a$, then the system of Bérenger's perfectly matched layer scheme is defined by

$$\left(\varepsilon\frac{\partial}{\partial t} + \sigma_2\right) E_{12} = \frac{\partial}{\partial x_2}(H_{31} + H_{32}),$$

$$\left(\varepsilon\frac{\partial}{\partial t} + \sigma_3\right) E_{13} = -\frac{\partial}{\partial x_3}(H_{23} + H_{21}),$$

$$\left(\varepsilon\frac{\partial}{\partial t} + \sigma_3\right) E_{23} = \frac{\partial}{\partial x_3}(H_{12} + H_{13}),$$

$$\left(\varepsilon\frac{\partial}{\partial t} + \sigma_1\right) E_{21} = -\frac{\partial}{\partial x_1}(H_{31} + H_{32}),$$

$$\left(\varepsilon\frac{\partial}{\partial t} + \sigma_1\right) E_{31} = \frac{\partial}{\partial x_1}(H_{23} + H_{21}),$$

$$\left(\varepsilon\frac{\partial}{\partial t} + \sigma_2\right) E_{32} = -\frac{\partial}{\partial x_2}(H_{12} + H_{13}),$$

$$\left(\mu\frac{\partial}{\partial t} + \sigma_2^*\right) H_{12} = -\frac{\partial}{\partial x_2}(E_{31} + E_{32}),$$

$$\left(\mu\frac{\partial}{\partial t} + \sigma_3^*\right) H_{13} = \frac{\partial}{\partial x_3}(E_{23} + E_{21}),$$

$$\left(\mu\frac{\partial}{\partial t} + \sigma_3^*\right) H_{23} = -\frac{\partial}{\partial x_3}(E_{12} + E_{13}),$$

$$\left(\mu\frac{\partial}{\partial t} + \sigma_1^*\right) H_{21} = \frac{\partial}{\partial x_1}(E_{31} + E_{32}),$$

$$\left(\mu\frac{\partial}{\partial t} + \sigma_1^*\right) H_{31} = -\frac{\partial}{\partial x_1}(E_{23} + E_{21}),$$

$$\left(\mu\frac{\partial}{\partial t} + \sigma_2^*\right) H_{32} = \frac{\partial}{\partial x_2}(E_{12} + E_{13}).$$

Let

$$s_k = 1 + i\frac{\sigma_k}{\varepsilon\omega} = 1 + i\frac{\sigma_k^*}{\mu\omega}.$$

We define Fourier transform with respect to t: $\tilde{E}(x, -\omega) = F(E(x,t))$, and $\tilde{H}(x, -\omega) = F(H(x,t))$. Define new independent variables $\tilde{x}_k = \int_0^{x_k} s_k\, dx_k$, and

$$\tilde{\nabla} = \left(\frac{\partial}{\partial\tilde{x}_1}, \frac{\partial}{\partial\tilde{x}_2}, \frac{\partial}{\partial\tilde{x}_3}\right).$$

Analogous to (5.50)(5.51) the equations can be written by

$$-i\omega\varepsilon\tilde{E} = \tilde{\nabla} \times \tilde{H}, \tag{5.59}$$

$$-i\omega\mu\tilde{H} = -\tilde{\nabla} \times \tilde{E}, \tag{5.60}$$

without splitting the field.

We turn now to the UPML scheme. Define $\hat{E}_k = s_k\tilde{E}_k$ and $\hat{H}_k = s_k\tilde{H}_k$. $\hat{E}_k$ and $\tilde{E}_k$, $\hat{E}_k$ and $\tilde{E}_k$, are the same if $s_k = 1$, but they are different in the PML. Replace the equations (5.59) and (5.60) by

$$-i\omega\varepsilon \begin{pmatrix} s_2 s_3 s_1^{-1} & 0 & 0 \\ 0 & s_3 s_1 s_2^{-1} & 0 \\ 0 & 0 & s_1 s_2 s_3^{-1} \end{pmatrix} \hat{E} = \nabla \times \hat{H}, \tag{5.61}$$

$$-i\omega\mu \begin{pmatrix} s_2 s_3 s_1^{-1} & 0 & 0 \\ 0 & s_3 s_1 s_2^{-1} & 0 \\ 0 & 0 & s_1 s_2 s_3^{-1} \end{pmatrix} \hat{H} = -\nabla \times \hat{E}. \tag{5.62}$$

In order to take inverse Fourier transform we introduce some new variables. For example for the first row we set $\hat{D}_1 = \varepsilon s_3 s_1^{-1} \hat{E}_1$, then the equations are

$$\frac{\partial D_1}{\partial t} + \sigma_2 D_1 = \frac{\partial H_3}{\partial x_2} - \frac{\partial H_2}{\partial x_3},$$

$$\frac{\partial D_1}{\partial t} + \frac{\sigma_1}{\varepsilon} D_1 = \varepsilon \frac{\partial E_1}{\partial t} + \sigma_3 E_1.$$

5.6   Helmholtz equations

Following the analysis by [Collino, F. and Monk, P.B. (1998)] we consider the exterior boundary value problem of the Helmholtz equations:

$$(\triangle + \omega^2)u = f, \qquad x \in \Omega, \tag{5.63}$$

with the boundary conditions,

$$u = 0, \qquad x \in \partial\Omega, \tag{5.64}$$

$$\frac{\partial u}{\partial r} - i\omega u = O(\frac{1}{r^2}), \tag{5.65}$$

in this section. We assume that $f \in L^2(\Omega)$, and it is compactly supported. Let R be large enough, so that $B(O, R) \supset \operatorname{supp} f$. The solution for $|x| > R$ is thus expressed by (2.16). Introduce a complex radius $\tilde{r} = \int_0^r \alpha(\rho)\, d\rho = r\beta(r)$, where $\alpha(r) = 1 + i\sigma(r)$, and the real function $\sigma(r)$ satisfies

$$\sigma(r) \begin{cases} = 0, r \leq R, \\ > 0, r > R, \end{cases} \quad \lim_{r \to \infty} \int_R^r \sigma(\rho)\, d\rho = \infty.$$

We replace r in (2.16) by $\tilde{r}$ then we get another function,

$$\tilde{u} = \sum_{n=-\infty}^{\infty} a_n H_n^{(1)}(\omega\tilde{r})e^{in\theta}, \tag{5.66}$$

for $r > R$. By (2.12) the asymptotic behavior of the Hankel function is

$$H_n^{(1)}(\omega\tilde{r}) = \sqrt{\frac{2}{\pi\omega\tilde{r}}} e^{i\left(\omega\tilde{r} - \frac{n\pi}{2} - \frac{\pi}{4}\right)} + O(|\omega\tilde{r}|^{-3/2}),$$

and $i\omega\tilde{r} = i\omega r - \omega\operatorname{Im}\tilde{r}$. So $H_n^{(1)}(\omega\tilde{r})$ decays to zero as $r \to \infty$.

Obviously $\tilde{u}$ satisfies the equation

$$\frac{1}{\tilde{r}}\left(\frac{\partial}{\partial \tilde{r}}\left(\tilde{r}\frac{\partial \tilde{u}}{\partial \tilde{r}}\right) + \frac{1}{\tilde{r}}\frac{\partial^2 \tilde{u}}{\partial \theta^2}\right) + \omega^2 \tilde{u} = 0. \tag{5.67}$$

Returning to the original coordinates (r, θ), we have the equation,

$$\frac{1}{r}\left(\frac{\partial}{\partial r}\left(\frac{\beta}{\alpha}r\frac{\partial \tilde{u}}{\partial r}\right) + \frac{\alpha}{r\beta}\frac{\partial^2 \tilde{u}}{\partial \theta^2}\right) + \alpha\beta\omega^2 \tilde{u} = 0.$$

In Cartesian coordinates the equation is

$$\nabla \cdot (A\nabla \tilde{u}) + \alpha\beta\omega^2 \tilde{u} = f, \quad x \in \Omega, \tag{5.68}$$

where

$$A = \begin{pmatrix} \frac{\beta}{\alpha}\cos^2\theta + \frac{\alpha}{\beta}\sin^2\theta & \cos\theta\sin\theta\left(\frac{\beta}{\alpha} - \frac{\alpha}{\beta}\right) \\ \cos\theta\sin\theta\left(\frac{\beta}{\alpha} - \frac{\alpha}{\beta}\right) & \frac{\beta}{\alpha}\sin^2\theta + \frac{\alpha}{\beta}\cos^2\theta \end{pmatrix}.$$

The function $\tilde{u}$ is the solution to the equation (5.68) and the boundary condition (5.64), and it satisfies

$$\tilde{u} \text{ bounded as } |x| \to \infty, \tag{5.69}$$

by noting the above asymptotic behavior of the Hankel functions. Obviously we have

Lemma 68. *The solution $\tilde{u}$ of (5.68),(5.64),(5.69) is equal to u of the problem (5.63),(5.64),(5.65) on $B(O,R)\bigcap\Omega$.*

Moreover we have the following

Lemma 69. *The solution $\tilde{u}$ to the problem (5.68),(5.64),(5.69) is unique.*

Proof. If $f = 0$, by the Green's formula on the domain $\Omega_1 = \Omega \cap B(O,R)$, we get

$$\int_{|x|=R}\left\{\overline{\tilde{u}}\frac{\partial \tilde{u}}{\partial r} - \tilde{u}\frac{\overline{\partial \tilde{u}}}{\partial r}\right\} ds = 0. \tag{5.70}$$

For $r > R$, $\tilde{u}$ is the solution to (5.67), so it can be expressed in terms of Hankel functions,

$$\tilde{u}(r,\theta) = \sum_{-\infty}^{\infty}(a_n H_n^{(1)}(\omega\tilde{r}) + b_n H_n^{(2)}(\omega\tilde{r}))e^{in\theta}.$$

Since we assume that σ is chosen so that Im $(\omega \tilde{r}) \to \infty$ as $r \to \infty$, the Hankel functions $H_n^{(2)}(\omega \tilde{r})$ are unbounded. Therefore $b_n = 0$. By the integral (5.70) we get

$$\omega \sum_{n=-\infty}^{\infty} |a_n|^2 (H_n^{(1)}(\omega R) \overline{H_n^{(1)'}(\omega R)} - \overline{H_n^{(1)}(\omega R)} H_n^{(1)'}(\omega R)) = 0.$$

The Wronskian of Hankel functions does not vanish, so $a_n = 0$ for all n. Then $\tilde{u} = 0$ for $r > R$. $\tilde{u}$ is an analytic function, so $\tilde{u} = 0$ for $r < R$, too.
□

Since the solutions to (5.68),(5.64),(5.69) decay to zero as $|x| \to \infty$, the domain can be truncated. We assume that the thickness of the perfectly matched layer be d and denote $\Omega_d = \Omega \cap B(O, R + d)$ and $\Omega_2 = B(O, R + d) \setminus \overline{B(O, R)}$, then we get the approximate problem:

$$\nabla \cdot (A \nabla u_h) + \alpha \beta \omega^2 u_h = f, \quad x \in \Omega_d, \tag{5.71}$$

$$u_h = 0, \quad x \in \partial\Omega, \tag{5.72}$$

$$u_h = 0, \quad |x| = R + d. \tag{5.73}$$

We are going to investigate the well-posedness of the problem (5.71)-(5.73).

Theorem 69. *For almost all $\omega^2 \in (0, \infty)$, except for a point spectrum set σ_p', the problem (5.71)-(5.73) admits a unique solution.*

Proof. We consider the operator $L = -\nabla \cdot (A\nabla) + \lambda$. We claim that for positive λ the corresponding sesqui-linear form,

$$a(u_h, v) = (A\nabla u_h, \nabla v) + (\lambda u_h, v),$$

is coercive. To see that, for any $u \in H_0^1(\Omega_d)$,

$$|(A\nabla u, \nabla u)| \geq \text{Re}(A\nabla u, \nabla u) = \int_{\Omega_1} |\nabla u|^2 \, dx + \int_{\Omega_2} \text{Re}(\nabla u \cdot A\nabla u) \, dx.$$

By the expression of A,

$$\int_{\Omega_2} \text{Re}(\nabla u \cdot A\nabla u) \, dx$$

$$= \int_R^{R+d} \int_0^{2\pi} r \left(\text{Re}\left(\frac{\beta}{\alpha}\right) \left|\frac{\partial u}{\partial r}\right|^2 + \text{Re}\left(\frac{\alpha}{\beta r^2}\right) \left|\frac{\partial u}{\partial \theta}\right|^2 \right) d\theta dr.$$

It suffices to show that

$$\min(\text{Re}(\alpha/\beta), \text{Re}(\beta/\alpha)) \geq \delta > 0,$$

where δ is independent of r. Let $\beta(r) = 1 + i\bar{\sigma}(r)$, then

$$\text{Re}\left(\frac{\alpha}{\beta}\right) = \frac{1 + \sigma\bar{\sigma}}{1 + \bar{\sigma}^2} \geq \frac{1}{1 + \max_{R < r < R+d} \bar{\sigma}^2}.$$

$\text{Re}(\beta/\alpha)$ can be estimated similarly. Therefore

$$|a(u, u)| \geq \text{Re}((A\nabla u, \nabla u) + \lambda(u, u)) \geq \delta|u|_1^2 + \lambda\|u\|_0^2 \geq \min(\delta, \lambda)\|u\|_1^2.$$

By the Lax-Milgram theorem there is an inverse operator of L. Let λ be a fixed number, the equation (5.71) can be written as:

$$u_h = L^{-1}((\lambda + \alpha\beta\omega^2)u_h - f) = (\lambda + \alpha\beta\omega^2)L^{-1}u_h - L^{-1}f.$$

L^{-1} is a bounded operator from $L^2(\Omega_1)$ to $H^1(\Omega_1)$, then by the compact embedding of Sobolev spaces, it is a compact operator from $L^2(\Omega_1)$ to $L^2(\Omega_1)$. Applying Theorem 4, the spectrum includes point spectrum σ_p only. Therefore if $\lambda + \alpha\beta\omega^2$ is not included in σ_p, the problem (5.71)-(5.73) admits a unique solution. We define $\sigma_p' = \{\omega^2 \in (0, \infty); \lambda + \alpha\beta\omega^2 \in \sigma_p\}$, then the conclusion follows. □

Generally speaking σ_p is not a null set. Because if $d = 0$, there is no unique solution when ω^2 is the eigenvalue of the Laplace operator. For small d the spectrum is a small perturbation of the spectrum of the Laplace operator. However for large d there exists a unique solution. See [Bramble, J.H. and Pasciak, J.E.].

Error estimates of this scheme can be found in [Chen, Z. and Liu, X.].[Bramble, J.H. and Pasciak, J.E.]. The analysis of this scheme for the time-harmonic Maxwell equations can be found in [Bramble, J.H. and Pasciak, J.E.][Bao, G. and Wu, H. (2005)].

Chapter 6

Spectral Method

The spectral method aims at using an orthogonal set of smooth basis functions to construct approximate solutions. The fast Fourier transform (FFT) is a powerful tool in scientific computing. One of the advantage of the spectral method is the application of FFT. Another advantage is that the method enjoys a theoretical spectral convergence rate. Sometimes the order of error estimates can be as high as desired provided the exact solution is smooth enough. According to the chosen of basis functions the method has the following ingredients: trigonometric functions, orthogonal polynomials, or orthogonal rational functions. For a comprehensive study of the spectral method we refer to the books [Guo, B. (1998)] and [Shen, J. and Tang, T. (2006)], and for the application of the spectral method to exterior problems we refer to the survey [Guo, B. (2002)], and references therein. Following the analysis of Guo, Shen, and their co-workers, we will restrict ourselves to some examples in this chapter to show this method in the simplest form.

6.1 Introduction

For simplicity we denote $\partial_x^k u = \frac{d^k}{dx^k} u(x)$ in this chapter. To show how the spectral method works for exterior problems, let us see a model problem first. Let $\Lambda = (0, \infty)$. We consider the following ordinary differential equation on Λ with a boundary condition at $x = 0$:

$$-\partial_x^2 u + u = e^{-x}, \qquad u(0) = 0. \tag{6.1}$$

The solution is

$$u = -\frac{1}{2}xe^{-x} + C(e^{-x} - e^x), \tag{6.2}$$

with an arbitrary constant C. Let us impose one boundary condition at the infinity by two means: either we assume that u is bounded, or we define the corresponding variational formulation: Find $u \in H_0^1(\Lambda)$, such that

$$a(u, v) = \int_0^\infty e^{-x} v(x)\, dx, \qquad \forall v \in H_0^1(\Lambda),$$

where $a(u, v) = \int_0^\infty (\partial_x u(x) \partial_x v(x) + u(x)v(x))\, dx$. Then $C = 0$ and the unique solution is $u = -\frac{1}{2} x e^{-x}$.

We look for a polynomial $p \in P_N(\Lambda)$ to approximate u. The first glance is that it is impossible, because all polynomials, except constants, tend to the infinity as $x \to \infty$, and the solution u decays exponentially. However it is still possible if we introduce some weights. Letting $\beta > 0$, multiplying the equation (6.1) by $e^{-\beta x} v$, and applying the Green's formula, we get

$$\int_0^\infty e^{-\beta x}(\partial_x u \partial_x v - \beta \partial_x u v + uv)\, dx = \int_0^\infty e^{-(\beta+1)x} v\, dx. \tag{6.3}$$

Thus we introduce a weight function, $\omega(x) = e^{-\beta x}$ and define a weighted L^2-space,

$$L_\omega^2(\Lambda) = \{u \in L_{loc}^2(\Lambda); \int_0^\infty \omega u^2\, dx < \infty\},$$

with norm $\|u\|_{0,\omega} = \|\omega^{\frac{1}{2}} u\|_0$.

For any non-negative integer m and weight χ, we define the weighted Sobolev space

$$H_\chi^m(\Lambda) = \{v; \partial_x^k v \in L_\chi^2(\Lambda), 0 \le k \le m\}$$

equipped with the following inner product, semi-norm and norm

$$(u, v)_{m,\chi,\Lambda} = \sum_{0 \le k \le m} (\partial_x^k u, \partial_x^k v)_{0,\chi,\Lambda}, \quad |v|_{m,\chi,\Lambda} = \|\partial_x^m v\|_{0,\chi,\Lambda},$$

$$\|v\|_{m,\chi,\Lambda} = (v, v)_{m,\chi,\Lambda}^{\frac{1}{2}}.$$

For any real $r > 0$, the space $H_\chi^r(\Lambda)$ and its norm $\|v\|_{r,\chi,\Lambda}$ are defined by space interpolation. [Adams, R.A. (1975)]

We return to the expression (6.3). With regard to the boundary condition, a subspace is defined by

$$H_{0,\omega}^1(\Lambda) = \{u \in H_\omega^1(\Lambda); u(0) = 0\}.$$

We define a new bilinear form,

$$a(u, v) = \int_0^\infty e^{-\beta x}(\partial_x u \partial_x v - \beta \partial_x uv + uv)\, dx, \tag{6.4}$$

then the new variational formulation is: Find $u \in H_{0,\omega}^1(\Lambda)$, such that

$$a(u, v) = \int_0^\infty e^{-(\beta+1)x} v\, dx, \qquad \forall v \in H_{0,\omega}^1(\Lambda). \tag{6.5}$$

We notice that

$$\left| \int_0^\infty e^{-\beta x} \beta \partial_x uu\, dx \right| \leq \frac{\beta}{2} \int_0^\infty e^{-\beta x}(\partial_x u)^2\, dx + \frac{\beta}{2} \int_0^\infty e^{-\beta x} u^2\, dx.$$

Therefore owing to the Lax-Milgram theorem if $\beta < 2$, the problem (6.5) admits a unique solution. The space $P_N(\Lambda)$ is a finite dimensional subspace of $H_\omega^1(\Lambda)$, so the following spectral scheme also admits a unique solution: Find $p \in P_N(\Lambda)$, such that $p(0) = 0$, and

$$a(p, v) = \int_0^\infty e^{-(\beta+1)x} v\, dx, \qquad \forall v \in P_N(\Lambda), v(0) = 0. \tag{6.6}$$

The condition $\beta < 2$ is natural, because if $\beta > 2$, there are infinitely many solutions (6.2) in the space $H_{0,\omega}^1(\Lambda)$.

To solve (6.6) numerically, one needs a basis of the space $P_N(\Lambda)$. This is the subject of the next section. One will see the Laguerre polynomials are suitable for this problem. Therefore applying the spectral method to solve an exterior problem, one needs a finite set of basis functions, a functional space which involves the analytic solution and the the given finite set, and a variational formulation in which the analytical problem and the approximate problem are both well posed.

The solution p to the problem (6.6) can be a good approximation to u near $x = 0$, but it is a poor approximation on the far field. To improve the result, one can make a transform of variables. Let $\tilde{u} = e^{x/2}u$, then it is the solution to

$$-\partial_x^2 \tilde{u} + \partial_x \tilde{u} + \frac{3}{4}\tilde{u} = e^{-x/2}.$$

We define a weight $\omega = e^{-x}$, and a new bilinear form,

$$a(\tilde{u}, v) = \int_0^\infty e^{-x}(\partial_x \tilde{u} \partial_x v + \frac{3}{4}\tilde{u}v)\, dx,$$

then it is easy to see that the corresponding variational problem admits a unique solution in $H^1_{0,\omega}(\Lambda)$ with $\omega = e^{-x}$, and the spectral scheme is: Find $p \in P_N(\Lambda)$, such that $p(0) = 0$, and

$$a(p, v) = \int_0^\infty e^{-3x/2} v \, dx, \qquad \forall v \in P_N(\Lambda), v(0) = 0, \qquad (6.7)$$

which also admits a unique solution in $P_N(\Lambda)$. Returning to the original variable u we have

$$\|u - e^{-x/2} p\|_1^2 = \|\partial_x(e^{-x/2}\tilde{u} - e^{-x/2}p)\|_0^2 + \|e^{-x/2}\tilde{u} - e^{-x/2}p\|_0^2$$

$$= \|\partial_x(\tilde{u} - p)\|_{0,\omega}^2 + \frac{5}{4}\|\tilde{u} - p\|_{0,\omega}^2,$$

so it is just the approximation in the space $H^1(\Lambda)$.

Another approach is to transfer the exterior domain into a finite one. Let

$$x = -2\ln(1 - \xi) + 2\ln 2,$$

then the domain is transferred to $\tilde{\Lambda} = (-1, 1)$ and the equation (6.1) is transferred to

$$-\partial_\xi^2 u + \frac{1}{1 - \xi}\partial_\xi u + \frac{4}{(1 - \xi)^2} u = 1.$$

We can define a bilinear form,

$$a(u, v) = \int_{-1}^1 ((1 - \xi)^2 \partial_\xi u \partial_\xi v - (1 - \xi)\partial_\xi u v + 4uv) \, d\xi,$$

and define a weighted Sobolev space $H^1_\chi(\tilde{\Lambda})$ in accordance with the inner product,

$$((u, v)) = \int_{-1}^1 ((1 - \xi)^2 \partial_\xi u \partial_\xi v + uv) \, d\xi.$$

Since

$$\left| \int_{-1}^1 (1 - \xi)\partial_\xi u u \, d\xi \right| \leq \frac{1}{2}\int_{-1}^1 ((1 - \xi)^2((\partial_\xi u)^2 + u^2) \, d\xi,$$

the corresponding variational problem admits a unique solution, that is: Find $u \in H^1_\chi(\tilde{\Lambda})$, such that $u(-1) = 0$, and

$$a(u, v) = \int_{-1}^1 v \, d\xi, \qquad \forall v \in H^1_\chi(\tilde{\Lambda}), v(-1) = 0.$$

The spectral scheme is: Find $p \in P_N(\tilde{\Lambda})$, such that $p(-1) = 0$, and

$$a(p, v) = \int_{-1}^{1} v \, d\xi, \qquad \forall v \in P_N(\tilde{\Lambda}), v(-1) = 0,$$

which also admits a unique solution. We will see that the Jacobi polynomials are suitable for an orthogonal basis with some weight.

The third approach is applying rational and irrational functions. Let us introduce a weight $\omega = \frac{1}{(x+1)^2}$ and define a bilinear form in analogous to (6.4),

$$a(u, v) = \int_0^\infty \frac{1}{(x+1)^2} (\partial_x u \partial_x v - \frac{2}{x+1} \partial_x uv + uv) \, dx.$$

The corresponding variational formulation is: Find $u \in H_{0,\omega}^1(\Lambda)$, such that

$$a(u, v) = \int_0^\infty \frac{1}{(x+1)^2} e^{-x} v \, dx, \qquad \forall v \in H_{0,\omega}^1(\Lambda).$$

Lemma 70. *There exists $\alpha > 0$, such that*

$$a(u, u) \geq \alpha \|u\|_{1,\omega}^2, \qquad \forall u \in H_{0,\omega}^1(\Lambda).$$

Proof. We may assume that $u \in C^1(\Lambda)$. Let $\varepsilon \in (0, 1)$, then

$$\left| \int_0^\infty \frac{2}{(x+1)^3} \partial_x uu \, dx \right|$$

$$\leq (1 - \varepsilon) \int_0^\infty \frac{1}{(x+1)^2} (\partial_x u)^2 \, dx + \frac{1}{1 - \varepsilon} \int_0^\infty \frac{1}{(x+1)^4} u^2 \, dx.$$

Let $c > 0$, then

$$a(u, u) \geq \varepsilon \int_0^\infty \frac{1}{(x+1)^2} (\partial_x u)^2 \, dx + \int_0^\infty \frac{1}{(x+1)^2} u^2 \, dx$$

$$+ \frac{1}{1 - \varepsilon} \int_0^\infty \frac{1}{(x+1)^4} u^2 \, dx$$

$$= \int_0^c \frac{1}{(x+1)^2} \left(\varepsilon(\partial_x u)^2 + u^2 - \frac{u^2}{(1 - \varepsilon)(x+1)^2} \right) dx$$

$$+ \int_c^\infty \frac{1}{(x+1)^2} \left(\varepsilon(\partial_x u)^2 + u^2 - \frac{u^2}{(1 - \varepsilon)(x+1)^2} \right) dx.$$

For the second term on the right, we take a large c, such that $\frac{1}{(1-\varepsilon)(c+1)^2} < 1$, that is $c > (1 - \varepsilon)^{-\frac{1}{2}} - 1$, i.e. for a positive constant $c_1 > 0$, $c > c_1 \varepsilon$. For the first term we have

$$\int_0^c \frac{1}{(x+1)^4} u^2\, dx = \int_0^c \frac{1}{(x+1)^4} \left(\int_0^x \partial_t u(t)\, dt \right)^2 dx$$

$$\leq \int_0^c \frac{x}{(x+1)^4} \int_0^x (\partial_t u(t))^2\, dt\, dx = \int_0^c (\partial_t u(t))^2\, dt \int_t^c \frac{x}{(x+1)^4}\, dx$$

$$\leq \frac{c^2}{2} \int_0^c (\partial_t u(t))^2\, dt.$$

We take a small c, such that $\frac{c^2}{2(1-\varepsilon)} < \frac{\varepsilon}{(c+1)^2}$, that is $c^2 < \varepsilon\frac{2(1-\varepsilon)}{(c+1)^2}$, i.e. for a positive constant $c_2 > 0$, $c < c_2\varepsilon^{1/2}$. Let ε be sufficient small, so that the chosen of c is possible. Thus $a(\cdot, \cdot)$ is coercive. □

Owing to the Lax-Milgram theorem, the variational problem admits a unique solution. Polynomials are no longer in this space $H^1_{0,\omega}(\Lambda)$. However it is possible to use rational functions. We define a finite dimensional space,

$$V_N = \left\{ R_l; R_l(x) = v\left(\frac{x-1}{x+1} \right), v \in P_N(-1,1) \right\},$$

then $V_{N0} = \{ u \in V_N; u(0) = 0 \}$ is a subspace of $H^1_{0,\omega}(\Lambda)$. The spectral scheme is: Find $u_N \in V_{N0}$, such that

$$a(u_N, v) = \int_0^\infty \frac{1}{(x+1)^2} e^{-x} v\, dx, \qquad \forall v \in V_{N0},$$

which also admits a unique solution. We will see the Legendre rational functions, $R_l(x) = \sqrt{2}L_l\left(\frac{x-1}{x+1} \right)$. generate an orthogonal basis in V_N with the weight ω. The definitions of Legendre functions L_l will be given in the next section.

Another approach is introducing the weight $\omega(x) = \frac{1}{\sqrt{x}(x+1)}$. We will see in the next section the rational functions generated by Chebyshev polynomials form an orthogonal basis of functions. Multiplying the weight ω and a test function v to the equation (6.1), and integrating by parts, we define a bilinear form

$$a(u,v) = \int_0^\infty \partial_x u \partial_x v\omega\, dx + \frac{1}{2} \int_0^\infty (v\partial_x u - u\partial_x v)\partial_x \omega\, dx$$

$$- \frac{1}{2} \int_0^\infty uv\partial_x^2\omega\, dx + \int_0^\infty uv\omega\, dx.$$

$$(6.8)$$

The corresponding variational formulation is: Find $u \in H^1_{0,\omega}(\Lambda)$, such that

$$a(u, v) = \int_0^\infty e^{-x} v\omega \, dx, \qquad \forall v \in H^1_{0,\omega}(\Lambda). \tag{6.9}$$

The point $x = 0$ is a singular point in the integrals in (6.8). To verify these integrals are well defined, we need the following lemma.

Lemma 71. *If $v \in H^1_{0,\omega}(\Lambda)$, then*

$$\int_\Lambda \frac{v(x)^2}{x^2} \frac{3x+1}{2(x+1)} \omega(x) \, dx \leq \int_\Lambda (\partial_x v(x))^2 \omega(x) \, dx.$$

Proof. It is not obvious that the left hand side is well defined either. We will prove the convergence of this integral in the process of estimation. It has no harm in assuming $v \in C_0^\infty(\Lambda)$ first. We notice that

$$\partial_x \omega = -\frac{3x+1}{2(x+1)x}\omega,$$

then we have

$$\frac{v^2(x)}{x}\omega(x) = \int_0^x \partial_t \left(\frac{v^2(t)}{t}\omega(t) \right) dt$$

$$= \int_0^x \left(2\frac{v(t)}{t}\partial_t v(t)\omega(t) - \frac{v^2(t)}{t^2}\omega(t) + \frac{v^2(t)}{t}\partial_t\omega(t) \right) dt$$

Using the Schwarz inequality we find

$$\frac{v^2(x)}{x}\omega(x) + \int_0^x \frac{v^2(t)}{t^2}\frac{5t+3}{2(t+1)}\omega(t) \, dt = \int_0^x 2\frac{v(t)}{t}\partial_t v(t)\omega(t) \, dt$$

$$\leq \int_0^x \frac{v^2(t)}{t^2}\omega(t) \, dt + \int_0^x (\partial_t v(t))^2 \omega(t) \, dt.$$

Letting $x \to \infty$, we get the desired inequality. Since $C_0^\infty(\Lambda)$ is dense in $H^1_{0,\omega}(\Lambda)$, the results hold for $v \in H^1_{0,\omega}(\Lambda)$. $\square$

Using the Schwarz inequality and the inequality

$$\partial_x^2 \omega = -\frac{15x^2 + 10x + 3}{4(x+1)^2 x^2}\omega \leq \frac{15}{4}\frac{\omega}{x^2}$$

we can estimate the terms in (6.8). For example

$$\int_0^\infty v\partial_x u\partial_x \omega \, dx \le \left(\int_0^\infty (\partial_x u)^2 \omega \, dx\right)^{\frac{1}{2}} \left(\int_0^\infty v^2 \frac{(\partial_x \omega)^2}{\omega} \, dx\right)^{\frac{1}{2}}$$

$$\le \left(\int_0^\infty (\partial_x u)^2 \omega \, dx\right)^{\frac{1}{2}} \left(\int_0^\infty \left(\frac{3x+1}{2(x+1)}\right)^2 \frac{v^2}{x^2} \omega \, dx\right)^{\frac{1}{2}}$$

$$\le \frac{3}{2} \left(\int_0^\infty (\partial_x u)^2 \omega \, dx\right)^{\frac{1}{2}} \left(\int_0^\infty (\partial_x v)^2 \omega \, dx\right)^{\frac{1}{2}}.$$

At the same time we have proved that $a(\cdot, \cdot)$ is bounded,

$$|a(u, v)| \le M\|u\|_{1,\omega}\|v\|_{1,\omega}.$$

To verify that the problem (6.9) admits a unique solution, we need to prove

$$a(u, u) \ge \alpha\|u\|_{1,\omega}^2, \qquad \alpha > 0,$$

which needs a little more careful investigation. Let us verify

$$\frac{15x^2 + 10x + 3}{8(x+1)^2} \le \frac{14}{27}x^2 + \frac{15}{32}, \qquad x \ge 0. \tag{6.10}$$

Let

$$f(x) = \frac{15x^2 + 10x + 3}{8(x+1)^2} - \frac{14}{27}x^2 - \frac{15}{32},$$

then we have

$$f'(x) = \frac{5x+1}{2(x+1)^3} - \frac{28}{27}x, \quad f'(\frac{1}{2}) = 0,$$

$$f''(x) = -\frac{5x+1}{(x+1)^4} - \frac{28}{27} < 0.$$

Hence $\frac{1}{2}$ is the only root of $f'(x)$ in $[0, \infty)$. Thus $f(x) \le f(\frac{1}{2}) < 0$, $\forall x \ge 0$, which implies (6.10). By (6.8) we get

$$a(u, u) = \int_0^\infty (\partial_x u)^2 \omega \, dx - \frac{1}{2}\int_0^\infty u^2 \partial_x^2 \omega \, dx + \int_0^\infty u^2 \omega \, dx$$

$$\ge \frac{1}{16}\int_0^\infty (\partial_x u)^2 \omega \, dx + \frac{13}{27}\int_0^\infty u^2 \omega \, dx \ge \frac{1}{16}\|u\|_{1,\omega}^2.$$

Owing to the Lax-Milgram theorem the problem (6.9) admits a unique solution. The finite dimensional space V_{N0} defined above is also a subspace of $H^1_{0,\omega}(\Lambda)$, and the spectral scheme is: Find $u_N \in V_{N0}$, such that

$$a(u_N, v) = \int_0^\infty e^{-x} v\omega \, dx, \qquad \forall v \in V_{N0},$$

which also admits a unique solution.

We will investigate the spectral method for more complicated exterior problems in the following sections.

6.2 Orthogonal systems of polynomials

Let $\Lambda = (-1, 1)$. The Legendre polynomial of degree l is

$$L_l(x) = \frac{(-1)^l}{2^l l!} \partial_x^l (1 - x^2)^l.$$

It is the l-th eigenfunction of the singular Sturm-Liouville problem,

$$\partial_x((1 - x^2)\partial_x v) + \lambda v = 0, \qquad x \in \Lambda,$$

related to the l-th eigenvalue $\lambda_l = l(l + 1)$. Clearly $L_0(x) = 1$, $L_1(x) = x$, and they satisfy the recurrence relations,

$$L_{l+1}(x) = \frac{2l + 1}{l + 1} x L_l(x) - \frac{l}{l + 1} L_{l-1}(x), \qquad l \geq 1,$$

$$(2l + 1)L_l(x) = \partial_x L_{l+1}(x) - \partial_x L_{l-1}(x), \qquad l \geq 1.$$

It can be checked that $L_l(1) = 1$, $L_l(-1) = (-1)^l$, $\partial_x L_l(1) = \frac{1}{2}l(l + 1)$, and $\partial_x L_l(-1) = \frac{(-1)^{l+1}}{2}l(l + 1)$. Moreover

$$|L_l(x)| \leq 1, \qquad \partial_x L_l(x) \leq \frac{1}{2}l(l + 1), \qquad x \in \Lambda.$$

The set of Legendre polynomials is the L^2-orthogonal system in Λ, i.e.,

$$\int_\Lambda L_l(x) L_m(x) \, dx = \left(l + \frac{1}{2}\right)^{-1} \delta_{lm}. \tag{6.11}$$

By integrating by parts, we derive that

$$\int_\Lambda (\partial_x L_l(x))^2 \, dx = l(l + 1).$$

The Legendre expansion of a function $v \in L^2(\Lambda)$ is

$$v(x) = \sum_{l=0}^{\infty} v_l L_l(x),$$

with

$$v_l = \left(l + \frac{1}{2}\right) \int_{\Lambda} v(x) L_l(x) \, dx.$$

For the rational functions $R_l(x) = \sqrt{2} L_l \left(\frac{x-1}{x+1}\right)$, defined in Section 6.1, we make a variable transform in (6.11) to obtain

$$\int_0^{\infty} \frac{1}{(x+1)^2} R_l(x) R_m(x) \, dx = \left(l + \frac{1}{2}\right)^{-1} \delta_{lm}.$$

Therefore the set of R_l is a weighted orthogonal system.

Let the weight $\omega(x) = (1 - x^2)^{-\frac{1}{2}}$. The Chebyshev polynomial of the first kind of degree l is

$$T_l(x) = \cos(l \arccos x).$$

It is the l-th eigenfunction of the singular Sturm-Liouville problem,

$$\partial_x \left((1 - x^2)^{\frac{1}{2}} \partial_x v\right) + \lambda (1 - x^2)^{-\frac{1}{2}} v = 0, \qquad x \in \Lambda,$$

related to the l-th eigenvalue $\lambda_l = l^2$. Clearly $T_0(x) = 1$, $T_1(x) = x$, and they satisfy the recurrence relations,

$$T_{l+1}(x) = 2x T_l(x) - T_{l-1}(x), \qquad l \geq 1,$$

$$2T_l(x) = \frac{1}{l+1} \partial_x T_{l+1}(x) - \frac{1}{l-1} \partial_x T_{l-1}(x), \qquad l \geq 1.$$

It can be checked that $T_l(1) = 1$, $T_l(-1) = (-1)^l$, $\partial_x T_l(1) = l^2$, and $\partial_x T_l(-1) = (-1)^{l+1} l^2$. Moreover

$$|T_l(x)| \leq 1, \qquad \partial_x T_l(x) \leq l^2, \qquad x \in \Lambda.$$

The set of Chebyshev polynomials is the L_{ω}^2-orthogonal system in Λ, i.e.,

$$\int_{\Lambda} T_l(x) T_m(x) \omega(x) \, dx = \frac{\pi}{2} c_l \delta_{lm},$$

where $c_0 = 2$ and $c_l = 1$ for $l \geq 1$. The Chebyshev expansion of a function $v \in L_\omega^2(\Lambda)$ is

$$v(x) = \sum_{l=0}^{\infty} v_l T_l(x),$$

with

$$v_l = \frac{2}{\pi c_l} \int_\Lambda v(x) T_l(x) \omega(x) \, dx.$$

We can define rational functions, $R_l(x) = T_l\left(\frac{x-1}{x+1}\right)$, then they generate a weighted L^2-orthogonal system on the interval $(0, \infty)$, i.e.,

$$\int_0^\infty R_l(x) R_m(x) \frac{1}{\sqrt{x}(x+1)} \, dx = \frac{\pi}{2} c_l \delta_{lm}.$$

Let the weight $\omega(x) = (1-x)^\alpha (1+x)^\beta$ for $\alpha, \beta > -1$, the Jacobi polynomial $J_l^{\alpha,\beta}(x)$ is the l-th eigenfunction of the singular Sturm-Liouville problem,

$$(1-x)^{-\alpha}(1+x)^{-\beta}\partial_x\left((1-x)^{\alpha+1}(1+x)^{\beta+1}\partial_x v\right) + \lambda v = 0, \qquad x \in \Lambda,$$

related to the l-th eigenvalue $\lambda_l = l(l+1+\alpha+\beta)$, and normalized by

$$J_l^{\alpha,\beta}(1) = \frac{\Gamma(l+\alpha+1)}{l!\Gamma(\alpha+1)}.$$

The Jacobi polynomials satisfy the recurrence relations,

$$J_0^{\alpha,\beta}(x) = 1, \quad J_1^{\alpha,\beta}(x) = \frac{1}{2}(\alpha+\beta+2)x + \frac{1}{2}(\alpha-\beta),$$

$$J_{l+1}^{\alpha,\beta}(x) = (a_l^{\alpha,\beta} x - b_l^{\alpha,\beta}) J_l^{\alpha,\beta}(x) - c_l^{\alpha,\beta} J_{l-1}^{\alpha,\beta}(x), \qquad l \geq 1,$$

where

$$a_l^{\alpha,\beta} = \frac{(2l+\alpha+\beta+1)(2l+\alpha+\beta+2)}{2(l+1)(l+\alpha+\beta+1)},$$

$$b_l^{\alpha,\beta} = \frac{(\beta^2-\alpha^2)(2l+\alpha+\beta+1)}{2(l+1)(l+\alpha+\beta+1)(2l+\alpha+\beta)},$$

$$c_l^{\alpha,\beta} = \frac{(l+\alpha)(l+\beta)(2l+\alpha+\beta+2)}{(l+1)(l+\alpha+\beta+1)(2l+\alpha+\beta)}.$$

The set of Jacobi polynomials is the L_ω^2-orthogonal system in Λ, i.e.,

$$\int_\Lambda J_l^{\alpha,\beta}(x) J_m^{\alpha,\beta}(x) \omega(x) \, dx = 0, \qquad l \neq m.$$

We also have the following derivative recurrence relation

$$\partial_x J_l^{\alpha,\beta}(x) = \frac{1}{2}(l + \alpha + \beta + 1) J_{l-1}^{\alpha+1,\beta+1}(x).$$

By integrating by parts, we deduce that

$$\int_\Lambda \partial_x J_l^{\alpha,\beta}(x) \partial_x J_m^{\alpha,\beta}(x)(1-x)^{\alpha+1}(1+x)^{\beta+1} \, dx = 0, \qquad l \neq m.$$

Both Legendre and Chebyshev polynomials are special cases of Jacobi polynomials, namely, the Legendre polynomials $L_l(x)$ correspond to $\alpha = \beta = 0$ with the normalization $L_l(1) = 1$ and the Chebyshev polynomials $T_l(x)$ correspond to $\alpha = \beta = -\frac{1}{2}$ with the normalization $T_l(1) = 1$. For $\alpha \leq -1$ or $\beta \leq -1$ the weight function $\omega(x)$ is not in $L_\omega^2(\Lambda)$, so the Jacobi polynomials are defined for $\alpha, \beta > -1$. However, the generalized Jacobi polynomials are defined for the cases where the indexes α and β can be negative integers.

$$J_l^{\alpha,\beta}(x) = \begin{cases} (1-x)^{-\alpha}(1+x)^{-\beta} J_{l-l_0}^{-\alpha,-\beta}(x), & \text{if } \alpha, \beta \leq -1, \\ (1-x)^{-\alpha} J_{l-l_0}^{-\alpha,\beta}(x), & \text{if } \alpha \leq -1, \beta > -1 \\ (1+x)^{-\beta} J_{l-l_0}^{\alpha,-\beta}(x), & \text{if } \alpha > -1, \beta \leq -1, \end{cases}$$

where $l \geq l_0$ with $l_0 = -\alpha - \beta, -\alpha, -\beta$ respectively for the above three cases. The generalized Jacobi polynomials $J_l^{\alpha,\beta} \in P_l$ and they are orthogonal with the generalized weight,

$$\int_\Lambda J_l^{\alpha,\beta}(x) J_m^{\alpha,\beta}(x) \omega(x) \, dx = 0, \qquad l \neq m.$$

The generalized Jacobi polynomials with negative integer indexes can be expressed in terms of Legendre polynomials,

$$J_l^{\alpha,\beta}(x) = \sum_{j=l+\alpha+\beta}^{l} a_j L_j(x), \qquad l \geq -\alpha - \beta.$$

Finally, by the definition the generalized Jacobi polynomials satisfy some zero boundary conditions,

$$\partial_x^i J_l^{\alpha,\beta}(1) = 0, \qquad i = 0, 1, \cdots, -\alpha - 1,$$

$$\partial_x^i J_l^{\alpha,\beta}(-1) = 0, \qquad i = 0, 1, \cdots, -\beta - 1.$$

Let $\Lambda = (0, \infty)$ and the weight $\omega(x) = e^{-x}$. The Laguerre polynomial of degree l is

$$\mathcal{L}_l(x) = \frac{1}{l!} e^x \partial_x^l (x^l e^{-x}).$$

It is the l-th eigenfunction of the singular Sturm-Liouville problem,

$$\partial_x (xe^{-x}\partial_x v) + \lambda e^{-x} v = 0, \qquad x \in \Lambda,$$

related to the l-th eigenvalue $\lambda_l = l$. Clearly $\mathcal{L}_0(x) = 1$, $\mathcal{L}_1(x) = 1 - x$, and they satisfy the recurrence relations,

$$(l+1)\mathcal{L}_{l+1}(x) = (2l+1-x)\mathcal{L}_l(x) - l\mathcal{L}_{l-1}(x), \qquad l \geq 1,$$

$$\mathcal{L}_l(x) = \partial_x \mathcal{L}_l(x) - \partial_x \mathcal{L}_{l+1}(x), \qquad l \geq 0.$$

It can be checked that $\mathcal{L}_l(0) = 1$, $\partial_x \mathcal{L}_l(0) = -l$ for $l \geq 1$, and

$$|\mathcal{L}_l(x)| \leq e^{\frac{x}{2}}, \qquad x \in \Lambda.$$

The set of Laguerre polynomials is the L_ω^2-orthogonal system in Λ, i.e.,

$$\int_\Lambda \mathcal{L}_l(x)\mathcal{L}_m(x)\omega(x)\, dx = \delta_{lm}.$$

By integrating by parts, we deduce that

$$\int_\Lambda \partial_x \mathcal{L}_l(x)\partial_x \mathcal{L}_m(x) x\omega(x)\, dx = l\delta_{lm}.$$

The Laguerre expansion of a function $v \in L_\omega^2(\Lambda)$ is

$$v(x) = \sum_{l=0}^\infty v_l \mathcal{L}_l(x),$$

with

$$v_l = \int_\Lambda v(x)\mathcal{L}_l(x)\omega(x)\, dx.$$

The generalized Laguerre polynomial of degree l is defined by

$$\mathcal{L}_l^{(\alpha,\beta)}(x) = \frac{1}{l!} x^{-\alpha} e^{\beta x} \partial_x^l (x^{l+\alpha} e^{-\beta x}), \qquad l = 0, 1, \cdots.$$

If $\beta = 1$ we drop the superscript β and denote the generalized Laguerre polynomial by $\mathcal{L}_l^{(\alpha)}(x)$.

It is straightforward to derive the following properties:

$$\mathcal{L}_l^{(\alpha,\beta)}(0) = \mathcal{L}_l^{(\alpha)}(0) = \frac{\Gamma(l+\alpha+1)}{\Gamma(\alpha+1)\Gamma(l+1)}, \quad l \geq 0,$$

$$\partial_x \mathcal{L}_l^{(\alpha,\beta)}(x) = -\beta \mathcal{L}_{l-1}^{(\alpha+1,\beta)}(x), \quad l \geq 1,$$

$$(l+1)\mathcal{L}_{l+1}^{(\alpha,\beta)}(x) = (2l+\alpha+1-\beta x)\mathcal{L}_l^{(\alpha,\beta)}(x) - (l+\alpha)\mathcal{L}_{l-1}^{(\alpha,\beta)}(x), \quad l \geq 1,$$

$$\begin{aligned}
\mathcal{L}_l^{(\alpha,\beta)}(x) &= \mathcal{L}_l^{(\alpha+1,\beta)}(x) - \mathcal{L}_{l-1}^{(\alpha+1,\beta)}(x) \\
&= \beta^{-1}(\partial_x \mathcal{L}_l^{(\alpha,\beta)}(x) - \partial_x \mathcal{L}_{l+1}^{(\alpha,\beta)}(x)), \quad l \geq 1.
\end{aligned}$$

Let $\omega_{\alpha,\beta}(x) = x^\alpha e^{-\beta x}$, $\alpha > -1, \beta > 0$. In particular, we denote $\omega_\alpha(x) = \omega_{\alpha,1}(x) = x^\alpha e^{-x}$. The set of $\mathcal{L}_l^{(\alpha,\beta)}(x)$ is the complete $L^2_{\omega_{\alpha,\beta}}(\Lambda)$-orthogonal system, namely,

$$(\mathcal{L}_l^{(\alpha,\beta)}, \mathcal{L}_m^{(\alpha,\beta)})_{0,\omega_{\alpha,\beta},\Lambda} = \gamma_l^{\alpha,\beta}\delta_{lm}, \quad \gamma_l^{\alpha,\beta} = \frac{\Gamma(l+\alpha+1)}{\beta^{\alpha+1}l!}. \tag{6.12}$$

6.3   Laguerre spectral methods

6.3.1   *Mixed Laguerre-Fourier spectral method*

First of all let us consider the mixed Laguerre-Fourier spectral method for the Helmholtz exterior problem (1.25)(1.26)

$$-(\triangle + \lambda)u = f, \tag{6.13}$$

$$u = 0, \quad x \in \partial\Omega, \tag{6.14}$$

with a real $\lambda < 0$, dim $(\Omega) = 2$, and $\Omega = \{x; |x| > 1\}$. Owing to Theorem 17, the problem (6.13)(6.14) admits a unique solution in $H_0^1(\Omega)$. Under the following polar transformation:

$$x_1 = (\rho+1)\cos\theta, \quad x_2 = (\rho+1)\sin\theta,$$

the equation (6.13) becomes

$$-\frac{1}{\rho+1}\partial_\rho((\rho+1)\partial_\rho u) - \frac{1}{(\rho+1)^2}\partial_\theta^2 u - \lambda u = f. \tag{6.15}$$

We use the following change of variables

$$\tilde{u} = (\rho + 1)^{-\frac{1}{2}} e^{\frac{\rho}{2}} u, \quad \tilde{f} = (\rho + 1)^{\frac{3}{2}} e^{\frac{\rho}{2}} f,$$

to transform (6.15) to

$$-(\rho+1)^2 \partial_\rho^2 \tilde{u} + (\rho^2 - 1)\partial_\rho \tilde{u} - \partial_\theta^2 \tilde{u} + (-\lambda(\rho+1)^2 + \frac{1}{2} + \frac{\rho}{2} - \frac{1}{4}\rho^2)\tilde{u} = \tilde{f}. \quad (6.16)$$

To solve the boundary value problem of (6.16) we introduce some spaces. Let $I = (0, 2\pi)$ and $H_p^m(I)$ be the subspace of $H^m(I)$ consisting of all functions whose derivatives of order up to $m-1$ are periodic with the period 2π. Let M be any positive integer, and $\tilde{V}_M(I) =$ span $\{e^{il\theta}; |l| \le M\}$. We denote by $V_M(I)$ the subset of $\tilde{V}_M(I)$ consisting of all real-valued functions. Let $\Lambda = (0, \infty)$, $D = \Lambda \times I$ and $L_\chi^2(D)$ be the weighted Sobolev space with the following inner product and norm,

$$(u, v)_{0,\chi} = \int_D u(\rho, \theta)v(\rho, \theta)\chi(\rho)\, d\rho d\theta, \qquad \|v\|_{0,\chi} = (v, v)_{0,\chi}^{\frac{1}{2}}.$$

The weighted Sobolev spaces $H_\chi^r(D)$ and its norm $\|v\|_{r\chi}$ and semi-norm $|v|_{r,\chi}$ are defined in the usual manner. In particular we set

$$H_{0,p,\chi}^1(D) = \{v \in H_\chi^1(D); v(\rho, \theta + 2\pi) = v(\rho, \theta), v(0, \theta) = 0, \forall \theta \in I, \rho \in \Lambda\}.$$

With the weights $\omega(\rho) = e^{-\rho}$ and $\eta(\rho) = (\rho + 1)^2 e^{-\rho}$ we define the non-isotropic space

$$H_{0,p,\eta,\omega}^1(D) = \{v \in L_{loc}^1(D); \|v\|_{1,\eta,\omega} < \infty\},$$

where

$$\|v\|_{1,\eta,\omega} = (|v|_{1,\eta,\omega}^2 + \|v\|_{0,\omega}^2)^{\frac{1}{2}}, \quad |v|_{1,\eta,\omega} = (\|\partial_\rho v\|_{0,\eta}^2 + \|\partial_\theta v\|_{0,\omega}^2)^{\frac{1}{2}}.$$

Let us denote $P_{0,N}(\Lambda) = \{v \in P_N(\Lambda); v(0) = 0\}$ and

$$V_{N,M}(D) = P_{0,N}(\Lambda; V_M(I)).$$

Let us define the variational formulation of the boundary value problems of (6.16). For this purpose we define a bilinear form,

$$a(u, v) = \int_D (\rho + 1)^2 e^{-\rho} \partial_\rho u \partial_\rho v \, d\rho d\theta + \int_D e^{-\rho} \partial_\theta u \partial_\theta v \, d\rho d\theta$$

$$+ \int_D e^{-\rho} \left(-\lambda(\rho + 1)^2 - \frac{1}{4}\rho^2 + \frac{1}{2}\rho + \frac{1}{2} \right) uv \, d\rho d\theta.$$

The variational formulation is: Find $\tilde{u} \in H^1_{0,p,\eta,\omega}(D)$, such that

$$a(\tilde{u}, v) = (\tilde{f}, v)_{0,\omega}, \qquad \forall v \in H^1_{0,p,\eta,\omega}(D). \tag{6.17}$$

To investigate the well-posedness of (6.17), we need some lemmas.

Lemma 72. *For any* $v \in H^1_{0,\omega_2}(\Lambda) \bigcap L^2_{\omega_0}$ *it holds that*

$$\|v\|^2_{0,\omega_2,\Lambda} \leq 8|v|^2_{1,\omega_2,\Lambda} + 8\|v\|^2_{0,\omega_0,\Lambda},$$

where ω_α *with* $\alpha = 0, 2$ *is defined in Section 6.2.*

Proof. We have

$$\omega_2(\rho)v^2(\rho) = \int_0^\rho \partial_\xi(\omega_2(\xi)v^2(\xi)) \, d\xi$$

$$= 2\int_0^\rho \omega_2(\xi)v(\xi)\partial_\xi v(\xi) \, d\xi + 2\int_0^\rho \omega_0(\xi)v^2(\xi) \, d\xi$$

$$- \int_0^\rho \omega_2(\xi)v^2(\xi) \, d\xi.$$

Letting $\rho \to \infty$ and using the Schwarz inequality we obtain

$$\|v\|^2_{0,\omega_2,\Lambda} \leq \frac{1}{2}\|v\|^2_{0,\omega_2,\Lambda} + 2\|\partial_\rho v\|^2_{0,\omega_2,\Lambda} + 2\|v\|^2_{0,\omega_1,\Lambda}.$$

Thus

$$\|v\|^2_{0,\omega_2,\Lambda} \leq 4\|\partial_\rho v\|^2_{0,\omega_2,\Lambda} + 4\|v\|^2_{0,\omega_1,\Lambda}. \tag{6.18}$$

An integration by parts yields

$$\|v\|^2_{0,\omega_1,\Lambda} = \int_\Lambda \rho e^{-\rho} v(\rho)\partial_\rho v(\rho) \, d\rho + \|v\|^2_{0,\omega_0,\Lambda}.$$

The Schwarz inequality yields

$$2\int_\Lambda \rho e^{-\rho}v(\rho)\partial_\rho v(\rho) \, d\rho \leq |v|^2_{1,\omega_2,\Lambda} + \|v\|^2_{0,\omega_0,\Lambda}.$$

Substituting the above into (6.18) we obtain the result. $\qquad\qquad \square$

Lemma 73. *For any* $u, v \in H^1_{0,p,\eta,\omega}(D)$,

$$a(v,v) \geq \int_D (\rho+1)^2 e^{-\rho}(\partial_\rho v)^2 \, d\rho d\theta + \int_D e^{-\rho}(\partial_\theta v)^2 \, d\rho d\theta$$

$$- (\lambda + \frac{1}{4})\int_D (\rho+1)^2 e^{-\rho}v^2 \, d\rho d\theta + \frac{3}{4}\int_D e^{-\rho}v^2 \, d\rho d\theta,$$

and

$$|a(u,v)| \leq C\|u\|_{1,\eta,\omega}\|v\|_{1,\eta,\omega}.$$

Proof. Obviously

$$-\frac{1}{4}\rho^2 + \frac{1}{2}\rho + \frac{1}{2} \geq -\frac{1}{4}(\rho+1)^2 + \frac{3}{4},$$

which leads to the first result, and the second result follows from Lemma 72. $\qquad\square$

Theorem 70. *If* $\lambda \leq -\frac{1}{4}$ *and* $(\rho+1)^{-1}\tilde{f} \in L_\omega^2(D)$, *then the problem (6.17) admits a unique solution with*

$$\|\tilde{u}\|_{1,\eta,\omega} \leq C\|(\rho+1)^{-1}\tilde{f}\|_{0,\omega}.$$

Proof. Due to $\lambda \leq -\frac{1}{4}$ and Lemma 73, $a(\cdot,\cdot)$ is coercive on $H_{0,p,\eta,\omega}^1(D) \times H_{0,p,\eta,\omega}^1(D)$. Moreover by Lemma 72,

$$|(\tilde{f},v)_{0,\omega}| \leq C\|v\|_{1,\eta,\omega}\|(\rho+1)^{-1}\tilde{f}\|_{0,\omega}.$$

Thus the conclusion follows from the Lax-Milgram theorem. $\qquad\square$

Next, we consider the mixed Laguerre-Fourier approximation for (6.17): Find $u_{N,M} \in V_{N,M}$, such that

$$a(u_{N,M},v) = (\tilde{f},v)_{0,\omega}, \qquad \forall v \in V_{N,M}. \tag{6.19}$$

The following result is a direct consequence of the Lax-Milgram theorem.

Theorem 71. *If* $\lambda \leq -\frac{1}{4}$ *and* $(\rho+1)^{-1}\tilde{f} \in L_\omega^2(D)$, *then the problem (6.19) admits a unique solution with*

$$\|u_{N,M}\|_{1,\eta,\omega} \leq C\|(\rho+1)^{-1}\tilde{f}\|_{0,\omega}.$$

We next describe the implementation for (6.19). For simplicity, we denote $\mathcal{L}_l^{(0)}$ by $\mathcal{L}$ and set $\psi_l(\rho) = \mathcal{L}_{l-1}(\rho) - \mathcal{L}_l(\rho)$, $1 \leq l \leq N$, and

$$v_{l,m}^{(1)}(\rho,\theta) = \psi_l(\rho)\cos m\theta, \quad 1 \leq l \leq N, 0 \leq m \leq M,$$

$$v_{l,m}^{(2)}(\rho,\theta) = \psi_l(\rho)\sin m\theta, \quad 1 \leq l \leq N, 1 \leq m \leq M.$$

Since $\psi_l(0) = 0$, $v_{l,m}^{(j)}$ can be used as basis functions for $V_{N,M}$. Hence we can expend $u_{N,M}$ as

$$u_{N,M}(\rho,\theta) = \sum_{l=1}^{N}\left(\sum_{m=0}^{M} u_{l,m}^1 v_{l,m}^{(1)}(\rho,\theta) + \sum_{m=1}^{M} u_{l,m}^2 v_{l,m}^{(2)}(\rho,\theta)\right).$$

On the other hand we write

$$\tilde{f}(\rho,\theta) = \sum_{l=0}^{\infty}\sum_{m=0}^{\infty}(f_{l,m}^1 \mathcal{L}_l(\rho)\cos m\theta + f_{l,m}^2 \mathcal{L}_l(\rho)\sin m\theta).$$

Let us denote

$$Z_M = \{(j,n); j=1, n=0,1,\cdots,M; j=2, n=1,2,\cdots,M\}.$$

Taking $v = v_{k,n}^j$ in (6.19) for $(j,n) \in Z_M$, we derive by using the orthogonality of the trigonometric functions that (6.19) is equivalent to the following $2M+1$ linear systems:

$$\sum_{l=1}^{N}\left(\int_{\Lambda}(\rho+1)^2 e^{-\rho}\partial_\rho\psi_l\partial_\rho\psi_k\,d\rho\right.$$
$$\left.+\int_{\Lambda} e^{-\rho}(-\lambda(\rho+1)^2 + n^2 - \frac{\rho^2}{4} + \frac{\rho}{2} + \frac{1}{2})\psi_l\psi_k\,d\rho\right)u_{l,m}^j = g_{l,m}^j,$$
$$1 \le k \le N,$$

where $g_{k,n}^j = f_{k-1,n}^j - f_{k,n}^j$, $1 \le k \le N$. Let the matrices $A = (a_{k,l})$, $B = (b_{k,l})$, $C = (c_{k,l})$ and $D = (D_{k,l})$ with the following entries:

$$a_{l,k} = \int_{\Lambda}(\rho+1)^2 e^{-\rho}\partial_\rho\psi_l(\rho)\partial_\rho\psi_k(\rho)\,d\rho, \quad b_{l,k} = \int_{\Lambda} e^{-\rho}\psi_l(\rho)\psi_k(\rho)\,d\rho,$$

$$c_{l,k} = \int_{\Lambda}\rho e^{-\rho}\psi_l(\rho)\psi_k(\rho)\,d\rho, \quad d_{l,k} = \int_{\Lambda}\rho^2 e^{-\rho}\psi_l(\rho)\psi_k(\rho)\,d\rho.$$

Furthermore, let

$$X_n^j = (u_{1,n}^j, u_{2,n}^j,\cdots,u_{N,n}^j), \quad G_n^j = (g_{1,n}^j, g_{2,n}^j,\cdots,g_{N,n}^j),$$

$$j = 1, 2.$$

We obtain the system of equations,

$$\left\{A + (-\lambda + n^2 + \frac{1}{2})B + (-2\lambda + \frac{1}{2})C - (\lambda + \frac{1}{4})D\right\}X_n^j = G_n^j, (j,n) \in Z_M.$$

Indeed, A and C are five-diagonal matrices, B is a three-diagonal matrix and D is a seven-diagonal matrix. Moreover all matrices are symmetric and positive definite.

6.3.2 *Spherical harmonic-generalized Laguerre spectral method*

We consider the problem (6.13),(6.14) in three space dimension and the spherical harmonic-generalized Laguerre spectral method to solve this problem. Let ϕ and θ be the longitude and the latitude respectively. Then the spherical surface is $S_3 = \{(\phi, \theta); 0 \le \phi \le 2\pi, -\frac{\pi}{2} \le \theta \le \frac{\pi}{2}\}$. The Laplace operator with spherical coordinates is given by

$$\triangle = \frac{1}{r^2}\partial_r(r^2\partial_r) + \frac{1}{r^2\cos\theta}\partial_\theta(\cos\theta\partial_\theta) + \frac{1}{r^2\cos^2\theta}\partial_\phi^2.$$

Being Analogous to the investigation of (6.13),(6.14) we make the variable transformation

$$r = \rho + 1, \tilde{u} = e^{\frac{\beta}{2}\rho}u, \tilde{f} = (\rho+1)^2 e^{\frac{\beta}{2}\rho}f.$$

Then the equation (6.13) is changed to

$$-(\rho+1)^2\partial_\rho^2\tilde{u} - (2(\rho+1) - \beta(\rho+1)^2)\partial_\rho\tilde{u}$$
$$- \frac{1}{\cos\theta}\partial_\theta(\cos\theta\partial_\theta\tilde{u}) - \frac{1}{\cos^2\theta}\partial_\phi^2\tilde{u} \tag{6.20}$$
$$+ (-\lambda(\rho+1)^2 + \beta(\rho+1) - \frac{\beta^2}{4}(\rho+1)^2)\tilde{u} = \tilde{f}.$$

We define some spaces. Let $D = \Lambda \times S_3$ and the weights be $\omega_{0,\beta}(\rho) = e^{-\beta\rho}$ and $\eta_\beta(\rho) = (\rho+1)^2 e^{-\beta\rho}$. The spaces $L_\chi^2(D)$ and $H_\chi^r(D)$ are defined in the usual way. We denote

$$H_{0,p,\omega_{0,\beta}}^1(D) = \{v \in H_{\omega_{0,\beta}}^1(D); v(\rho, \phi+2\pi, \theta) = v(\rho, \phi, \theta), v(0, \phi, \theta) = 0\},$$

and

$$H_{0,p,\eta_\beta,\omega_{0,\beta}}^1(D) = \{v \in H_{0,p,\omega_{0,\beta}}^1(D); \|v\|_{1,\eta_\beta,\omega_{0,\beta},D} < \infty\},$$

where

$$|v|_{1,\eta_\beta,\omega_{0,\beta},D} = (\|\partial_\rho v\|_{0,\eta_\beta,D}^2 + \|\frac{1}{\cos\theta}\partial_\phi v\|_{0,\omega_{0,\beta},D}^2 + \|\partial_\theta v\|_{0,\omega_{0,\beta},D}^2)^{\frac{1}{2}}.,$$

$$\|v\|_{1,\eta_\beta,\omega_{0,\beta},D} = (|v|_{1,\eta_\beta,\omega_{0,\beta},D}^2 + \|v\|_{0,\omega_{0,\beta},D}^2)^{\frac{1}{2}}.$$

Let $Y_{m,l}(\phi, \theta)$, $0 \le m < \infty$, $|l| \le m$, be the normalized spherical harmonic functions. They form the complete normalized $L^2(S_3)$-orthogonal system.

Let M be any positive integer, and define the finite dimensional space $\hat{V}_M(S_3)$ as

$$\hat{V}_M(S_3) = \text{span}\{Y_{m,l}(\phi,\theta); m \le M, |l| \le m\}.$$

Denote by $V_M(S_3)$ the subset of $\hat{V}_M(S_3)$ containing all real-valued functions. Let $V_{N,M} = P_{0,N}(\Lambda; V_M(S_3))$.

For ease of notations, let

$$\nabla_s v = \left(\frac{1}{\cos\theta}\partial_\phi v, \partial_\theta v\right)^T.$$

We define a bilinear form

$$a(\tilde{u}, v) = \int_D (\rho+1)^2 e^{-\beta\rho}\partial_\rho\tilde{u}\partial_\rho v \, dsd\rho + \int_D e^{-\beta\rho}\nabla_s\tilde{u}\nabla_s v \, dsd\rho$$
$$+ \int_D (-\lambda(\rho+1)^2 + \beta(\rho+1) - \frac{\beta^2}{4}(\rho+1)^2)e^{-\beta\rho}\tilde{u}v \, dsd\rho.$$

Obviously

$$|a(\tilde{u}, v)| \le \max(-\lambda + \frac{\beta^2}{2}, 1)\|\tilde{u}\|_{1,\eta_\beta,\omega_{0,\beta},D}\|v\|_{1,\eta_\beta,\omega_{0,\beta},D}, \qquad (6.21)$$

and for $\lambda \le -\frac{\beta^2}{4}$

$$a(v, v) \ge (-\lambda - \frac{\beta^2}{4})\|v\|^2_{1,\eta_\beta,\omega_{0,\beta},D}. \qquad (6.22)$$

The weak formulation of the problem is: Find $\tilde{u} \in H^1_{0,p,\eta_\beta,\omega_{0,\beta}}(D)$, such that

$$a(\tilde{u}, v) = (\tilde{f}, v)_{0,\omega_{0,\beta},D}, \qquad \forall v \in H^1_{0,p,\eta_\beta,\omega_{0,\beta}}(D). \qquad (6.23)$$

If $\lambda < -\frac{\beta^2}{4}$ and $f \in \left(H^1_{0,p,\eta_\beta,\omega_{0,\beta}}(D)\right)'$, then by (6.21),(6.22) and the Lax-Milgram theorem, (6.23) admits a unique solution.

The mixed spectral scheme for (6.23) is: Seek $u_{N,M} \in V_{N,M}$, such that

$$a(u_{N,M}, v) = (\tilde{f}, v)_{0,\omega_{0,\beta},D}, \qquad \forall v \in V_{N,M}. \qquad (6.24)$$

Under the same conditions the problem (6.24) also admits a unique solution.

We next describe the implementation for (6.24). Set $\psi_k(\rho) = \mathcal{L}^{(0,\beta)}_{k-1}(\rho) - \mathcal{L}^{(0,\beta)}_k(\rho)$ and

$$Z^1_{m,l}(\phi,\theta) = \frac{1}{\sqrt{2\pi}}\sin(l\phi)\mathcal{L}(\sin\theta), \quad Z^2_{m,l}(\phi,\theta) = \frac{1}{\sqrt{2\pi}}\cos(l\phi)\mathcal{L}(\sin\theta).$$

Moreover, let $v_{k,m,l}^j(\rho,\phi,\theta) = \psi_k(\rho)Z_{m,l}^j(\phi,\theta)$, $j = 1,2$, and

$$u_{N,M}(\rho,\phi,\theta) = \sum_{j=1}^{2}\sum_{k=1}^{N}\sum_{m=0}^{M}\sum_{l=-m}^{m} u_{k,m,l}^j v_{k,m,l}^j(\rho,\phi,\theta),$$

$$\tilde{f}(\rho,\phi,\theta) = \sum_{j=1}^{2}\sum_{m=0}^{M}\sum_{l=-m}^{m} f_{m,l}^j(\rho)Z_{m,l}^j(\phi,\theta).$$

Take $v = v_{k,m,l}^j(\rho,\phi,\theta)$, $j = 1,2$, in (6.24). Then we derive a linear system for the coefficients $u_{k,m,l}^j$. More precisely, let the matrices $A = (a_{n,k})$, $B = (b_{n,k})$, $C = (c_{n,k})$ and $D = (D_{n,k})$ with the following entries:

$$a_{n,k} = \int_\Lambda \rho^2 e^{-\beta\rho}\partial_\rho\psi_n(\rho)\partial_\rho\psi_k(\rho)\,d\rho, \quad b_{n,k} = \int_\Lambda \rho^2 e^{-\beta\rho}\psi_n(\rho)\psi_k(\rho)\,d\rho,$$

$$c_{n,k} = \int_\Lambda \rho e^{-\beta\rho}\psi_n(\rho)\psi_k(\rho)\,d\rho, \quad d_{n,k} = \int_\Lambda e^{-\beta\rho}\psi_n(\rho)\psi_k(\rho)\,d\rho.$$

Furthermore, let

$$X_{m,l}^j = (u_{1,m,l}^j, u_{2,m,l}^j, \cdots, u_{N,m,l}^j), \quad G_{m,l}^j = (g_{1,m,l}^j, g_{2,m,l}^j, \cdots, g_{N,m,l}^j),$$

$$j = 1,2,$$

where

$$g_{k,m,l}^j = \int_\Lambda e^{-\beta\rho}f_{m,l}^j(\rho)\psi_k(\rho)\,d\rho.$$

We obtain the system of equations,

$$\left\{A - (\lambda + \frac{\beta^2}{4})B + \beta C + m(m+1)D\right\}X_{m,l}^j = G_{m,l}^j, \qquad j = 1,2.$$

Indeed, A and C are five-diagonal matrices, B is a seven-diagonal matrix and D is a three-diagonal matrix. Moreover all matrices are symmetric and positive definite.

6.3.3 *Generalized Laguerre pseudo-spectral method*

We consider the spherically symmetric solution of the equation (6.20), which is reduced to

$$-(\rho+1)^2\partial_\rho^2\tilde{u} - (2(\rho+1) - \beta(\rho+1)^2)\partial_\rho\tilde{u}$$
$$+ (-\lambda(\rho+1)^2 + \beta(\rho+1) - \frac{\beta^2}{4}(\rho+1)^2)\tilde{u} = \tilde{f}. \qquad (6.25)$$

Let $\sigma_{\alpha,\beta}(\rho) = (\rho + 1)^{\alpha} e^{-\beta\rho}$, and denote $H^1_{0,\sigma_{\alpha,\beta}}(\Lambda) = \{v \in H^1_{\sigma_{\alpha,\beta}}(\Lambda); v(0) = 0\}$. The variational formulation of the boundary value problems of (6.25) is: Find $\tilde{u} \in H^1_{0,\sigma_{2,\beta}}(\Lambda) \bigcap L^2_{\sigma_{1,\beta}}(\Lambda)$, such that

$$a(\tilde{u}, v) = (\tilde{f}, v)_{0,\omega_{0,\beta}}, \qquad \forall v \in H^1_{0,\sigma_{2,\beta}}(\Lambda) \bigcap L^2_{\sigma_{1,\beta}}(\Lambda), \qquad (6.26)$$

where the bilinear form is defined by

$$a(\tilde{u}, v) = (\partial_\rho \tilde{u}, \partial_\rho v)_{0,\sigma_{2,\beta}} - (\lambda + \frac{\beta^2}{4})(\tilde{u}, v)_{0,\sigma_{2,\beta}} + \beta(\tilde{u}, v)_{0,\sigma_{1,\beta}}.$$

One can verify readily that

$$|a(\tilde{u}, v)| \le C((1+\beta)\|\tilde{u}\|_{1,\sigma_{2,\beta}} + \beta^{\frac{1}{2}}\|\tilde{u}\|_{0,\sigma_{1,\beta}})((1+\beta)\|v\|_{1,\sigma_{2,\beta}} + \beta^{\frac{1}{2}}\|v\|_{0,\sigma_{12,\beta}}),$$

and for $\lambda < -\frac{\beta^2}{4}$,

$$|a(v, v)| \ge C(\|v\|^2_{1,\sigma_{2,\beta}} + \beta\|v\|^2_{0,\sigma_{2,\beta}}).$$

Hence, if $\tilde{f} \in \left(H^1_{0,\sigma_{2,\beta}}(\Lambda) \bigcap L^2_{\sigma_{1,\beta}}(\Lambda)\right)'$, then (6.26) admits a unique solution.

To define the generalized Laguerre pseudo-spectral scheme for (6.26) we need to introduce the generalized Laguerre-Gauss-Radan interpolation. Let $\xi^{(\alpha,\beta)}_{N,j}$ be the zeros of $x\partial_x \mathcal{L}^{(\alpha,\beta)}_{N+1}(x)$. They are arranged in ascending order. Denote the corresponding weights, $\omega^{(\alpha,\beta)}_{N,j}$, $0 \le j \le N$, such that

$$\int_\Lambda \phi(x)\omega_{\alpha,\beta}(x)\,dx = \sum_{j=0}^N \phi(\xi^{(\alpha,\beta)}_{N,j})\omega^{(\alpha,\beta)}_{N,j}, \qquad \forall \phi \in P_{2N}. \qquad (6.27)$$

The weights can be evaluated as

$$\omega^{(\alpha,\beta)}_{N,j} = \begin{cases} \frac{(\alpha+1)\Gamma^2(\alpha+1)\Gamma(N+1)}{\beta^{\alpha+1}\Gamma(N+\alpha+2)}, & j = 0, \\[2mm] \frac{\Gamma(N+\alpha+1)}{\beta^\alpha\Gamma(N+2)} \frac{1}{\mathcal{L}^{(\alpha,\beta)}_{N+1}(\xi^{(\alpha,\beta)}_{N,j})\partial_x \mathcal{L}^{(\alpha,\beta)}_N(\xi^{(\alpha,\beta)}_{N,j})}, & 1 \le j \le N. \end{cases}$$

The discrete inner product and norm are defined by

$$(u, v)_{0,\omega_{\alpha,\beta},N} = \sum_{j=0}^N u(\xi^{(\alpha,\beta)}_{N,j})v(\xi^{(\alpha,\beta)}_{N,j})\omega^{(\alpha,\beta)}_{N,j}, \qquad \|v\|_{0,\omega_{\alpha,\beta},N} = (v, v)^{\frac{1}{2}}_{0,\omega_{\alpha,\beta},N}.$$

By the exactness of (6.27)

$$(u, v)_{0,\omega_{\alpha,\beta},N} = (u, v)_{0,\omega_{\alpha,\beta}}, \qquad \forall uv \in P_{2N}. \qquad (6.28)$$

In particular

$$\|u\|_{0,\omega_{\alpha,\beta},N} = \|u\|_{0,\omega_{\alpha,\beta}}, \qquad \forall u \in P_N.$$

The generalized Laguerre-Gauss-Radan interpolation $I_{N,\alpha,\beta}v \in P_N$ is defined by

$$I_{N,\alpha,\beta}v(\xi_{N,j}^{(\alpha,\beta)}) = v(\xi_{N,j}^{(\alpha,\beta)}), \qquad 0 \le j \le N.$$

The generalized Laguerre pseudo-spectral scheme for (6.26) is: Seek $u_N \in P_{0,N} = \{u \in P_N; u(0) = 0\}$, such that

$$\tilde{a}(u_N, v) = (\tilde{f}, v)_{0,\omega_{0,\beta},N}, \qquad \forall v \in P_{0,N}, \tag{6.29}$$

where

$$\tilde{a}(u, v) = (\partial_x u, \partial_x v)_{0,\omega_{2,\beta},N} + 2(\partial_x u, \partial_x v)_{0,\omega_{1,\beta},N} + (\partial_x u, \partial_x v)_{0,\omega_{0,\beta},N}$$
$$- (\lambda + \frac{1}{4}\beta^2)(u, v)_{0,\omega_{2,\beta},N} - (2\lambda + \frac{1}{2}\beta^2 - \beta)(u, v)_{0,\omega_{1,\beta},N}$$
$$- (\lambda + \frac{1}{4}\beta^2 - \beta)(u, v)_{0,\omega_{0,\beta},N}.$$

According to (6.28), the formulation (6.29) is equivalent to

$$\tilde{a}(u_N, v) = (I_{N,0,\beta}\tilde{f}, v)_{0,\omega_{0,\beta}}, \qquad \forall v \in P_{0,N}.$$

6.3.4 *Nonlinear equations*

As an example, we consider the spectral method for the initial boundary value problem of the Burgers equation,

$$\partial_t u + \frac{1}{2}\partial_x(u^2) = \mu\partial_x^2 u + f(x,t), \qquad x \in (0,\infty), \tag{6.30}$$

$$u|_{x=0} = 0, \tag{6.31}$$

$$u|_{t=0} = u_0(x), \tag{6.32}$$

where $\mu > 0$ is the viscosity and the conditions on f and u_0 will be specified later on. Let $\tilde{u} = e^{x/2}u$, $\tilde{f} = e^{x/2}f$, and $\tilde{u}_0 = e^{x/2}u_0$, then the equation and the initial condition are transferred to,

$$\partial_t\tilde{u} + \frac{1}{2}e^{x/2}\partial_x(e^{-x}\tilde{u}^2) = \mu(\partial_x^2\tilde{u} - \partial_x\tilde{u} + \frac{1}{4}\tilde{u}) + \tilde{f},$$

and

$$\tilde{u}|_{t=0} = \tilde{u}_0.$$

Let the weight be $\omega = e^{-x}$, and let $\Pi_N : H^1_{0,\omega}(\Lambda) \to P_{0,N}(\Lambda)$ be an orthogonal projection operator, defined by: $\Pi_N u \in P_{0,N}(\Lambda)$,

$$(\partial_x(\Pi_N u - u), \partial_x v)_{0,\omega} + (\Pi_N u - u, v)_{0,\omega} = 0, \qquad \forall v \in P_{0,N}(\Lambda).$$

We assume that $\tilde{u}_0 \in H^1_{0,\omega}(\Lambda)$ and $\tilde{f} \in L^\infty((0,T); L^2_\omega(\Lambda))$, then multiply the equation by ωv, and take integral on $\Lambda = (0,\infty)$ to get

$$\partial_t(\tilde{u}, v)_{0,\omega} + \frac{1}{2}\int_\Lambda e^{-x/2}\partial_x(e^{-x}\tilde{u}^2)v\,dx + \mu(\partial_x\tilde{u}, \partial_x v)_{0,\omega} - \frac{\mu}{4}(\tilde{u}, v)_{0,\omega} = (\tilde{f}, v)_{0,\omega}. \tag{6.33}$$

Thus the Laguerre spectral scheme is: Find $u_N(t) \in P_N(\Lambda)$ for $t > 0$, such that $u_N|_{x=0} = 0$, and

$$\partial_t(u_N, v)_{0,\omega} + \frac{1}{2}\int_\Lambda e^{-x/2}\partial_x(e^{-x}u_N^2)v\,dx + \mu(\partial_x u_N, \partial_x v)_{0,\omega}$$
$$- \frac{\mu}{4}(u_N, v)_{0,\omega} = (\tilde{f}, v)_{0,\omega}, \quad \forall v \in P_N(\Lambda), v|_{x=0} = 0, \tag{6.34}$$

and

$$u_N|_{t=0} = \Pi_N\tilde{u}_0. \tag{6.35}$$

We develop the function u_N into a series in terms of Laguerre polynomials:

$$u_N(t) = \sum_{l=0}^N c_l(t)\mathcal{L}_l(x),$$

then the system (6.34),(6.35) is an initial value problem of ordinary differential equations for $c_l(t)$. The function v is also developed in the same way, then thanks to the orthogonal property of Laguerre polynomials, the matrix associated with $(u_N, v)_{0,\omega}$ is the identity. Therefore the system (6.34),(6.35) at least admits a local solution. Next let us prove global existence of (6.34),(6.35). Set $v = u_N$ in (6.34), then we get

$$\partial_t\|u_N\|^2_{0,\omega} + \frac{1}{2}\int_\Lambda e^{-x/2}\partial_x(e^{-x}u_N^2)u_N\,dx + \mu|u_N|^2_{1,\omega} - \frac{\mu}{4}\|u_N\|^2_{0,\omega} = (\tilde{f}, u_N)_{0,\omega}.$$

We have

$$\frac{1}{2}\int_\Lambda e^{-x/2}\partial_x(e^{-x}u_N^2)u_N\,dx = \frac{1}{3}\int_\Lambda \partial_x\left(e^{-3x/2}u_N^3\right)dx = 0,$$

and

$$(\tilde{f}, u_N)_{0,\omega} \le \frac{1}{2}\|\tilde{f}\|_{0,\omega}^2 + \frac{1}{2}\|u_N\|_{0,\omega}^2.$$

Therefore it holds that

$$\partial_t \|u_N\|_{0,\omega}^2 + \mu|u_N|_{1,\omega}^2 \le \left(\frac{\mu}{4} + \frac{1}{2}\right)\|u_N\|_{0,\omega}^2 + \frac{1}{2}\|\tilde{f}\|_{0,\omega}^2. \tag{6.36}$$

Integrating (6.36) over $(0, t)$, $t \le T$, gives

$$\|u_N(t)\|_{0,\omega}^2 + \mu \int_0^t |u_N|_{1,\omega}^2 \, dt$$
$$\le \|u_N(0)\|_{0,\omega}^2 + \left(\frac{\mu}{4} + \frac{1}{2}\right)\int_0^t \|u_N\|_{0,\omega}^2 \, dt + \frac{1}{2}\int_0^t \|\tilde{f}\|_{0,\omega}^2 \, dt. \tag{6.37}$$

Set $E(t) = \|u_N(t)\|_{0,\omega}^2$. Thanks to the Gronwall inequality (Theorem 67) with $a = \frac{\mu}{4} + \frac{1}{2}$ and $b = \|u_N(0)\|_{0,\omega}^2 + \frac{1}{2}\int_0^T \|\tilde{f}\|_{0,\omega}^2 \, dt$, it holds that

$$E(t) \le \left(\|u_N(0)\|_{0,\omega}^2 + \frac{1}{2}\int_0^T \|\tilde{f}\|_{0,\omega}^2 \, dt\right) e^{(\frac{\mu}{4} + \frac{1}{2})t}.$$

Let $t = T$ then

$$E(T) \le \left(\|u_N(0)\|_{0,\omega}^2 + \frac{1}{2}\int_0^T \|\tilde{f}\|_{0,\omega}^2 \, dt\right) e^{(\frac{\mu}{4} + \frac{1}{2})T}.$$

Since T is arbitrary, we replace T by t and obtain finally that

$$E(t) \le \left(\|u_N(0)\|_{0,\omega}^2 + \frac{1}{2}\int_0^t \|\tilde{f}\|_{0,\omega}^2 \, dt\right) e^{(\frac{\mu}{4} + \frac{1}{2})t}. \tag{6.38}$$

$E(t)^{1/2}$ plays the role of a norm on $P_N(\Lambda)$, so the solution to (6.34),(6.35) is bounded uniformly with respect to t. The solution can be extended continuously to T, which proves the global existence.

6.4 Jacobi spectral methods

As an example we consider the spectral method for the initial-boundary value problem of the Klein-Gordon equation,

$$\partial_t^2 u + u^3 = \partial_x^2 u + f(x, t), \qquad x \in (0, \infty), \tag{6.39}$$

$$u|_{x=0} = 0, \tag{6.40}$$

$$u|_{t=0} = u_0(x), \quad \partial_t u|_{t=0} = u_1(x), \tag{6.41}$$

where the conditions on f, u_1 and u_1 will be specified later on. We introduce a new variable ξ and set $x = -2\ln(1 - \xi) + 2\ln 2$, then the equation (6.39) is changed to

$$\partial_t^2 u + u^3 = \frac{1}{4}(1 - \xi)\partial_\xi((1 - \xi)\partial_\xi u) + f, \quad \xi \in (-1, 1).$$

Letting $\Lambda = (-1, 1)$ and $\omega = (1-\xi)^2$, we define $V = \{u \in L^4(\Lambda); \|\partial_\xi u\|_{0,\omega} < \infty, u(-1) = 0\}$, then the weak formulation is: Find $u(t) \in V$ such that $u|_{t=0} = u_0$, $\partial_t u|_{t=0} = u_1$, and

$$(\partial_t^2 u, v) + (u^3, v) + \frac{1}{4}((1 - \xi)\partial_\xi u, \partial_\xi((1 - \xi)v)) = (f, v), \quad \forall v \in V,$$

where $(\cdot, \cdot)$ is the L^2-inner product. The Jacobi spectral scheme is: Find $u_N(t) \in P_N(\Lambda)$ such that $u_N(t)|_{\xi=-1} = 0$, $u_N|_{t=0} = u_{0N}$, $\partial_t u_N|_{t=0} = u_{1N}$, and

$$(\partial_t^2 u_N, v) + (u_N^3, v) + \frac{1}{4}((1-\xi)\partial_\xi u_N, \partial_\xi((1-\xi)v)) = (f, v), \quad \forall v \in P_N, v(-1) = 0$$

where $u_{0N}, u_{1N} \in P_N(\Lambda)$ are some approximations of u_0 and u_1. Being analogous to the investigation of (6.34),(6.35), to prove global existence we only need to verify a global estimate. Let $v = \partial_t u_N$, then we get

$$\int_\Lambda \left\{ \frac{1}{2}\partial_t(\partial_t u_N)^2 + \frac{1}{4}\partial_t(u_N^4) \right\} d\xi + \frac{1}{4}((1-\xi)\partial_\xi u_N, \partial_\xi((1-\xi)\partial_t u_N)) = (f, \partial_t u_N).$$

Let us estimate the terms as follows:

$$\frac{1}{4}((1 - \xi)\partial_\xi u_N, \partial_\xi((1 - \xi)\partial_t u_N))$$

$$= \frac{1}{4}\int_\Lambda (1 - \xi)^2 \partial_\xi u_N \partial_\xi \partial_t u_N \, d\xi - \frac{1}{4}\int_\Lambda (1 - \xi)\partial_\xi u_N \partial_t u_N \, d\xi$$

$$= \frac{1}{8}\int_\Lambda (1 - \xi)^2 \partial_t(\partial_\xi u_N)^2 \, d\xi - \frac{1}{4}\int_\Lambda (1 - \xi)\partial_\xi u_N \partial_t u_N \, d\xi$$

$$\geq \frac{1}{8}\int_\Lambda (1 - \xi)^2 \partial_t(\partial_\xi u_N)^2 \, d\xi - \frac{1}{8}\int_\Lambda (1 - \xi)^2(\partial_\xi u_N)^2 \, d\xi - \frac{1}{8}\int_\Lambda (\partial_t u_N)^2 \, d\xi.$$

$$(f, \partial_t u_N) \leq \frac{1}{2}\int_\Lambda f^2 \, d\xi + \frac{1}{2}\int_\Lambda (\partial_t u_N)^2 \, d\xi.$$

We set

$$E(t) = \int_\Lambda \left\{ \frac{1}{2}(\partial_t u_N)^2 + \frac{1}{4}(u_N^4) + \frac{1}{8}(1-\xi)^2(\partial_\xi u_N)^2 \right\} d\xi,$$

then it follows that

$$\partial_t E(t) \le 2E(t) + \frac{1}{2}\int_\Lambda f^2 \, d\xi.$$

By integrating over the interval $(0,t)$ it yields

$$E(t) \le 2\int_0^t E(\tau) \, d\tau + E(0) + \frac{1}{2}\int_0^t \int_\Lambda f^2 \, d\xi.$$

We assume that $E(0) < \infty$ and $\int_0^T \int_\Lambda f^2 \, d\xi < \infty$, then apply the Gronwall inequality to obtain the boundedness of $E(t)$, then global existence follows.

6.5   Rational and irrational spectral methods

The rational spectral method can be applied to the initial boundary value problem of the Klein-Gordon equation (6.39)-(6.41). The weight is $\omega = \frac{1}{(x+1)^2}$, and we set $\Lambda = (0,\infty)$ and $V = H^1_{0,\omega}(\Lambda) \bigcap L^4_\omega(\Lambda)$. We multiply the equation (6.39) by ωv, then integrate the equation over the interval Λ, thus the weak formulation of the problem is derived as: Find $u(t) \in V$, such that $u|_{t=0} = u_0$, $\partial_t u|_{t=0} = u_1$, and

$$\int_\Lambda \frac{1}{(x+1)^2} \left\{ \partial_t^2 uv + u^3 v - \frac{2}{x+1}\partial_x uv + \partial_x u \partial_x v \right\} dx$$
$$= \int_\Lambda \frac{1}{(x+1)^2} fv \, dx, \quad \forall v \in V.$$

We define a finite dimensional subspace of V:

$$V_N = \left\{ u = \sqrt{2}v\left(\frac{x-1}{x+1}\right) ; v \in P_N(\tilde{\Lambda}), v(-1) = 0 \right\},$$

where $\tilde{\Lambda} = (-1,1)$. The rational spectral spectral scheme is: Find $u_N(t) \in V_N$, such that $u_N|_{t=0} = u_{0N}$, $\partial_t u_N|_{t=0} = u_{1N}$, and

$$\int_\Lambda \frac{1}{(x+1)^2} \left\{ \partial_t^2 u_N v + u_N^3 v - \frac{2}{x+1}\partial_x u_N v + \partial_x u_N \partial_x v \right\} dx$$
$$= \int_\Lambda \frac{1}{(x+1)^2} fv \, dx, \quad \forall v \in V_N,$$

where $u_{0N}, u_{1N} \in V_N$ are the orthogonal projection of u_0 and u_1, namely,

$$(\partial_x(u_{0N} - u_0), \partial_x v)_{0,\omega} + (u_{0N} - u_0, v)_{0,\omega} = 0, \quad \forall v \in V_N,$$

and

$$(\partial_x(u_{1N} - u_1), \partial_x v)_{0,\omega} + (u_{1N} - u_1, v)_{0,\omega} = 0, \quad \forall v \in V_N.$$

Being analogous to the investigation of (6.34),(6.35), to prove global existence we only need to verify a global estimate. Let $v = \partial_t u_N$, then we get

$$\int_\Lambda \frac{1}{(x+1)^2} \left\{ \partial_t^2 u_N \partial_t u_N + u_N^3 \partial_t u_N - \frac{2}{x+1} \partial_x u_N \partial_t u_N + \partial_x u_N \partial_x \partial_t u_N \right\} dx$$
$$= \int_\Lambda \frac{1}{(x+1)^2} f \partial_t u_N \, dx,$$

where

$$\partial_t^2 u_N \partial_t u_N = \frac{1}{2} \partial_t (\partial_t u_N)^2, \quad u_N^3 \partial_t u_N = \frac{1}{4} \partial_t (u_N^4),$$

$$\left| \int_\Lambda \frac{2}{(x+1)^3} \partial_x u_N \partial_t u_N \, dx \right| \leq \int_\Lambda \frac{1}{(x+1)^2} (\partial_x u_N)^2 \, dx + \int_\Lambda \frac{1}{(x+1)^2} (\partial_t u_N)^2 \, dx,$$

$$\partial_x u_N \partial_x \partial_t u_N = \frac{1}{2} \partial_t (\partial_x u_N)^2,$$

and

$$\left| \int_\Lambda \frac{1}{(x+1)^2} f \partial_t u_N \, dx \right| \leq \int_\Lambda \frac{1}{(x+1)^2} (\partial_t u_N)^2 \, dx + \int_\Lambda \frac{1}{4(x+1)^2} f^2 \, dx.$$

We set

$$E(t) = \int_\Lambda \frac{1}{2(x+1)^2} \left\{ (\partial_t u_N)^2 + \frac{1}{2}(u_N^4) + (\partial_x u_N)^2 \right\} dx,$$

then it follows that

$$\partial_t E(t) \leq 4E(t) + \frac{1}{2} \int_\Lambda f^2 \, dx.$$

By integrating over the interval $(0, t)$ it yields

$$E(t) \leq 4 \int_0^t E(\tau) \, d\tau + E(0) + \int_0^t \int_\Lambda \frac{1}{2(x+1)^2} f^2 \, d\xi.$$

We assume that $E(0) < \infty$ and $\int_0^T \int_\Lambda \frac{1}{(x+1)^2} f^2 \, d\xi < \infty$, then apply the Gronwall inequality to obtain the boundedness of $E(t)$, then global existence follows.

6.6 Error estimates

The error estimates of the spectral method rely on the approximation theory of orthogonal sets of functions. Here we study the error estimates for the Laguerre spectral methods as examples.

To begin with, we prove some properties of weighted Sobolev spaces. For definiteness we denote $\Lambda = (0, \infty)$, $\omega(x) = e^{-\beta x}$ and $\omega_{\alpha,\beta}(x) = x^\alpha e^{-\beta x}$ in this section.

Lemma 74. *We assume that $\beta \in (0, 2)$, then it holds that*

$$\|e^{-\frac{\beta x}{2}} v\|_{0,\infty} \le \|v\|_{1,\omega}, \qquad \forall v \in H_{0,\omega}^1(\Lambda). \tag{6.42}$$

Proof. Thanks to the embedding theorem , $v \in C(\Lambda)$. We have

$$
\begin{aligned}
(e^{-\frac{\beta x}{2}} v(x))^2 &= \int_0^x \partial_t (e^{-\frac{\beta t}{2}} v(t))^2 \, dt \\
&= \int_0^x e^{-\beta t} 2 v(t) \partial_t v(t), dt - \beta \int_0^x e^{-\beta t} v^2(t) \, dt \\
&\le \int_0^x e^{-\beta t} (\partial_t v(t))^2 \, dt + (1 - \beta) \int_0^x e^{-\beta t} v^2(t) \, dt.
\end{aligned}
$$

Thus the proof is complete. □

Lemma 75. *Let $P_{0,N} = \{p \in P_N; p(0) = 0\}$, then*

$$\inf_{v_N \in P_{0,N+1}} \|v_N - v\|_{1,\omega} \le C N^{-\frac{r}{2}} \|\partial_x^{r+1} v\|_{0,\omega_{r,\beta}},$$

$$\forall v \in H_{0,\omega_{r,\beta}}^1(\Lambda) \bigcap H_{\omega_{r,\beta}}^{r+1}(\Lambda), r > 0 \tag{6.43}$$

for $N > r$, where the constant C depends on β only.

Proof.   We have

$$\int_0^\infty e^{-\beta x} v^2(x)\, dx = \int_0^\infty e^{-\beta x} \int_0^x 2v(t)\partial_t v(t)\, dt$$

$$= \int_0^\infty 2v(t)\partial_t v(t)\, dt \int_t^\infty e^{-\beta x}\, dx = 2\beta^{-1} \int_0^\infty e^{-\beta t} v(t)\partial_t v(t)\, dt$$

$$\leq 2\beta^{-1} \left(\int_0^\infty e^{-\beta t} v^2(t)\, dt \right)^{\frac{1}{2}} \left(\int_0^\infty e^{-\beta t}(\partial_t v(t))^2\, dt \right)^{\frac{1}{2}}.$$

Therefore

$$\left(\int_0^\infty e^{-\beta t} v^2(t)\, dt \right)^{\frac{1}{2}} \leq 2\beta^{-1} \left(\int_0^\infty e^{-\beta t}(\partial_t v(t))^2\, dt \right)^{\frac{1}{2}},$$

that is

$$\|v\|_{0,\omega} \leq 2\beta^{-1}\|\partial_x v\|_{0,\omega}. \tag{6.44}$$

We develop the function $\partial_x v$ in terms of generalized Laguerre polynomials:

$$\partial_x v(x) = \sum_{l=0}^\infty c_l \mathcal{L}_l^{(0,\beta)}(x),$$

where

$$c_l = \beta \int_0^\infty e^{-\beta x} \partial_x v(x) \mathcal{L}_l^{(0,\beta)}(x)\, dx.$$

Let

$$w_N(x) = \sum_{l=0}^N c_l \mathcal{L}_l^{(0,\beta)}(x),$$

then

$$w_N(x) - \partial_x v(x) = - \sum_{l=N+1}^\infty c_l \mathcal{L}_l^{(0,\beta)}(x).$$

Owing to the orthogonal property, it holds that

$$\|w_N - \partial_x v\|_{0,\omega}^2 = \sum_{l=N+1}^\infty c_l^2 \gamma_l^{0,\beta},$$

where $\gamma_l^{0,\beta}$ are given in (6.12). On the other hand

$$\|\partial_x^{r+1}v\|_{0,\omega_{r,\beta}}^2 = \|\partial_x^r \sum_{l=0}^{\infty} c_l \mathcal{L}_l^{(0,\beta)}\|_{0,\omega_{r,\beta}}^2 = \|\sum_{l=r}^{\infty} c_l \mathcal{L}_{l-r}^{(r,\beta)}\|_{0,\omega_{r,\beta}}^2 = \sum_{l=r}^{\infty} c_l^2 \gamma_{l-r}^{r,\beta}.$$

Thanks to (6.12) we obtain

$$\frac{\gamma_l^{(0,\beta)}}{\gamma_{l-r}^{(r,\beta)}} = \frac{\beta^r(l-r)!}{l!} \leq \left(\frac{\beta}{N}\right)^r.$$

Therefore

$$\|w_N - \partial_x v\|_{0,\omega}^2 \leq \left(\frac{\beta}{N}\right)^r \|\partial_x^{r+1}v\|_{0,\omega_{r,\beta}}^2$$

Letting $v_N(x) = \int_0^x w_N(x)\,dx$, we apply (6.44) for $v_N - v$, then (6.43) follows. $\square$

Having proved the above primary results, the error estimates for the schemes (6.6) and (6.7) are routine. Because

$$\alpha\|u - p\|_{1,\omega} \leq a(u-p, u-p) = a(u-p, u-q), \quad \forall q \in P_{0,N},$$

and

$$a(u-p, u-q) < M\|u-p\|_{1,\omega}\|u-q\|_{1,\omega},$$

we obtain that

$$\|u-p\|_{1,\omega} \leq \frac{M}{\alpha} \inf_{q \in P_{0,N}} \|u-q\|_{1,P_{0,N}} \leq CN^{-\frac{r}{2}}\|\partial_x^{r+1}u\|_{0,\omega_{r,\beta}},$$

provided the right hand side is bounded.

We turn now to the error estimate of the scheme (6.34),(6.35) for the problem (6.30)-(6.32). The assumptions on the functions u_0 and f will be given in the process of estimation. We first assume that u_0 and f are bounded functions, then by the maximum principle of parabolic equations the exact solution u is bounded, $|u| \leq M$. Then it yields that $|e^{-\frac{x}{2}}\tilde{u}| \leq M$. As $N \to \infty$ $\Pi_N \tilde{u}_0$ converges to $\tilde{u}_0$ in H_ω^1 with $\beta = 1$. Owing to Lemma 75 and Lemma 74 with $\beta = 1$, $|e^{-\frac{x}{2}}(\Pi_N\tilde{u}_0 - \tilde{u}_0)| < M$ if N is large enough. We notice that there are only finite terms in the spectral expression of u_N, hence by continuity there exists $T^* > 0$, such that $|e^{-\frac{x}{2}}(u_N - \tilde{u})| < M$ for $t \in [0, T^*]$. Let us derive the energy estimate on the interval $t \in [0, T^*]$ first.

Let $w_N = \Pi_N \tilde{u}$, then being analogous to the estimates for (6.6),(6.7) we get

$$\|\tilde{u} - w_N\|_{1,\omega} \leq CN^{-\frac{r}{2}} \|\partial_x^{r+1} \tilde{u}\|_{0,\omega_{r,1}}, \qquad (6.45)$$

provided the right hand side is bounded. It remains to estimate $e_N = u_N - w_N$. By (6.33) and (6.34) we take $v = e_N$ to get

$$(\partial_t(u_N - \tilde{u}), e_N)_{0,\omega} + \frac{1}{2}\int_\Lambda e^{-x/2}\{\partial_x(e^{-x}u_N^2) - \partial_x(e^{-x}\tilde{u}^2)\}e_N\, dx$$

$$+ \mu(\partial_x(u_N - \tilde{u}), \partial_x e_N)_{0,\omega} - \frac{\mu}{4}(u_N - \tilde{u}, e_N)_{0,\omega} = 0.$$

We note that $|e^{-\frac{x}{2}}(u_N + \tilde{u})| < 3M$, so the nonlinear term can be estimated as the following:

$$\left| \frac{1}{2}\int_\Lambda e^{-x/2}\{\partial_x(e^{-x}u_N^2) - \partial_x(e^{-x}\tilde{u}^2)\}e_N\, dx \right|$$

$$= \left| -\frac{1}{2}\int_\Lambda e^{-x/2}\{e^{-x}u_N^2 - e^{-x}\tilde{u}^2\}\partial_x e_N\, dx \right.$$

$$\left. + \frac{1}{4}\int_\Lambda e^{-x/2}\{e^{-x}u_N^2 - e^{-x}\tilde{u}^2\}e_N\, dx \right|$$

$$= \int_\Lambda e^{-3x/2}(-\frac{1}{2}\partial_x e_N + \frac{1}{4}e_N)(u_N^2 - \tilde{u}^2)\, dx$$

$$\leq 3M\int_\Lambda e^{-x}|-\frac{1}{2}\partial_x e_N + \frac{1}{4}e_N| \cdot |u_N - \tilde{u}|\, dx$$

$$\leq \varepsilon\int_\Lambda e^{-x}(-\frac{1}{2}\partial_x e_N + \frac{1}{4}e_N)^2\, dx + \frac{9M^2}{4\varepsilon}\int_\Lambda e^{-x}(u_N - \tilde{u})^2\, dx,$$

where $\varepsilon > 0$ is a constant to be determined. Noting that $u_N - \tilde{u} = e_N + (w_N - \tilde{u})$ we get

$$\frac{1}{2}\partial_t\|e_N\|_{0,\omega}^2 + \mu|e_N|_{1,\omega}^2 - \frac{\mu}{4}\|e_N\|_{0,\omega}^2 = R,$$

where

$$|R| \leq \varepsilon|e_N|_{1,\omega}^2 + C\{\|\partial_t(w_N - \tilde{u})\|_{0,\omega}^2 + (1 + \varepsilon^{-1})(\|e_N\|_{0,\omega}^2 + \|w_N - \tilde{u}\|_{1,\omega}^2)\}.$$

Being analogous to (6.45) we have the estimate

$$\|\partial_t(\tilde{u} - w_N)\|_{1,\omega} \leq CN^{-\frac{r}{2}} \|\partial_t\partial_x^{r+1}\tilde{u}\|_{0,\omega_{r,1}}, \qquad (6.46)$$

provided the right hand side is bounded. Let $\varepsilon = \frac{\mu}{2}$, then we obtain

$$\frac{1}{2}\partial_t \|e_N\|_{0,\omega}^2 + \frac{\mu}{2}|e_N|_{1,\omega}^2 \leq C\|e_N\|_{0,\omega}^2 + CN^{-r}.$$

Integrate over the interval $(0, t)$ to get

$$\frac{1}{2}\|e_N\|_{0,\omega}^2 + \frac{\mu}{2}\int_0^t |e_N|_{1,\omega}^2 \, dt \leq C\int_0^t \|e_N\|_{0,\omega}^2 \, dt + \frac{1}{2}\|e_N(0)\|_{0,\omega}^2 + CtN^{-r}.$$

The inequality (6.45) implies

$$\|e_N(0)\|_{0,\omega}^2 \leq CN^{-r}.$$

Applying the Gronwall inequality we get an estimate on $(0, T^*)$,

$$\frac{1}{2}\|e_N\|_{0,\omega}^2 + \frac{\mu}{2}\int_0^t |e_N|_{1,\omega}^2 \, dt \leq Ce^{Ct}(1+t)N^{-r}.$$

Along with (6.45) we obtain

$$\frac{1}{2}\|u_N - \tilde{u}\|_{0,\omega}^2 + \frac{\mu}{2}\int_0^t |u_N - \tilde{u}|_{1,\omega}^2 \, dt \leq Ce^{Ct}(1+t)N^{-r}. \tag{6.47}$$

Then we take $v = \partial_t e_N$ in (6.33) and (6.34) to get

$$(\partial_t(u_N-\tilde{u}), \partial_t e_N)_{0,\omega} + \frac{1}{2}\int_\Lambda e^{-x/2}\{\partial_x(e^{-x}u_N^2) - \partial_x(e^{-x}\tilde{u}^2)\}\partial_t e_N \, dx$$
$$+ \mu(\partial_x(u_N - \tilde{u}), \partial_x\partial_t e_N)_{0,\omega} - \frac{\mu}{4}(u_N - \tilde{u}, \partial_t e_N)_{0,\omega} = 0. \tag{6.48}$$

The linear terms can be estimated as follows:

$$(\partial_t(u_N - \tilde{u}), \partial_t e_N)_{0,\omega} - \|\partial_t e_N\|_{0,\omega}^2 - (\partial_t(\tilde{u} - w_N), \partial_t e_N)_{0,\omega}$$
$$\geq \|\partial_t e_N\|_{0,\omega}^2 - (\|\partial_t(\tilde{u} - w_N)\|_{0,\omega}^2 + \frac{1}{4}\|\partial_t e_N\|_{0,\omega}^2)$$
$$\geq \frac{3}{4}\|\partial_t e_N\|_{0,\omega}^2 - CN^{-r}\|\partial_t\partial_x^{r+1}\tilde{u}\|_{0,\omega_{r,1}}^2,$$

$$\mu(\partial_x(u_N - \tilde{u}), \partial_x\partial_t e_N)_{0,\omega}$$
$$= \mu(\partial_x(u_N - \tilde{u}), \partial_x\partial_t(u_N - \tilde{u}))_{0,\omega} + \mu(\partial_x(u_N - \tilde{u}), \partial_x\partial_t(\tilde{u} - w_N))_{0,\omega}$$
$$\geq \frac{\mu}{2}\partial_t|u_N - \tilde{u}|_{1,\omega}^2 - \frac{\mu}{2}|u_N - \tilde{u})|_{1,\omega}^2 - \frac{\mu}{2}|\partial_t(\tilde{u} - w_N)|_{1,\omega}^2$$
$$\geq \frac{\mu}{2}\partial_t|u_N - \tilde{u}|_{1,\omega}^2 - \frac{\mu}{2}|u_N - \tilde{u})|_{1,\omega}^2 - CN^{-r}\|\partial_t\partial_x^{r+1}\tilde{u}\|_{0,\omega_{r,1}}^2,$$

$$-\frac{\mu}{4}(u_N - \tilde{u}, \partial_t e_N)_{0,\omega} \geq -Ce^{Ct}(1+t)N^{-r} - \frac{1}{4}\|\partial_t e_N\|_{0,\omega}^2,$$

where we employ the estimates (6.46) and (6.47). The nonlinear term can be estimated as follows:

$$\left|\frac{1}{2}\int_\Lambda e^{-x/2}\{\partial_x(e^{-x}u_N^2) - \partial_x(e^{-x}\tilde{u}^2)\}\partial_t e_N\,dx\right|$$

$$=\left|\frac{1}{2}\int_\Lambda e^{-3x/2}\{\tilde{u}^2 - u_N^2 + 2u_N\partial_x u_N - 2\tilde{u}\partial_x\tilde{u}\}\partial_t e_N\,dx\right|$$

$$\leq\frac{1}{2}\|\partial_t e_N\|_{0,\omega}^2 + \frac{1}{8}\int_\Lambda e^{-2x}\{\tilde{u}^2 - u_N^2 + 2u_N\partial_x u_N - 2\tilde{u}\partial_x\tilde{u}\}^2\,dx \cong I.$$

We assume that the exact solution $\tilde{u}$ satisfies $|e^{-x/2}\partial_x\tilde{u}| < M_1$, then

$$e^{-2x}\{\tilde{u}^2 - u_N^2 + 2u_N\partial_x u_N - 2\tilde{u}\partial_x\tilde{u}\}^2$$

$$=e^{-2x}\{(\tilde{u} + u_N)(\tilde{u} - u_N) + 2\partial_x\tilde{u}(u_N - \tilde{u}) + 2u_N\partial_x(u_N - \tilde{u})\}^2$$

$$\leq 3e^{-x}\{(3M)^2(\tilde{u} - u_N)^2 + 4M_1^2(u_N - \tilde{u})^2 + 4(2M)^2(\partial_x(u_N - \tilde{u}))^2\}.$$

Then

$$I \leq \frac{1}{2}\|\partial_t e_N\|_{0,\omega}^2 + C\|u_N - \tilde{u}\|_{1,\omega}^2.$$

Substituting these estimates into (6.48) gives

$$\frac{\mu}{2}\partial_t|u_N - \tilde{u}|_{1,\omega}^2 \leq C|u_N - \tilde{u}|_{1,\omega}^2 + CN^{-r}.$$

Integrate over the interval $(0,t)$ to get

$$\frac{\mu}{2}|u_N - \tilde{u}|_{1,\omega}^2 \leq \frac{\mu}{2}|(u_N - \tilde{u})(0)|_{1,\omega}^2 + C\int_0^t |u_N - \tilde{u}|_{1,\omega}^2\,dt + CtN^{-r}.$$

Applying the Gronwall inequality we get another estimate on $(0, T^*)$,

$$|u_N - \tilde{u}|_{1,\omega}^2 \leq Ce^{Ct}(1+t)N^{-r}. \tag{6.49}$$

Thanks to Lemma 74 it holds that

$$|e^{-x/2}(u_N - \tilde{u})|^2 \leq Ce^{CT}(1+T)N^{-r}, \tag{6.50}$$

for $t < T^*$.

Let N be large enough so that $|e^{-x/2}(u_N - \tilde{u})| \leq \frac{M}{2}$ on $[0, T^*]$, then by the definition of T^* this interval can be extended. Since the constant C in (6.50) is independent of T^*, we can always extend this interval up to T, then the global error estimates (6.47) and (6.49) are valid for all $t \in [0, T]$.

Bibliography

Abarbanel, S. and Gottlieb, D. (1997). A mathematical analysis of the PML method, *J. Comput. Phys.*, 134, pp. 357-363.

Abarbanel, S. and Gottlieb, D. (1998). On the construction and analysis of absorbing layers in CEM, *Appl. Numer. Math.*, 27, pp. 331-340.

Abarbanel, S., Gottlieb, D. and Hesthaven, J.S. (2002). Long time behavior of the perfectly matched layer equations in computational electromagnetics, *J Sci. Comp.*, 17,4, pp. 405-422.

Adams, R.A. (1975). *Sobolev Spaces*, Academic Press, New York.

Andrew, W., Banalis, C. and Tirkas, P. (1995). A comparison of the Berenger perfectly matched layer and the Lindman higher-order ABC for the FDTD method, *IEEE Microwave Guided Wave Lett.*, 5, pp. 192-194.

Bao, G. and Wu, H. (2005). Convergence analysis of the perfectly matched layer problems for time-harmonic Maxwell's equations, *SIAM J. Numer. Anal.*, 43,5, pp. 2121-2143.

Bayliss, A. and Turkel, E. (1980). Radiation boundary conditions for wave-like equations, *Comm. Pure Appl. Math.*, 33,6, pp. 707-725.

Ben-Porat, G. and Givoli, D. (1995). Solution of unbounded domain problems using elliptic artificial boundary, *Comm. Numer. Meth. Engng.*, 11, pp. 735-741.

Berenger, J.P. (1994). A perfectly matched layer for the absorption of electromagnetic waves, *J. Comput. Phys.*, 114, pp. 185-200.

Berenger, J.P. (1996). Perfectly matched layer for the FDTD solution of wave-structure interaction problems, *IEEE Trans.Anttennas Propagat.*, 44, pp. 110-117.

Berenger, J.P. (1996). Three dimensional perfectly matched layer for the absorption of electromagnetic waves, *J. Comput. Phys.*, 127, pp. 363-379.

Berenger, J.P. (2002). Numerical reflection from FDTD-PMLs: a comparison of the split PML with the unsplit and CFS PMLs , *IEEE Trans.Anttennas Propagat.*, 50,3, pp. 258-265.

Bönisch,S., Heuveline, V. and Wittwer, P. (2005). Adaptive boundary conditions for exterior flow problems, *J. Math. Fluid Mech.*, 7, pp. 85-107.

Bossavit, A. and Verite, J.C. (1982). A mixed FEM-BEM method to solve 3-D eddy current problems, *IEEE Tran. Magnetics*, 18, pp. 431-435.

Boyd, J.P. (1987). Orthogonal rational functions on a semi-infinite interval, *J. Comput. Phys.*,70, pp. 63-88.

Boyd, J.P. (1987). Spectral methods using rational basis on an infinite interval, *J. Comput. Phys.*,69, pp. 112-142.

Bramble, J.H. and Pasciak, J.E. Analysis of a finite PML approximation for the three dimensional time-harmonic Maxwell and acoustic scattering problems, *Math. Comp.*.

Bruneau, C.H. (2000). Boundary conditions on artificial frontiers for incompressible and compressible Navier-Stokes equations, *RAIRO Model Math. Anal. Numer.*, 34, pp. 303-314.

Bruneau, C.H. and Fabrie, P. (1996). New efficient Boundary conditions for incompressible Navier-Stokes equations: a well-posedness result, *RAIRO Model Math. Anal. Numer.*, 30, pp. 815-840.

Bruno, O.P. and Reitich, F. (1996). Calculation of electromagnetic scattering via boundary variations and analytic continuation, *Appl. Comput. Electromagn. Soc. J.*, 11,1, pp. 17-31.

Cai, M. and Zhou, H. (1995). Proof of perfectly matched layer conditions in three dimensions, *Electron Lett.*, 31, pp. 1675-1676.

Calderon, A.P. and Zygmund, A. (1957). Singular integral operators and differential equations, *Amer. J. Math.*, 79, pp. 901-921.

Chen, B., Fang, D.G. and Zhou, B.H. (1995). Modified Berenger PML absorbing boundary condition for FDTD meshes, *IEEE Microwave Guided Wave Lett.*, 5, pp. 399-401.

Chen, Z. and Liu, X., An adaptive perfectly matched layers technique for time-harmonic scattering problems, *SIAM J. Numer. Anal.*, (to appear).

Chen, Z. and Wu, H. (2003). An adaptive finite element method with perfectly matched absorbing layers for the wave scattering by periodic structures, *SIAM J. Numer. Anal.*, 41, pp. 799-826.

Chew, W.C. and Weedon, W.H. (1994). A 3D perfectly matched medium from modified Maxwell's equations with stretched coordinates, *IEEE Microwave and Guided Wave Letters*, 7, pp. 599-604.

Christov, C.J. (1982) A complete orthogonal system of functions in $L^2(-\infty,\infty)$ space, *SIAM J. Appl. Math.*, 42, pp. 1337-1344.

Ciarlet, P.G. (1978). *The Finite Element Method for Elliptic Problems*, North-Holland, Amsterdam, New York, Oxford.

Ciarlet Jr, P. and Zou, J. (1997). Finite element convergence for the Darwin model to Maxwell's equations, *Math. Modelling Numer. Anal.*, 31,2, pp. 213-250.

Clement, P. (1975). Approximation by finite element functions using local regularization, *Rev. Francaise Automat. Informat. Researche Operationnelle Ser. Rouge Anal. Numer.*, R.2 pp. 77-84.

Collino, F. (1997). Perfectly matched layer for the paraxial equation, *J. Comput. phys.*, 131, pp. 164-170.

Collino, F. and Monk, P.B. (1998). The perfectly matched layer in curvilinear coordinates, *SIAM J. Sci. Comput.*, 19, pp. 2061-2090.

Collino, F. and Monk, P.B. (1998). Optimizing the perfectly matched layer, *Com-*

put. Methods Appl. Mech. Engng., 164, pp. 157-171.

Colton, D. and Kress, R. (1983). *Integral Equation Methods in Scattering Theory*, John Wiley and Sons.

Colton, D. and Kress, R. (1992). *Inverse Acoustic and Electromagnetic Scattering Theory*, Applied Mathematical Sciences, vol.93, Springer-Verlag.

Coulaud, O., Funaro, D. and Kavian, O. (1990). Laguerre spectral approximation of elliptic problems in exterior domains, *Comp. Mech. Appl. Mech. Engng.*, 80, pp. 451-458.

Degond, P. and Raviart, P.A. (1992). An analysis of the Darwin model of approximation to Maxwell's equations, *Forum Math.*, 4, pp. 13-44.

DeMoerloose, J. and Stuchly, M.a. (1995). Behavior of Berenger absorbing boundary condition for evanescent wave, *IEEE Microwave Guided Wave Lett.*, 5, pp. 344-346.

Deuring, P. (1997). Finite element methods for the Stokes system in three dimensional exterior domains, *Math. Methods Appl. Sci.*, 20, pp. 245-269.

Deuring, P. (1998). A stable mixed finite element method on truncated exterior domains, *RAIRO Model Math. Anal. Numer.*, 32, pp. 283-305.

Deuring, P. and Kracmar, S (2000). Artificial boundary conditions for the Oseen system in 3D exterior domains, *Analysis*, 20, pp. 65-90.

Deuring, P. and Kracmar, S (2004). Exterior stationary Navier-Stokes flows in 3D with non-zero velocity at infinity: approximation by flows in bounded domains, *Math. Nachr.*, 269-270, 1, pp. 86-115.

Du, Q. and Yu, D. (2000). Natural integral equation of parabolic initial boundary value problem and its numerical implementation, *Chinese J. Numer. Math. Appl.*, 22,1, pp. 88-101.

Du, Q. and Yu, D. (2001). On the natural integrl equation for two dimensional hyperbolic equation, *Acta Math. Appl. Sinica*, 24,1, pp. 17 26.

Du, Q. and Yu, D. (2003). Dirichlet-Neumann alternating algorithm based on the natural boundary reduction for time-dependent problems over an unbounded domain, *Applied Numerical Mathematics*, 44, pp. 471-486.

Engquist, B. and Majda, A. (1977). Absorbing boundary conditions for numerical simulation of waves, *Math. Comp.*, 31,139, pp. 629-651.

Engquist, B. and Majda, A. (1979). Radiation boundary conditions for acoustic and elastic calculations, *Comm. Pure Appl. Math.*, 32, pp. 313-357.

Engquist, B. and Zhao, H.-K. (1998). Absorbing boundary conditions for domain decomposition, *Applied Numerical Mathematics*, 27, pp. 341-365.

Ernst, O.G. (1996). A finite element capacitance matrix method for exterior Helmholtz problems, *Numer. Math.*, 75.

Fan, G.X. and Liu, Q.H. (2003). A strongle well-posed PML in lossy media, *IEEE Antennas and Wireless Propagation Letters, 2,7, pp. 97-100.*

Fang, J.Y. amd Mei, K.K. (1988). A super-absorbing boundary algorithm for numerical solving electromagnetic problems by time-domain finite difference method, *IEEE AP-S International Symposium*, Syracuse, pp. 427-475.

Fang, J. and Wu, Z. (1995). Generalized perfectly matched layer-An extension of Berebger perfectly matched layer boundary condition, *IEEE Microwave Guided Wave Lett.*, 5, pp. 451-453.

Farwig, R. (1992). A variational approach in weighted Sobolev spaces to the operator $-\triangle \, \partial/\partial X_1$ in exterior domains of $\mathbb{R}^3$, *Math. Z.*, 210, pp. 449-464.

Farwig, R. (1992). The stationary exterior 3D problem of Oseen and Navier-Stokes equations in anisotropically weighted Sobolev spaces, *Math. Z.*, 211, pp. 409-447.

Farwig, R. (1998). The stationary Navier-Stokes equations in a 3D exterior domain, in *Recent Topics on Mathematical Theory of Viscous Incompressible Fluid, ed. H. Kozono and Y. Shibata, Lecture Notes in Applied and Numerical Analysis*, Vol.16, Kinokyniya, Tokyo, pp. 53-115.

Farwig, R. and Sohr, H.(1998). Weighted estimates for the Oseen equations and Navier-Stokes equations in exterior domains, in Theory of the Navier-Stokes Equations, ed. J.G. Heywood, K. Masuda, R. Rautmann, and V.A. Solonnikov, Series on Advances in Mathematics in Applied Sciences, Vol.47, World Scientific, Singapore, pp. 11-31.

Feistauer, M. and Schwab, C. (1999). Coupled problems for viscous incompressible flow in exterior domains, in *Applied Nonlinear Analysis, ed. A. Sequeira et al.*, Kluwer/Plenum, New York, pp. 97-116.

Feistauer, M. and Schwab, C. (2001). Coupling of an interior Navier-Stokes problem with an exterior Oseen problem, *J. Math. Fluid Mech.*, 3, pp. 1-17.

Feng, K. (1980). Differential versus integral equations and finite versus infinite elements, *Math. Numer. Sinica*, 2,1, pp. 100-105.

Feng, K. (1982). Canonical boundary reduction and finite element method, *Proc. of Symposium on Finite Element Methods, (Hefei,1981)*, Science Press, Beijing, pp. 330-352.

Feng, K. (1983). Finite element method and natural boundary reduction, *Proceedings of the International Congress of Mathematicians*, Warszawa, pp.1439-1453.

Feng, K. (1984). Asymptotic radiation conditions for reduced wave equations, *J. Comp. Math.*, 2,2, pp. 130-138.

Feng, K. (1994). Natural boundary reduction and domain decomposition, *Collected Works of Feng Kang*, Defence Industry Press, Beijing, pp. 367-371.

Feng, K. and Yu, D. (1983). Canonical integral equations of elliptic boundary value problems and their numerical solutions, *Proceedings of China-France Symposium on the Finite Element Method (Feng, K. and Lions, J.L. ed.)*, Science Press and Gordon Breach, Beijing, New York, pp.211-252.

Feng, K. and Yu, D. (1994). A theorem for the natural integral operator of harmonic equation, *J. Math. Numer. Sinica*, 16,2, pp. 221-226.

Finn, R. (1965). On the exterior stationary problem for the Navier-Stokes equations, and associated perturbation problems, *Arch. Rational Mech. Anal.*, 19, pp. 363-406.

Friedlander, F.G. (1962). On the radiation field of pulse solutions of the wave equation, *Proc. Royal Soc. London, Ser.A*, 269, pp. 53-69.

Funaro, D. and Kavian, O. (1991). Approximation of some diffusion evolution equations in unbounded domains by Hermite functions, *Math. Comp.*, 57, pp. 597-619.

Galdi, G.P. (1994). *An Introduction to the Mathematical Theory of the Navier-*

Stokes equations, vol.I, vol.II, Springer-Verlag, New York.

Gatica, G.H. and Hsiao G.C. (1992). On the coupled BEM and FEM for a nonlinear exterior Dirichlet problem in R^2, *Numer. Math.*, 61, pp. 171-214.

Gatica, G.H. and Hsiao G.C. (1996). The uncoupling of boundary and finite element methods for nonlinear boundary value problems, *J.Math.Anal.Appl.*, 189,1, pp. 442-461.

Gatica, G.H. and Hsiao, G.C. (2000). A dual-dual formulation for the coupling of mixed FEM and BEM to hyperelascity, *SIAM J. Numer. Anay.*, 38,2, pp. 380-400.

Gendey, S.D. (1996). An anisotropic perfectly matched layer absorbing media for the truncation of FDTD lattices, *IEEE Trans. Antennas Propagat.*, 44, pp. 1630-1639.

Gendey, S.D. (1996). An anisotropic PML absorbing media for the FDTD simulation of fields in lossy and dispersive media, *Electromagnetics*, 16,4, pp. 399-415.

Gilbarg, D. and Trudinger, N.S. (1977). *Elliptic Partial Differential Equations of Second Order*, Springer-Verlag, Heidelberg, New York.

Girault, V. and Raviart, P.A. (1988). *Finite Element Methods for Navier-Stokes Equations*, Springer, Berlin.

Girault and Sequeira, (1991). *Arch.Ration.Mech.Anal.*, 114, pp. 303-333.

Givoli, D. (1991). Nonreflecting boundary conditions, *J. Comput. Phys.*, 94, pp. 1-29.

Givoli, D. (1992). *Numerical Methods for Problems in Infinite Domains*, Elsevier, Amsterdam.

Goldstein, C.I. (1981). The finite element method with nonuniform mesh sizes for unbounded domains, *Math. Comp.*, 36, pp. 387-404.

Goldstein, C.I. (1993). Multigrid methods for elliptic problems in unbounded domains, *SIAM J. Numer. Anal.*, 336, pp. 159-183.

Grosch, C.E. and Orszag, S.A. (1977). Numerical solution of problems in unbounded regions: coordinates transforms, *J. Comput. Phys.*, 25, pp. 273-296.

Grote, M.J. and Keller, J.B. (1995). Exact non-reflecting boundary conditions for the dependent wave equation, *SIAM J. Appl. Math.*, 55,2, pp. 280-297.

Grote, M.J. and Keller, J.B. (1996). Non-reflecting boundary conditions for the dependent scatering problems, *J. Comput. Phys.*, 127,1, pp. 52-65.

Guidotti, P. (2004). Elliptic and parabolic problems in unbounded domains, *Math. Nachr.*, 272,1, pp. 32-45.

Guirguis, G.H. (1987). On the coupling of boundary integral and finite element methods for the exterior Stokes problem in three dimensions, *Math. Comp.*, 49, pp. 370 380.

Guirguis, G.H. and Gunzberger, M.D. (1987). On the approximation of the exterior Stokes problem in three dimensions, *RAIRO Model Math. Anal. Numer.*, 21, pp. 445-464.

Guo, B. (1998). *Spectral Methods and Their Applications*, World Scientific, Singapore.

Guo, B. (1998). Gegenbauer approximation and its applications to differential equations on the whole line, *J. Math. Anal. Appl.*, 226, pp. 180-206.

Guo, B. (2000). Jacobi spectral approximation and its applications to differential equations on the half line, *J. Comp. Math.*, 18, pp. 95-112.

Guo, B. (2002). Some developments in spectral methods for nonlinear partial differential equations in unbounded domains, in *Differential Geometry and Related Topics, ed. Gu Chaohao, Hu Hesheng and Li Tatsien*, World Scientific, pp. 68-90.

Guo, B. and Ma, H. (2001). Composite Legendre-Laguerre approximation in unbounded domains, *J. Comp. Math.*, 19, pp. 101-112.

Guo, B. and Shen, J. (2000). Laguerre-Galerkin method for nonlinear partial differential equations on a semi-infinite interval, *Numer. Math.*, 86, pp. 635-654.

Guo, B. and Shen, J. (2001). On spectral approximations using modified Legendre rational functions: application to the Korteweg-de Vries equation on the half line, *Indiana Univ. Math. J.*, 50, pp. 181-204.

Guo, B. and Shen, J. (2006). Irrational approximations and their applications to partial differential equations in exterior domains, *Adv. comput. Math.*.

Guo, B., Shen, J. and Wang, Z. (2000). A rational approximation and its applications to differential equations on the half line, *J. Sci. Comp.*, 15, pp. 117-147.

Guo, B., Shen, J. and Wang, Z. (2002). Chebyshev rational spectral and pseudospectral methods on a semi-infinite interval, *Int. J. Numer. Meth. Engng.*, 53, pp. 65-84.

Guo, B., Shen, J. and Xu, C. (2005). Generalized Laguerre approximation and its applications to exterior problems, *J. Comp. Math.*, 23,2, pp. 113-130.

Guo, B. and Wang, L. (2006). Modified Laguerre pseudospectral method refined by multidomain Legendre pseudospectral approximation, *J. Comp. Appl. Math.*, 190, pp. 304-324.

Guo, B. Wang, L. and Wang, Z. (2006). Generalized Laguerre interpolation and pseudospectral method for unbounded domains, *SIAM J. Numer. Anal.*, 43,6, pp. 2567-2589.

Guo, B. and Zhang, X. (2005). A new generalized Laguerre spectral approximation and its applications, *Journal of Computational and Applied Mathematics*, 181, pp. 342-363.

Halpern, L. and Schatzmann, M. (1989). Artificial boundary conditions for incompressible viscous flows, *SIAM J. Math. Anal.*, 20, pp. 308-353.

Halpern, L. and Trefethen, L.N. (1988). Wide-angle one-way wave equations, *J. Acoust. Soc. Am.*, 84,4, pp. 1397-1404.

Han, H. and Bao, W. (1999). The discrete artificial boundary contition on a polygonal artificial boundary for the exterior problem of Poisson equation by using the direct method of lines, *Comp. Meth. Appl. Mech. Engng.*, 179, pp. 345-360.

Han, H. and Huang, Z. (2002). Exact and approximating boundary conditions for the parabolic problems on unbounded domains, *Comp. Math. Appl.*, 44, pp. 655-666.

Han, H. and Huang, Z. (2002). A class of artificial boundary conditions for heat equation in unbounded domains, *Comp. Math. Appl.*, 43, pp. 889-900.

Han, H. and Wu, X. (1985). Approximation of infinite boundary condition and its application to finite element methods, *J. Comp. Math.*, 33, pp. 179-192.

Han, H. and Wu, X. (1992). The approximation of the exact boundary conditions at an artificial boundary for linear elastic equations and its application, *Math. Comp.*, 59, pp. 21-37.

Han, H. and Ying, L. (1980). Huge element and local finite element method, *Acta Math. Appl. Sinica*, 3,3, pp. 237-249.

Hewett, D.W. and Nielson, C. (1978). A multidimensional quasineutral plasma simulation model, *J.Comput. Phys.*, 29, pp. 219-236.

Heywood, J.G., Rannacher, R. and Turek, S. (1992). Artificial boundaries and flux and pressure conditions for the incompressible Navier-Stokes equations, *Int. J. Numer. Meth. Fluids*, 22, pp. 325-352.

Higdon, R.L. (1986). Absorbing boundary conditions for difference approximations to the multi-dimensional wave equation, *Math. Comp.*, 47,176, pp. 437-459.

Higdon, R.L. (1987). Numerical absorbing boundary conditions for the wave equation, *Math. Comp.*, 49,179, pp. 65-90.

Hsiao, G. (1988). The coupling of BEM and FEM-a brief review, *Boundary Elements X*, Computational Mechanics Pub., Southampton, pp. 431-445.

Hsiao, G. and Wendland, W. (1983). On a boundary integral method for some exterior problems in elasticity, *Dokl.Akad.Nauk SSSR*.

Hsiao, G. and Wendland, W. (1997). A finite element method for some integral equations of the first kind, *J. Math. Anal. Appl.*,58,3, pp. 449-483 .

Hsiao, G. and Zhang, S. (1994). Optimal order multigrid methods for solving exterior boundary value problems, *SIAM J. Numer. Anal.* 31, pp. 680-694.

Hu, Q. and Yu, D. (2001). A solution method for a certain nonlinear interface problem in unbounded domains, *Computing*, 47, pp. 119-140.

Hu, Q. and Yu, D. (2001). A coupling of FEM-BEM for a kind of Signorini contact problem, *Science in China (Series A)*, 44,7, pp. 895-906.

Hu, Q. and Yu, D. (2001). Solving singularity problems in unbounded domains by coupling of natural BEM and composite grid FEM, *Applied Numerical Mathematics*, 37, pp. 127-143.

Hu, Q. and Yu, D. (2002). A preconditioner for coupling system of natural boundary element and composite grid finite element, *J. Comp. Math.*, 20,2, pp. 165-174.

Huan, R. and Thompson, L.L. (2000). Accurate radiation boundary conditions for the time-dependent wave equation on unbounded domain, *Inter. J. Numer. Meth. Engng.*, 47,9, pp. 1569-1603.

Jia, Z., Wu, J. and Yu, D. (2001). The coupled natural boundary-finite element method for solving 3-D exterior Helmholtz problem, *Math. Numer. Sinica*, 23,3, pp. 357-368.

Johnson, C. and Nedelec, J.C. (1980).On the coupling of boundary integral and finite element methods, *Math. Comp.*, 35, pp. 1063-1079.

Joseph, R.M., Hagness, S.C. and Taflove, A. (1991). Direct time integration of Maxwell's equations in linear dispersive media with absorption for scatering and propagation of femtosecond electromagnetic pulse, *Opt. Lett.*, 16, pp. 1412-1414.

Karp, S.N. (1961). A convergent far-field expansion for two dimensional radiation functions, *Comm. Pure Appl. Math.*, 14, pp. 427-434.

Katz, D.S., Thiele, E.T. and Taflove, A. (1994). Validation and extension to three-dimensions of the Berenger PML absorbing boundary condition for FDTD meshes, *IEEE Microwave and Guided wave Letters*, 4, pp. 268-270.

Keller, J.B. and Givoli, D. (1989). Exact non-reflecting boundary conditions, J. Comput. Phys., 82, pp. 172-192.

Kopriva, D.A., Woodruff, S.L. and Hussaini, M.Y. (2002). Computation of electromagnetic scatering with a non-conforming discontinuous spectral element method, *Int. J. Numer. Meth. Engng.*, 53, pp. 105-122.

Ladyzhenskaya, O.A. (1969). *The Mathematical Theory of Viscous Incompressible Flow, Second edition*, Gordon and Breach, New York, English Translation.

Lai, M., Wenston, P. and Ying, L. (2002). Bivariate C^1 cubic splines for exterior biharmonic equations, *Approx. Th X, (Charles K.Chui, Larry L. Schumaker, and Joachim Stöckler eds.)* Vanderbilt University Press, pp. 385-404.

Lassas, M. and Somersalo, E. (1998). On the existence and convergence of the solution of PML equations, *Computing*, 60, pp. 229-241.

Lax, P.D. and Philips, R.S. (1967). *Scatering Theory.* Academic Press, New York, London.

Leis, R. (1986). *Initial Boundary Value Problems in Mathematical Physics*, John Wiley and Sons.

Li, T.-t. and Qin, T. (1997). *Physics and Partial Differential Equations*, Higher Education Press, Beijing.

Liao, Z., Wong, H., Yang, B. and Yuan, Y. (1984). A transmitting boundary for transient wave analysis, *Scientia Sinica, Ser. A*, 26, pp. 1063-1076.

Liao, Z. and Wong, H. (1984). A transmitting boundary for the numerical simulation of elastic wave propagation, *Soil Dynamics and Earthquake Engineering*, 3,4, pp. 174-183.

Lindman, E.L. (1975). Free space boundaries for the scalar wave equation, *J. Comp. Phys.*, 18, pp. 66-78.

Liu, Q. (1997). A FDTD algorithm with perfectly mathched layer for conductivity media, *Mocrowave Opt. Thchnol. Lett.*, 127,2, pp. 134-137.

Liu, Q. and Fan, G. (2000). A FDTD algorithm with perfectly mathched layers for general dispersive media, *IEEE Trans. Antennas Propagat.*, 48,5, pp. 637-646.

Lizorkin, P.I. (1963). (L^p, L^q) multipliers of Fourier integrals, *Dokl. Akad. Nauk. SSSR*, 152, pp. 808-811, English Transl.: *Soviet Math. Dokl.*, 4, 1963, pp. 1420-1424.

Lu, T., Zhang, P. and Cai, W. (2004). Discontinuous Galerkin methods for dispersive and lossy Maxwell's equation and PML boundary conditions, *J. Comput. Phys.*, 200, pp. 549-580.

Lysmer, J. and Kuhlemeyer, R.L. (1969). Finite dynamic model for infinite media, *Journal of the Engineering Mechanics Division, Proceedings of the American Society of Civil engineers*, 95(EM4), pp. 859-877.

Ma, H. and Guo, B. (2001). Composite Legendre-Laguerre pseudospectral approximation in unbounded domains, *IMA J. Numer. Anal.*, 21, pp. 587-602.

McLean, W. (2000). *Strongly Elliptic Systems and Boundary Integral Equations*, Cambridge University Press, Cambridge.

Mohammadia, A.H., Shankar, V. and Hall, W.F. (1991). Computation of electromagnetic scatering and radiation using a time domain finite volume discretization procedure, *Comput. Phys. Comm.*, 68, pp. 175-196.

Moore, T.G., Blaschak, J.G., Taflove, A. and Kriegsmann, G.A. (1988). Theory and application of radiation boundary operators, *IEEE Trans. Antennas Propagat.*, 36, pp. 1797-1812.

Mur, G. (1981). Absorbing boundary condition for ther finite difference approximation of the time-domain electromagnetic field equations, *IEEE Trans. Electromagnetic Compatibility*, 23, pp. 377-382.

Nazarov, S.A. and Pileckas, K. (1999). Asymptotics of solutions to the Navier-Stokes equations in the exterior of a bounded body, *Doklady Mathematics*, 60, pp. 133-135.

Nazarov, S.A. and Pileckas, K. (2000). On steady Stokes and Navier-Stokes problems with zero velocity at infinity in a three dimensional exterior domain, *J. Math. Kyoto Univ.*, 40, pp. 475-492.

Nazarov, S.A. and Specovius-Neugebauer, M. (1996). Approximation of exterior problems. Optimal conditions for the Laplacian, *Analysis*, 16, pp. 305-324.

Nazarov, S.A. and Specovius-Neugebauer, M. (1997). Approximation of exterior boundary value problems for the Stokes system, *Asymptotic Anal.*, 14, pp. 223-255.

Nazarov, S.A. and Specovius-Neugebauer, M. (2000). Artificial boundary conditions for the exterior spatial Navier-Stokes problem, *C. R. Acad. Sci. Paris*, Serie IIb 328, pp. 863-867.

Nazarov, S.A. and Specovius-Neugebauer, M. (2003). Nonlinear artificial boundary conditions with pointwise error estimates for the exterior three dimensional Navier-Stokes problem, *Math. Nachr.*, 252,1, pp. 86-105.

Nazarov, S.A., Specovius-Neugebauer, M. and Videman, J.H. (2004). Nonlinear artificial boundary conditions for the Navier-Stokes equations in an aperture domain, *Math. Nachr.*, 265,1, pp. 24-67.

Nicholis, D.P. and Reitich, F. (2003). Analytic continuation of Dirichlet-Neumann operators, *Numer. Math.*, 94,1, pp. 107-146.

Nicholis, D.P. and Reitich, F. (2004). Shape deformations in rough surface scattering: Cancellations, conditioning, and convergence, *J. Opt. Soc. Am. A.*, 21,4, pp. 590-605.

Nicholis, D.P. and Reitich, F. (2004). Shape deformations in rough surface scattering: Improved algorithms, *J. Opt. Soc. Am. A.*, 21,4, pp. 606 621.

Nicholis, D.P. and Shen, J. A stable high-order method for two dimensional bounded-obstacle scattering, (preprint).

Qi, Q. and Geers, T.L. (1998). Evaluation of the perfectly matched layer for computational acoustics, *J. Comput. Phys.*, 139, pp. 166-183.

Roden, J.A. and Gedney, S.D. (1997). Efficient implementation of the uniaxial based PML media in three-dimensional non-orthogonal coordinates using the FDTD technique, *Microwave Opt. Technol. Lett.*, 14, pp. 71-75.

Roden, J.A. and Gedney, S.D. (2000). Convolutional PML (CPML): an efficient FDTD implementation of the CFS-PML for arbitrary media, *Microwave Opt. Technol. Lett.*, 27,5, pp. 334-339.

Sacks, Z.S., Kingsland, D.M., Lee, R. and Lee, J.-F. (1995). A perfectly matched anisotropic absorber for use as an absorbing boundary condition, *IEEE Trans. Antennas Propagat.*, 43,12, pp. 1460-1463.

Safjan, A.J. (1998). Highly accurate non reflecting boundary conditions for finite element simulations of transient acoustics problems, *Comput. Methods Appl. Mech. Engng.*,152, pp. 175-193.

Sequeira, A. (1983). The coupling of boundary integral and finite element methods for the bidimensional exterior steady Stokes problem, *Math. Methods Appl. Sci.*, 5, pp. 356-375.

Sequeira, A. (1986). On the computer implementation of a coupled boundary and finite element method for the bidimensional exterior steady Stokes problem, *Math. Methods Appl. Sci.*, 8, pp. 117-133.

Shao, X., Chapter 13. Finite element methods for the wave equations, in *Handbook of the Finite element Method*, Science Press, Beijing, (to appear).

Shao, X. and Lan, Z., A kind of discrete non reflecting boundary conditions for varieties of wave equations (preprint).

Sheen,D. (1993). Second order absorbing boundary conditions for the wave equation in a rectangular domain, *Math. Comp.*, 61,204, pp. 595-606.

Shen, J. (2000). Stable and efficient spectral methods in unbounded domains using Laguerre functions, *SIAM J. Numer. Anal.*, 38, pp. 1113-1133.

Shen, J. and Tang, T. (2006). *High Order Numerical Methods and Algorithms* (preprint).

Sochacki, J., Kubichek, R., George, J., Fletcher, W.R. and Smithson, S. (1987). Absorbing boundary conditions and surface waves, *Geophysics*, 52,1, pp. 60-71.

Taflove, A. and Hagness, S.C. (2000). *Computational Electromagnetics: The Finite Difference Time-Domain Method, 2nd ed.*, Artech House, Boston, London.

Taylor, M. (1974). *Pseudo Differential Operators*, Lecture Notes in Math., 416, Springer-Verlag.

Teixeira, F.L. and Chew, W.C. (1997). Perfectly matched layer in cylindrical coordinates, *Proc. IEEE Antennas and Propagation Society International Symposium, vol.3*, Montreal, Canada, pp. 1908-1911.

Temam, R. (1984). *Navier-Stokes Equations, Theory and Numerical Analysis, Third edition*, North Holland, Netherlands.

Teng, Z. (2003). Exact boundary condition for time-dependent wave equation based on boundary integral, *J. Comput. Phys.*, 190, pp. 398-418.

Thatcher, R.W. (1976). The use of infinite grid refinements at singularities in the solution of Laplace's equations, *Numer. Math.*, 15, pp. 163-178.

Thatcher, R.W. (1978). On the finite element method for unbounded regions, *SIAM J. Numer. Anal.*, 15, pp. 466-477.

Thompson, K.V. (1990). Time dependent boundary conditions for hyperbolic systems II, *J. Comput. Phys.*, 89, pp. 439-461.

Thompson, K.V. (1992). Time dependent boundary conditions for nonlinear hyperbolic systems, *J. Comput. Phys.*, 61, pp. 171-214.

Trefethen, L.N. and Halpern, L. (1979). Well posedness of one way wave equation and absorbing boundary condition, *Math. Comp.*, 47, pp. 421-436.

Turkel, E. and Yefet, A. (1998). Absorbing PML boundary layers for wave-like equations, *Appl. Numer. Math.*, 27, pp. 533-557.

Varga, R.S. (1962). *Matrix Iterative Analysis*, Prentice-Hall, Englewood Cliffs, N.J..

Wang, L. and Guo, B. (2006). Stair Laguerre pseudospectral method for differential equations on the half line, *Adv. Comput. Math.*, 25, pp. 305-322.

Watson, G.N. (1946). *Treatise on the Theory of Bessel Functions*, Combridge.

Wilcox, C.H. (1956). An expansion theorem for electromagnetic fields, *Comm. Pure Appl. Math.*, 9, pp. 115-134.

Wittwer, P. (2002). On the structure of stationary solutions of the Navier-Stokes equations, *Comm. Math. Phys.*, 226, pp. 455-474.

Wu, J. and Yu, D. (2000). The overlapping domain decomposition method for harmonic equation over exterior three-dimensional domain, *J. Comp. Math.*, 18,1, pp. 83-94.

Xu, C. and Guo, B. (2002). Mixed Laguerre-Legendre spectral method for incompressible flow in an infinite strip, *Adv. Comput. Math.*, 16, pp.77-96.

Yee, K.S. (1966). Numerical solution of initial boundary value problems involving Maxwell's equations in isotropic media, *IEEE Trans. Antennas Propagat.*, 14, pp. 302-307.

Ying, L. (1978). The infinite similar element method for calculating stress intensity factors, *Scientia Sinica*, 21,1, pp. 19-43.

Ying, L. (1979). The convergence of the infinite similar element method, *Acta Math. Appl. Sinica*, 2,2, pp. 149-166.

Ying, L. (1982). The infinite element method, *Advances Math.*, 11,4, pp. 269-272.

Ying, L. (1986). Infinite element approximation to axial symmetric Stokes flow, *J. Comp. Math.*, 4,2, pp. 111-120.

Ying, L. (1991). Infinite element method for elliptic problems, *Science in China, Series A*, 34,12, pp. 1438-1447.

Ying, L. (1992). An introduction to the infinite element method, *Math. in Practice and Theory*, 2, pp. 69-78.

Ying, L. (1995). *Infinite Element Methods*, Peking University Press, Beijing, and Vieweg Publishing, Braunschweig/Wiesbaden.

Ying, L. (2000). Infinite element method for the exterior problems of the Helmholtz equations, *J. Comp. Math.*, 18,6, pp. 657-672.

Ying, L. (2001). Infinite element method for problems on unbounded and multiply connected domains, *Acta Mathematica Scientia*, 21,4, pp. 440-452.

Ying, L. and Han, H. (1980). On the infinite element method for unbounded domain and nonhomogeneous problems, *Acta Math. Sinica*, 23,1,pp. 118-127.

Ying, L. and Li, F. (2003). Exterior problem of the Darwin model and its numerical computation, *Mathematical Modelling and Numerical Analysis*, 37,3, pp. 515-532.

Ying, L. and Wei, W. (1993). Infinite element approximation to axial symmetric Stokes flow II, *Chinese J. Num. Math. Appl.*, 15,3, pp. 55-72.

Yosida, K. (1974). *Functional Analysis*, Springer-Verlag, Berlin, Heideberg, New York.

Yu, D. (1982). Canonical integral equations of biharmonic elliptic boundary value problems, *Math. Numer. Sinica*, 4,3, pp. 330-336.

Yu, D. (1983). Numerical solutions of harmonic and biharmonic canonical integral equations in interior or exterior domains, *J. Comp. Math.*, 1,1, pp. 52-62.

Yu, D. (1983). Coupling canonical boundary element method with FEM to solve harmonic problem over cracked domain, *J. Comp. Math.*, 1,3, pp. 195-202.

Yu, D. (1984). Canonical boundary element method for plane elasticity problems, *J. Comp. Math.*, 2,2, pp. 180-189.

Yu, D. (1985). Approximation of boundary conditions at infinity for a harmonic equation, *J. Comp. Math.*, 3,3, pp. 219-227.

Yu, D. (1991). A direct and natural coupling of BEM and FEM ,*Boundary Elements XIII, Brebbia C.A., Gipson G.S. eds., Southampton, Computational Mechanics Publications*, pp. 995-1004.

Yu, D. (1994). A domain decomposition method based on natural boundary reduction over unbounded domain, *Math. Numer. Sinica*, 16,4, pp. 448-459.

Yu, D. (1996). Discretization of non-overlapping domain decomposition method for unbounded domains and its convergence, *Math. Numer. Sinica*, 18,3, pp. 328-336.

Yu, D. (2002)., *Natural Boundary Integral Method and Ita Applioations*, Kluwer Academic Publisher.

Yu, D. (2004). Natural boundary integral method and its new development, *J. Comp. Math.* 22,2, pp.309-318.

Yu, D. and Du, Q. (2003). The coupling of natural boundary element and finite element method for 2D hyperbolic equations, *J. Comp. Math.*, 21,5, pp. 585-594.

Yu, D. and Jia, Z. (2002). Natural integral operator on elliptic boundary and the coupling method for an anisotropic problem, *Math. Numer. Sinica*, 24,3, pp. 375-384.

Yu, D. and Wu, J. (2001). A non-overlapping domain decomposition method for exterior 3-D problem, *J. Comp. Math.*, 19,1, pp. 77-86.

Yu, D. and Zhao, L. (2004). Natural boundary integral method and related numerical methods, *Engineering Analysis with Boundary Elements*, 28, pp. 937-944.

Zhang, G. (1981). Integrated solutions of ordinary differential equation system and two-point boundary value problems, Part I. Integrated solution method, *Math. Numer. Sinica*, 3, pp.'245-254.

Zhang, G. (1982). Integrated solutions of ordinary differential equation system and two-point boundary value problems, Part II. Stability, *Math. Numer. Sinica*, 4, pp.'329-339.

Zhang, G. (1985). High order approximation of one way wave equation, *J. Comp. Math.*, 3,1, pp. 90-97.

Zhang, G. (1993). Square root operator and its application in numerical solution of some wave problems, *Proceedings of the First China-Japan Seminar on Numerica Mathematics (eds. Shi, Z. and Ushijima, T.)*, World Scientific, Singapore, pp.'204-214.

Zhang, G. and Wei, S.(1998). Stability analysis of absorbing boundary conditions for acoustic wave equation, *Math. Numer. Sinica*, 20,1, pp. 103-120.

Zhang, X. and Guo, B. (2006). Spherical harmonic-generalized Laguerre spectral method for exterior problems, *J. Sci. Comp.*, DOI:10.1007/s10915-005-9056-6.

Zhu, J. (1991). *The Boundary Element Analysis for Elliptic Boundary Value Problems*, Science Press, Beijing (in Chinese).

Ziolkowski, R. (1997). Time derivative Lorentz model based absorbing boundary condition, *IEEE Trans. Antennas Propagat.*, 45,10, pp. 1530-1535.

Index